WINE
MICROBIOLOGY

WINE MICROBIOLOGY

Practical Applications and Procedures

Kenneth C. Fugelsang
Charles G. Edwards

Second edition

 Springer

Kenneth C. Fugelsang
Department of Viticulture
    and Enology
California State University,
    Fresno
Fresno, CA 93740-8003
Kennethf@csufresno.edu

Charles G. Edwards
Department of Food Science and
    Human Nutrition
Washington State University
Pullman, WA 99164-6376
edwardsc@wsu.edu

ISBN-13: 978-1-4419-4121-3

e-ISBN-10: 0-387-33349-5
e-ISBN-13: 978-0-387-33349-6

Printed on acid-free paper.

9 8 7 6 5 4 3 2 1

springer.com

CONTENTS

Section II: Vinification and Winery Processing

PREFACE TO THE SECOND EDITION

Organization of the large volume of material that needed to be included in *Wine Microbiology* was a difficult task as evidenced by the number of approaches attempted. The second edition is divided into three parts; *Grape and Wine Microorganisms* (Chapters 1 to 4), *Vinification and Winery Processing* (Chapters 5 to 11), and *Laboratory Procedures and Protocols* (Chapters 12 to 19). As subject areas frequently cross section or chapter boundaries, every effort was made to cross-reference related topics as a means to reduce difficulties in finding information.

Section I, *Grape and Wine Microorganisms*, describes those microorganisms found in grape must, juice, and wines; namely yeasts, lactic and acetic acid bacteria, and molds. Here, taxonomy, metabolism, nutritional requirements, and potential impacts on wine quality are areas of focus.

Section II, *Vinification and Winery Processing*, addresses on those microbiological issues of practical importance to the winemaker. Included here is a general discussion of microbial management followed by in-depth examination of microbial ecology. This section also describes general principles of sanitation (Chapter 9), implementation of a quality control program (Chapter 10), and specific wine spoilage issues (Chapter 11).

Section III, *Laboratory Procedures and Protocols*, begins with an introduction to the use of the microscope (Chapter 12) and follows with methodologies used to enumerate and identify wine microorganisms (Chapters 13 through 16). Because organic and inorganic precipitates found in wine are often confused with microorganisms, methods of identification as well as photomicrographs of typical precipitates are included in Chapter 17. Chapters 18 and 19 provide insight into designing a wine microbiological laboratory and related safety issues. The section ends with a glossary of terms commonly used by microbiologists.

ACKNOWLEDGMENTS

The authors would like to thank several individuals and organizations for their assistance in preparation of this book. First, we wish to extend thanks to California State University (Fresno, CA) and to Washington State University (Pullman, WA) for their support of this project. The authors also thank W.D. Edinger (Canandaigua Wine Company, Madera, CA), R. Morenzoni (Art of Winemaking, Modesto, CA), and B. Watson (Chemeketa Community College, Salem, OR) for technical review and S. Safren (Springer Science and Business Media) for her editorial assistance. Special thanks are extended to R.H. Dougherty and B.A. Rasco (Washington State University, Pullman, WA) for providing HACCP materials and technical advice as well as to Eiji Akaboshi, Peter Gray, Cheryl Mitchell, Henry Moore Jr., and Roy Thornton involved in obtaining and optimizing photomicrographs. Appreciation is given to the *American Journal of Enology and Viticulture*, American Public Health Association, *Australian Journal of Grape and Wine Research*, Blackwell Publishing, Elsevier Ltd., John Wiley & Sons Ltd., *Journal of Bacteriology*, Invitrogen Corporation, R. Pawsey, Springer Science and Business Media, WineBugs LLC, and *Wines and Vines* for granting copyright permissions to use figures and tables.

The authors convey their gratitude to the wine industry, commercial suppliers, and university colleagues around the world. Without continual input, advice, and innovative ideas from many individuals, this project would not have been possible.

Finally, special thanks are extended to undergraduate and graduate students, research associates, technicians, and other staff with whom the authors have worked. Through their efforts performing research and developing protocols, a much better understanding of wine microbiology has been gained by all.

INTRODUCTION

The winemaking community worldwide continues to be a study of philosophical contrasts. On the one hand, there are those winemakers and wineries that emphasize the "scientific" segment of winemaking through adoption of new research findings and technologies. On the other hand, others prefer to embrace Old World traditions and thereby accentuate the "artistic" aspects associated with wine production. In writing this book, the objective was not to debate the relative merits and deficiencies of either philosophy but, rather, to create a reference that was useful to enologists as well as to researchers and students globally.

Since publication of the first edition of *Wine Microbiology* in 1997, the volume of new information and concepts has dramatically increased. Perhaps one of the most intriguing developments in the past decade has been application of "real-time" molecular methods. Based on similarities at the gene level, these methods have evolved beyond esoteric laboratory exercises to the point where real-world problems can be solved through rapid identification of microorganisms. Another relatively new application has been the use of starter cultures of non-*Saccharomyces* yeasts, which yield wines that differ not only in flavor and aroma profiles but also in structure.

Winemakers are also increasingly facing spoilage issues associated with *Brettanomyces, Lactobacillus, Pediococcus,* and *Zygosaccharomyces,* some of these being consequences of changes in viticultural practices (e.g., increased so-called hang-time).

Even with the tremendous increase in available information, a comprehensive understanding regarding the role of individual microorganisms toward wine quality as well as the impact of complicated interactions between microorganisms and processing techniques is lacking. A good example would be *Brettanomyces,* probably the most enigmatic and controversial microorganism in the wine industry. Although initially thought of as a major threat, some winemakers are beginning to view *Brettanomyces* as a potential ally in the vintages of the new millennium. Hopefully, additional research and experience will provide winemakers with better microbiological control during vinification, which, in turn, will lead to a continued increase in wine quality.

We sincerely hope that you find the second edition of *Wine Microbiology* informative and useful in your winery, laboratory, or classroom. If you have any feedback for the authors (potential errors, ideas for the third edition, and the like), please feel free to write or e-mail us. Cheers!

Kenneth C. Fugelsang and Charles G. Edwards
February 14, 2006

SECTION I

GRAPE AND WINE MICROORGANISMS

CHAPTER 1

YEASTS

1.1 INTRODUCTION

Yeasts represent the most important group of microorganisms to wine-makers, because without *Saccharomyces*, producing quality wine would be impossible. Besides *Saccharomyces*, however there are many other genera and species present during vinification that ultimately impact quality, both positively and negatively (Fleet and Heard, 1993; Fugelsang et al., 1993; Deak and Beuchat, 1996; Loureiro and Malfeito-Ferreira, 2003).

1.2 REPRODUCTION

For yeasts, classification to the genus level requires demonstration of the presence or absence of a sexual phase in the life cycle. Sexual spores, called ascospores, are produced and upon germination yield the veget-ative budding yeast. The absence of sexual spores during the life cycle results in the yeast being classified as the anamorph or asexual ("imper-fect") form, whereas success in demonstrating the presence of sexual spores indicates the teleomorph or sexual form ("perfect"). Based on criteria other than spore formation (e.g., utilization of specific nitrogen compounds, fermentation of specific sugars, etc.), anamorphs and teleomorphs are identical.

1.2.1 Sexual Reproduction

Although presumably occurring in nature, the presence of a sexual cycle normally requires growing the yeast isolate on a specialized medium (Section 15.3.2). Complete identification of a particular isolate requires that an effort be made to verify the occurrence of a sexual phase. However, classification based on the failure to demonstrate ascopore formation may or may not be correct because many factors influence sporulation. First, conjugation cycles may be cyclical in nature; that is, most activity occurs over a relatively short time and laboratory personnel may fail to examine cultures at the right time, or sporulation may be ephemeral with ascospores rapidly germinating to yield the vegetative yeast. Second, the isolate may be heterothallic and haploid and thus require a compatible mating type that may not be present in the culture. Finally, expression of the sexual cycle may be prevented due to chemical (e.g., high concentrations of glucose) or physical (e.g., temperature) conditions.

1.2.2 Asexual Reproduction

Budding represents the most frequently encountered form of reproduction and, in the case of wet mounts from juice or fermenting wine, the only form that will be seen. Budding of the mother cell initially yields a bud and, eventually, a daughter cell after separation. Under optimal conditions, a single mother cell may bud many times during the cultivation period. However, under the restrictive conditions of fermentation, a single yeast will bud only three to four times. Here, the availability of oxygen, which is required for the formation of cell membrane precursors, limits further replication.

Yeasts exhibit several types of budding, which may be of diagnostic value. Among those isolated from fermenting and aging wine, multilateral and restricted polar budding are most frequently observed. Occasionally, yeasts that reproduce by fission through the formation of cross-walls can be isolated.

1.2.2.1 Multilateral Budding

Multilateral budding occurs on the "shoulder" area of the yeast where each bud arises at a location separate from others. As seen in the electron micrograph of *Saccharomyces* (Fig. 1.1), the mother cell is left with bud scars upon separation of daughter cells. The budding and separation cycle creates an increasing problem for yeasts during their fermentative phase of growth in that each cycle depletes the mother cell's membrane by approximately one-half. Because cell membrane synthesis occurs under aerobic conditions, this limits asexual replicative cycles. As a result, most fermentations are completed by yeasts in the stationary (nonbudding) phase.

Budding patterns vary depending on culture age, physiological status, and the physical/chemical conditions of culture. For example, *Dekkera/Brettanomyces* are described as reproducing asexually through multilateral budding. However, older cells may exhibit a cell shape suggestive of

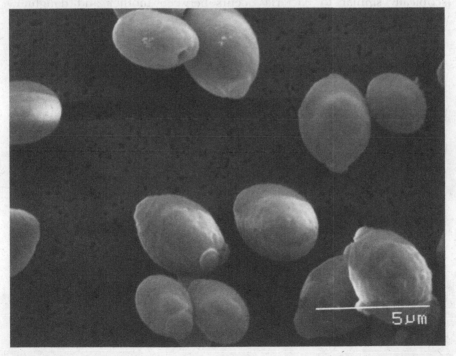

Figure 1.1.   *Saccharomyces cerevisiae* showing characteristic bud scars left upon separation of daughter cells. Photograph provided by A. Dumont with the kind permission of Lallemand Inc.

restricted polar budding. Thus, within a microscopic field, some cells can appear "box" or "rectangular" shaped, whereas others may be seen as "ogival," "boat-shaped," or "gothic arches." Furthermore, a pseudomycelium may also be produced.

1.2.2.2 Polar Budding

In contrast to multilateral budding, some native species found in the early stages of winemaking replicate by repeated budding at the same site(s). Known as polar budding, it can occur at one or two poles. In the case of *Kloeckera* and *Saccharomycodes*, budding occurs at either pole, giving rise to an older population of characteristic apiculate "lemon-shaped" yeasts (Fig. 1.2).

1.2.2.3 Fission

Fission reproduction is characteristic of the genus *Schizosaccharomyces*, where formation of the daughter cell occurs without the constriction seen in the above-described types. In this case, formation of a cross-wall microscopically similar to bacteria occurs between mother and daughter, followed by separation.

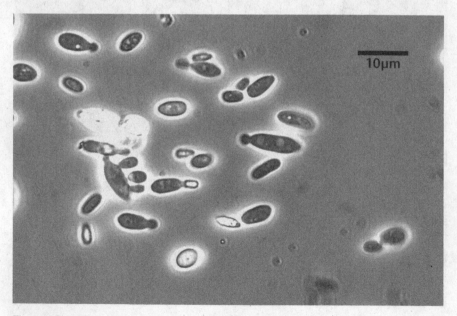

Figure 1.2.   *Kloeckera apiculata* as viewed with phase-contrast microscopy at a magnification of 1000×. Photograph provided with the kind permission of WineBugs LLC.

1.2.2.4 Pseudomycelium

Every winemaker has had occasion to note formation of a film (or pellicle) on the wine's surface. The film arises as a consequence of repeated budding of oxidatively growing yeast and failure of daughters to separate from the mother cells. With time, the previously defined constriction between mother and daughter widens, giving the impression of mold-like multi-cellular filamentation rather than a chain of yeast. The extent of pseudo-mycelial development varies from several adhering cells with apparent constriction between mother and daughter cells to elaborate arrays in which elongate stem cells visually appear distinct from ovoidal side buds. The latter may themselves elongate, which leads to further branching. Proliferation results from not only growth of cells in the main chain but also budding and branching of side chains leading to film formation that may rapidly cover the surface of wine.

The formation of a pseudomycelium is occasionally of diagnostic value, although several species of wine yeast including *Saccharomyces* are capable of this type of growth. Formation may be demonstrated by the use of slide cultures described in Section 12.6.

1.3 TAXONOMY

To distinguish yeasts that can produce ascospores from those that do not, mycologists use a dual taxonomy for classification. Unfortunately, the nomenclature for anamorph/teleomorph combinations is frequently different. Examples of sexual/asexual or teleomorph/anamorph yeasts include *Dekkera/Brettanomyces*, *Metschnikowia pulcherrima/Candida pulcher-rima*, *Hanseniaspora uvarum/Kloeckera apiculata*, and *Torulaspora delbrueckii/Candida colliculosa*. Some yeasts only exist in the anamorphic form because sporulation has yet to be demonstrated (e.g., *Candida vini*).

Whereas yeasts are differentiated by taxonomists through assignment of various genera/species names, many winemakers use an informal system to group yeasts based on their morphology or other characteristics. For example, *Kloeckera apiculata* is often referred to as an "apiculate yeast" due to its lemon-shaped morphology. This yeast as well as others present in grape musts, such as *Candida, Cryptococcus, Debaryomyces, Hansenula, Issatch-enkia, Kluyveromyces, Metschnikowia, Pichia*, and *Rhodotorula* (Section 6.2.1), are also called "native," "natural," or "wild" yeasts because they originate in the vineyard or winery. Some enologists have argued against the use of these terms because of the implication that other yeasts not included in this group are somehow "non-native" or "unnatural" (e.g., *Saccharomyces* or

Brettanomyces). Rather, the term "non-*Saccharomyces* yeasts" is now more commonly used to describe those yeasts commonly present in grape musts that are not of the genus *Saccharomyces*.

Yeasts found during and after alcoholic fermentation are also assigned to informal groups. For instance, yeasts that conduct the primary fermentation (*Saccharomyces*) are sometimes called "fermentative yeasts" based on their metabolism. During the aging of wine, some yeasts (*Candida, Hansenula,* and *Pichia*) can grow on the surface in the presence of oxygen. As a film on the wine surface is commonly associated with this spoilage, these yeasts are collectively referred to as "film yeasts" (Section 11.2.3).

The so-called black yeasts (*Aureobasidium pullulans*) are occasionally isolated from grape and wine environments. Pfaff et al. (1978) considered *Aureobasidium* among related "yeast-like" organisms. Microscopically, the asexual vegetative cell is ellipsoidal to apiculate in appearance. Budding is multilateral, and there is a strong tendency to form a mycelium. On agar, early growth is seen as light-cream to tan. With age, colonies turn olive-green (suggestive of mold contamination) and eventually become black. At all stages, colonies appear mucoidal, with the edges often fringe-like.

The following descriptions come from various sources, but the abilities to ferment or oxidatively assimilate specific sugars were obtained from the most current edition of *The Yeasts* (Kurtzman, 1998a; 1998b; 1998c; Meyer et al., 1998; Miller and Phaff, 1998a; 1998b; Smith, 1998a; 1998b; 1998c; Vaughan-Martini and Martini, 1998a; 1998b). These descriptions were not complete in that the ability of a given yeast to ferment some sugars important in grape musts/wines (e.g., arabinose and fructose) were not reported.

1.3.1 *Candida*

The anamophic genus *Candida* represents a very broad group with a number of species found in wines. The perfect or teleomorphic forms of *Candida* species are represented by a number of different genera including *Issatchenkia, Kluyveromyces, Pichia, Metschnikowia, Saccharomyces, Torulaspora,* and *Zygosaccharomyces* (Deak and Beuchat, 1996). As such, *Candida* represents a group with a wide range of physiological characteristics. For instance, cells can appear microscopically as being globose, ellipsoidal, cylindrical, or elongate (Meyer et al., 1998).

In general, reproduction in the case of *Candida* is accomplished through multilateral budding. Various sugars may be fermented and nitrate may be assimilated depending on species. Of the sugars tested, *Candida stellata* ferments and assimilates glucose, sucrose, and raffinose. Another species (*C. pulcherrima*) ferments glucose but assimilates glucose, galactose, L-sorbose, sucrose, maltose, cellobiose, trehalose, melezitose, D-xylose, *N*-acetyl-D-glucosamine, ethanol, glycerol, D-mannitol, D-glucitol, α-methyl-D-glucoside salicin, D-gluconate, succinate, and hexadecane (Meyer et al., 1998).

1.3.2 *Dekkera*

Many winemakers consider *Brettanomyces* and its sporulating equivalent *Dekkera* to be a threat to wine quality (Section 11.2.2). Worldwide, economic losses due to these yeasts are very high, not only from overtly spoiled and unmarketable wines but also from wines of diminished quality that do not command their expected market price. Despite the traditionally negative connotations surrounding *Dekkera/Brettanomyces*, some winemakers question whether some degree of infection can be beneficial in certain styles of wine. In these cases, the yeasts are thought to play a positive role in sensorial complexity as well as imparting aged characters in the case of some young red wines.

Brettanomyces was first isolated from French, South African, and Italian wines in the 1950s (Sponholz, 1993). In early reports, van der Walt and van Kerken (1959; 1961) described the occurrence of *B. intermedius* and *B. schanderlii* in wines from South Africa. Although a number of different species of *Dekkera/Brettanomyces* have been previously described, *D. anomala* (anamorph: *B. anomalus*), *D. bruxellensis* (anamorph: *B. bruxellensis*), *B. custersianus*, *B. naardenensis*, and *B. nanus* (Smith, 1998b) are currently accepted species (Smith, 1998a; 1998b). *B. intermedius*, *B. lambicus*, and *B. schanderlii* are considered to be synonyms of *D. bruxellensis* (Smith, 1998a).

Upon isolation, most laboratories rely on microscopic evaluation of cell morphology for identification. Although cell shape certainly plays a role in identification, caution must be applied when using this criterion. As previously noted, yeasts exhibit variable cell morphology depending on age, culture medium, and environmental stress. For example, *Brettanomyces* grown on solid agar substrate may appear considerably different from *Brettanomyces* isolated in barrel-aged wine. Classically, *Brettanomyces* exhibits cell shapes described as boat-shaped or ogival (Smith, 1998a). Reminiscent of gothic arches, ogival cell morphology is commonly seen in older cultures (Fig. 1.3). Furthermore, less than 10% of the cells in a given population may exhibit this shape.

When grown on solidified media (Section 13.5), colonies appear white to yellowish and may be glistening, moist and smooth, or dull and

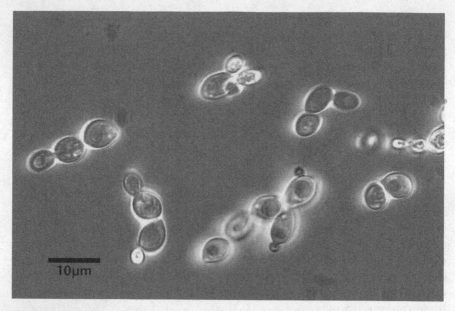

Figure 1.3. *Brettanomyces* as viewed with phase-contrast microscopy at a magnification of 1000×. Photograph provided with the kind permission of WineBugs LLC.

wrinkled. Grown on malt agar containing 2% w/v calcium carbonate, the colonies appear white to cream-colored, ranging from shiny to dull. Edges are entire or lobate (*D. anomala*) or entire and undulating (*D. bruxellensis*). Ascospores appear hat-shaped or somewhat spherical with tangential brims. All species ferment glucose, and other carbohydrates such as galactose, sucrose, maltose, and trehalose may be fermented depending on strain. The species found in wines, *B. anomalus* and *B. bruxellensis*, can be separated on the basis of lactose fermentation (most strains of *B. anomalus* ferment the sugar whereas *B. bruxellensis* does not) and succinate assimilation (most strains of *B. anomalus* assimilate the acid whereas *B. bruxellensis* does not). Both species can also assimilate nitrate and some *Brettanomyces* can use ethanol as a sole carbon and energy source (Silva et al., 2004).

Species of *Dekkera/Brettanomyces* produce large amounts of acetic acid when grown on glucose (Freer, 2002). In fact, acetic acid production can be sufficient to inhibit and eventually kill cultures maintained on unbuffered media. Thus, routine laboratory maintenance media contains 2% w/v calcium carbonate to buffer against the acids produced (Section 13.5). However, *Brettanomyces* tends to a slow growing yeast, often requiring several days for colonies to appear on solidified media.

Probably the most significant difficulty in successful routine laboratory identification of *Dekkera* and *Brettanomyces* lies in the fundamental requirement to demonstrate the presence (or absence) of a sexual phase. *Dekkera* requires a sporulation medium that includes augmentation with several vitamins (Section 15.3.2.4). However, Ilagan (1979) noted that even under ideal conditions, relatively poor sporulation (<1%) can be observed. As the yeast is not known to form spores in wine, suspect isolates are reported as *Dekkera/Brettanomyces* or, simply, "Brett-like."

1.3.3 *Hanseniaspora*

Hanseniaspora forms ovoid or spherical (young cultures) or apiculate or lemon-shaped (older cultures) cells as shown in Fig. 1.2. Vegetative reproduction is by budding at both poles (Smith, 1998c). Ascospores (one to four per ascus) are spherical and can become hat- or saturn-shaped. The species most commonly found in grapes, *H. uvarum* (anamorph: *Kloeckera apiculata*) ferments only glucose and assimilates glucose, cellobiose, 2-keto-D-gluconate, and salicin. The microorganism will also grow in the presence of 100 mg/L cycloheximide. Species of *Hanseniaspora* tend to be fructophilic in that most prefer fructose over glucose (Ciani and Fatichenti, 1999).

1.3.4 *Issatchenkia*

Species within this genus exhibit multilateral budding as well as pseudomycelia. One to four ascospores that appear roughened are formed. Glucose is fermented while nitrate is not assimilated. One species found in grape juice or wines, *I. orientalis* (anamorph: *Candida krusei*) appears as ovoidal to elongated cells. *I. orientalis* assimilates glucose, N-acetyl-D-glucosamine, ethanol, glycerol, DL-lactate, and succinate (Kurtzman, 1998a).

1.3.5 *Metschnikowia*

Like some other yeasts, *Metschnikowia* also forms multilateral buds as well as pseudohyphae. Asci produce one to two needle-shaped ascospores without any terminal appendages. A species found in grape musts or wines, *M. pulcherrima* (anamorph: *Candida pulcherrima*), ferments glucose and can assimilate a number of compounds including glucose, galactose, L-sorbose, sucrose, maltose, cellobiose, trehalose, melezitose, D-xylose, N-acetyl-D-glucosamine, ethanol, glycerol, D-mannitol, D-glucitol, α-methyl-D-glucose, salicin, D-gluconate, succinate, and hexadecane but not nitrate (Miller and

Phaff, 1998a). The species can assimilate various nitrogen sources including cadaverine, L-lysine, ethylamine, and tolerates 10 mg/L cycloheximide but is completely inhibited by 100 mg/L. Some species produce pulcherrimin, a brown/red pigment (Pallmann et al., 2001)

1.3.6 *Pichia*

A number of different species of *Pichia* are recognized including two found in wines, *P. anomala* (anamorph: *Candida pelliculosa*) and *P. membranifaciens* (anamorph: *Candida valida*). Another species, *P. guilliermondii* (anamorph: *Candida guilliermondii*) has also been recovered from grape musts and from winery equipment in contact with grape musts but not wines (Dias et al., 2003a). The cells microscopically appear as ovoid, ellipsoidal, or cylindrical and reproduce vegetatively by multilateral budding. A pseudomycelium may be poorly developed or absent. Colonies on solid media are white or cream, dull, and usually wrinkled. Ascospores (one to four per ascus) can appear spherical, hemispherical, hat- or saturn-shaped.

Depending on species, various carbohydrates can be fermented and nitrate may be assimilated. *P. anomala* ferments glucose and sucrose and assimilates glucose, sucrose, maltose, cellobiose, trehalose, raffinose, melezitose, soluble starch, ethanol, glycerol, erythritol, D-mannitol, D-glucitol, α-methyl-D-glucoside, salicin, DL-lactate, succinate, citrate, and nitrate (Kurtzman, 1998b). *P. membranifaciens* weakly ferments glucose and assimilates far fewer compounds than *P. anomala* (glucose, N-acetyl-D-glucosamine, and ethanol).

P. anomala, formerly *Hansenula anomala* (Deak and Beuchat, 1996), has limited fermentative abilities but can grow oxidatively as a film yeast. When growing fermentatively, *P. anomala* is capable of producing 0.2% to 4.5% v/v alcohol along with potentially large amounts of acetic acid, ethyl acetate, and isoamyl acetate (Shimazu and Watanabe, 1981). Acid utilization by this yeast may also be substantial, resulting in decreased titratable acidity and upward pH shifts (Sponholz, 1993).

1.3.7 *Saccharomyces*

As described by Vaughan-Martini and Martini (1998a), *Saccharomyces* appear microscopically as globose or ovoidal cells with multilateral budding and possibly pseudohyphae (Fig. 1.4). The yeast forms one to four ascospores, which are smooth and ellipsoidal. Colonies appear smooth, usually flat, and occasionally raised and opaque. The two primary species found in wines, *S. bayanus* and *S. cerevisiae* (anamorph: *Candida robusta*), ferment glucose, sucrose, and raffinose and assimilate glucose, sucrose, maltose, raffinose, and ethanol but not nitrate. *Saccharomyces* can not utilize five-carbon sugars (e.g., pentoses).

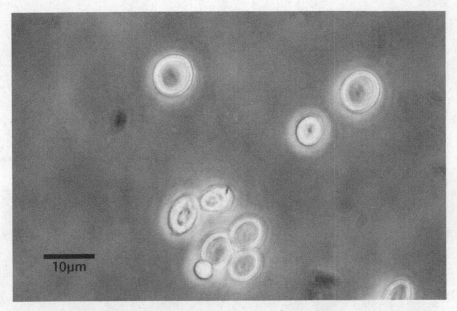

Figure 1.4.   *Saccharomyces* as viewed with phase-contrast microscopy at a magnification of 1000×. Photograph provided with the kind permission of WineBugs LLC.

Over the years, there have been numerous changes to the taxonomy of the genus *Saccharomyces* (Vaughan-Martini and Martini, 1995). In fact, Vaughan-Martini and Martini (1998a) listed a total of 97 synonyms for *S. cerevisiae*. As an example, *S. carlsbergensis* was initially reclassified as *S. uvarum*, a species that later became *S. cerevisiae* race *uvarum*, which then became *S. pastorianus*. Currently, *S. cerevisiae* and *S. bayanus* are thought to be either two separate species (Vaughan-Martini and Martini, 1998a) or the same species that differ slightly so as to be different races (Boulton et al., 1996).

1.3.8 *Saccharomycodes*

This genus is represented by a single species, *Saccharomycodes ludwigii*, which appears as lemon-shaped cells with blunt tips, sausage-shaped, curved, or elongated with a swelling in the middle (Fig. 1.5). At times, cells are single or appear in pairs or groups of three (Miller and Phaff, 1998b). Asexual reproduction is by bipolar budding. *Saccharomycodes* produces one to four smooth, spheroid ascospores with a small subequatorial ledge. Sugars fermented include glucose, sucrose, and raffinose while compounds assimilated are glucose, sucrose, raffinose, glycerol,

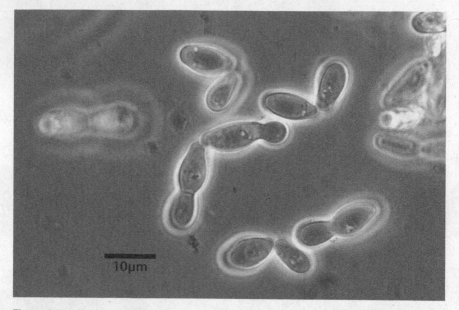

Figure 1.5.   *Saccharomycodes ludwigii* as viewed with phase-contrast microscopy at a magnification of 1000×. Photograph provided with the kind permission of WineBugs LLC.

cadaverine, and ethylamine but not nitrate. Growth is apparent in the presence of 1 mg/L cycloheximide but is inhibited by 10 mg/L.

1.3.9 *Schizosaccharomyces*

Cells of *Schizosaccharomyces* may be cylindrical, ovoid, or even spherical (Vaughan-Martini and Martini, 1998b). Of the yeasts found in grape juice or wine, this genus uniquely reproduces by fission. Mycelia may form and asci produce two to eight spherical or ellipsoidal ascospores. The primary species found in grape musts or wines, *S. pombe*, ferments glucose, sucrose, and maltose and can assimilate glucose, sucrose, maltose, raffinose, and D-gluconate. This species cannot use ethanol as a sole carbon source or nitrate as a nitrogen source.

1.3.10 *Zygosaccharomyces*

Zygosaccharomyces comprises nine species (Kurtzman, 1998c), of which *Z. bailii*, *Z. bisporous*, *Z. rouxii*, and *Z. florentinus* have been isolated from grape musts or wines. *Saccharomyces rouxii* and *Zygosaccharomyces barkeri* are synonyms of *Z. rouxii* (Deak and Beuchat, 1996).

Table 1.1. Inhibitory concentrations of various compounds against *Zygosaccharomyces*.

Compound	Concentration
Acetic acid	>2.5% v/v
Benzoic acid	>1000 mg/L
Ethanol	>20% v/v
Inhibitory pH	<2 and >7
Sorbic acid	>800 mg/L
Sugar	>70% w/v
Sulfur dioxide	>3 mg/L molecular

Adapted from Thomas and Davenport (1985) with the kind permission of Elsevier Ltd.

Zygosaccharomyces microscopically appears as spherical, ellipsoidal, or elongate cells with multilateral budding and possibly pseudohyphae (Kurtzman, 1998c). Ascospores are smooth, spherical, and ellipsoidal with one to four per ascus. A number of sugars are fermented depending on the species, and nitrate is not assimilated. *Z. rouxii* ferments glucose and maltose and assimilates glucose, trehalose, glycerol, D-mannitol, and D-glucitol.

Colony and cell morphologies vary with isolation media but *Z. bailii* can microscopically appear ovoidal to cylindrical in shape. The yeast exhibits multilateral budding leading to formation of a simple pseudomycelium. *Zygosaccharomyces* is haploid and heterothallic, meaning that sporulation requires the union (conjugation) of two compatible mating types prior to sporulation. Asci appear shaped as clubs (*Z. bailii*) or dumb-bells (*Z. bisporus*). Each conjugant produces two smooth, round ascosores.

Unlike many other yeasts, *Zygosaccharomyces* can grow in high-solute environments such as 60% w/w glucose (Thomas, 1993). Furthermore, the yeast is extraordinarily resistant to common preservatives used by the juice, concentrate, and wine industries (Table 1.1). Resistance to SO_2 is believed to be due to synthesis of extracellular sulfite-binding compounds such as acetaldehyde as well as other unidentified mechanisms (Deak and Beuchat, 1996). *Zygosaccharomyces* are also known to be extremely tolerant to alcohol and can grow in wines containing 18% v/v (Thomas and Davenport, 1985).

1.4 NUTRITIONAL REQUIREMENTS

Like other microorganisms, yeasts require a number of nutrients for growth including carbon (sugars), nitrogen (ammonia and/or amino acids), and various growth and survival factors such as vitamins and minerals.

1.4.1 Nitrogen

Of all the nutrients important for *Saccharomyces* to conduct alcoholic fermentation of grape musts, perhaps the most important is nitrogen (Section 8.2). Yeasts can assimilate nitrogen from organic (amino acids) or inorganic (ammonia or NH_4^+) sources. Inorganic nitrogen is "fixed" into organic forms through reaction with α-keto-glutarate to yield glutamate by glutamate dehydrogenase (Fig. 1.6). Glutamate can then be used by the cell to produce other amino acids important for metabolism.

Arginine is quantitatively the most important amino acid utilizable by *Saccharomyces* in grapes and, subsequently, unfermented juice/must (Spayd and Andersen-Bagge, 1996; Stines et al., 2000). This amino acid is rapidly

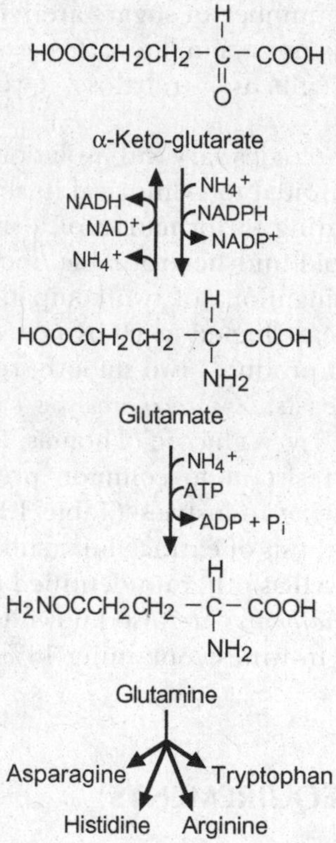

Figure 1.6. Reaction of ammonia with α-keto-glutarate to incorporate inorganic forms of nitrogen by *Saccharomyces*.

incorporated by yeast at the start of fermentation and subsequently released back into the wine during autolytic cycles. Catabolically utilized in production of intermediates useful to the cell and energy in the form of ATP, one by-product of the pathway is urea, a compound that reacts with ethanol to yield ethyl carbamate (Section 11.3.2).

1.4.2 Growth and Survival Factors

On a dry weight basis, *Saccharomyces* contains 3% to 5% phosphate, 2.5% potassium, 0.3% to 0.4% magnesium, 0.5% sulfur, and trace amounts of calcium, chlorine, copper, iron, zinc, and manganese (Monk, 1994; Walker, 1998). Yeast must be supplied with a source of phosphate, which is incorporated into nucleic acids, phospholipids, adenosine-5'-triphosphate (ATP), and other compounds. Potassium is necessary for uptake of phosphate, and a deficiency may be linked to sluggish alcoholic fermentations (Kudo et al., 1988). Other minerals needed by *Saccharomyces* during fermentation have a variety of functions but are used primarily as enzyme activators.

Besides minerals, yeasts require various vitamins such as thiamin, riboflavin, pantothenic acid, pyridoxine, nicotinamide, biotin, and inositol depending on species and specific growing conditions (Monk, 1994; Ough et al., 1989a). In general, practically all strains of *Saccharomyces* require biotin and pantothenic acid while some also need inositol and/or thiamin (Walker, 1998). Biotin is involved in carboxylation of pyruvic acid and the synthesis of nucleic acids, proteins, and fatty acids. Pantothenic acid is an essential part of coenzyme A, a molecule required for sugar and lipid metabolism. A deficiency of pantothenic acid can also lead to H_2S (Section 8.5.2). Though involved in oxoacid decarboxylations, thiamin may or may not be required because some strains of yeast can synthesize the vitamin. Sulfur dioxide can cleave thiamin making the vitamin unavailable. Finally, nicotinic acid is used in the synthesis of NAD^+ and $NADP^+$, and inositol is required for cell division. Nutrient requirements for yeasts other than *Saccharomyces* vary widely.

Like nitrogen, strains of commercial wine yeast vary in their demands for oxygen (Julien et al., 2000). Although *Saccharomyces* exhibits growth under anaerobic conditions, viability in the absence of oxygen is finite. Oxygen is required for the synthesis of certain metabolites, specifically lanosterol, ergosterol, and unsaturated fatty acyl coenzyme A esters (Walker, 1998; Ribéreau-Gayon et al., 2000), collectively termed "survival factors" (Lafon-Lafourcade et al., 1979). Addition of ergosterol to clarified grape musts can encourage completion of alcoholic fermentation (Houtman et al., 1980b). Sterols are necessary for membrane permeability and so will increase the viability of yeasts and prolong their fermentative activity (Lafon-Lafourcade et al., 1979). Sterols also affect the synthesis of volatile odor and flavor compounds, depending on the presence of oxygen (Mauricio et al., 1997). Addition of yeast hulls ("ghosts") to fermenting musts can be a source of survival factors during fermentation (Munoz and Ingledew, 1989a; 1989b).

Grapes supply a portion of the lipids needed by yeast during anaerobic growth. Oleanolic acid composes up to two-thirds of the cuticular waxes in some grape varieties and can replace the yeast's requirement for ergosterol supplementation under anaerobic conditions (Brechot et al., 1971).

Most non-*Saccharomyces* yeasts have greater requirements for oxygen than do *Saccharomyces*, with concentrations of a few parts per million needed to support growth. Reflecting their oxidative metabolism, many of these microorganisms are usually found growing on the surface of wines (Section 11.2.3).

1.5 METABOLISM

1.5.1 Glucose

Yeasts metabolize sugars (e.g., glucose) to produce energy in the form of ATP. When ATP is hydrolyzed to yield ADP (adenosine-5′-diphosphate) and P_i (inorganic phosphate), the energy released is used by the cell for various reactions and transformations.

Glucose is metabolized in a series of steps known as the Emben–Meyerhof–Parnas pathway, or glycolysis (Fig. 1.7). Through a series of biochemical reactions, metabolism of glucose eventually yields two three-carbon compounds, glyceraldehyde-3-phosphate and dihydroxyacetone-phosphate. The "trioses" are transformed into two molecules of pyruvate, which are subsequently decarboxylated and reduced to yield two carbon dioxide and alcohol. In summary, glycolysis produces two molecules each of ATP and reduced coenzyme (NADH) per molecule of glucose metabolized.

Under some conditions, dihydroxyacetone formed during glycolysis can be reduced to glycerol through glycerol-3-phosphate (Fig. 1.8), resulting in the oxidation of an additional molecule of NADH to NAD^+. This reaction is important under fermentation conditions when NAD^+ is in short supply within the cell. The formation of glycerol is favored if acetaldehyde is not available to be reduced to ethanol as would be the case in formation of "bound SO_2" by reaction of acetaldehyde and bisulfite (Section 5.2.1).

Under aerobic conditions, pyruvate produced by glycolysis will enter the Krebs, or tricarboxylic acid cycle (Fig. 1.9). Initially, pyruvate is either

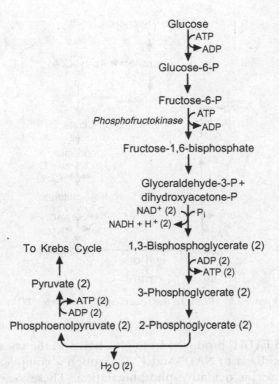

Figure 1.7. Utilization of glucose by *Saccharomyces* through Embden–Meyerhof–Parnas (EMP) pathway or glycolysis.

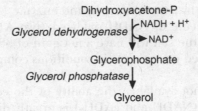

Figure 1.8. Reduction of dihydroxyacetone to glycerol and the subsequent oxidation of NADH to NAD^+.

carboxylated to yield oxaloacetate or can be directly used to produce acetyl CoA. Both oxaloacetate and acetyl CoA will react to form the first product, citric acid. With one "turn" of the cycle, carbon entering will be lost as CO_2 and additional molecules of reduced coenzyme (NADH and $FADH_2$) are produced in addition to energy in the form of guanosine triphosphate (GTP).

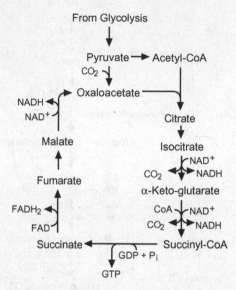

Figure 1.9. Krebs or the tricarboxylic acid cycle (TCA).

NADH and $FADH_2$ produced from glycolysis and the tricarboxylic acid cycle are reoxidized to NAD^+ and FAD through a complicated series of reactions known as oxidative phosphorylation. These reactions involve the transfer of electrons through cytochromes with the ultimate electron acceptor being oxygen to form H_2O. Because of the need of oxygen, these reactions are active only under aerobic conditions. ATP will be produced from these reactions by a membrane-bound enzyme (ATPase) at a rate of three ATP per molecule of NADH oxidized (two ATP per molecule of $FADH_2$). Therefore, the cell will have a net gain of 38 ATP per molecule of glucose metabolized under aerobic conditions compared with only two from glycolysis.

When oxygen is not available, the ability of the cell to reoxidize the reduced coenzymes (NADH and $FADH_2$) is greatly diminished. To compensate, the biochemistry of the cell is altered such that pyruvate is reduced to acetaldehyde and then to ethanol (fermentation), the latter step requiring NADH (Fig. 1.10). Thus, the formation of ethanol allows the cell to reoxidize the NADH that was produced in earlier steps in glycolysis.

Although not completely functional, certain enzymes Krebs cycle are, active under anaerobic conditions (Fig. 1.11). These additional pathways are very important to *Saccharomyces* during fermentation because NADH can be reoxidized and other precursors important for cellular functions synthesized (e.g., α-keto-glutarate involved in NH_4^+ assimilation).

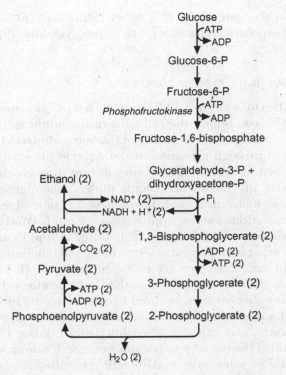

Figure 1.10. Utilization of glucose by *Saccharomyces* under anaerobic (fermentative) conditions.

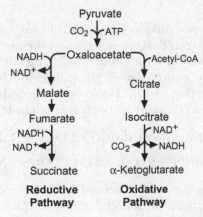

Figure 1.11. Operational portions of Krebs cycle during fermentation by *Saccharomyces*. Adapted from Boulton et al. (1999) with the kind permission of Springer Science and Business Media.

Depending on the state of the cell, either "reductive" (fermentation) or "oxidative" (respiration) pathways are active, yielding different final products.

1.5.1.1 Pasteur and Crabtree Effects

Because only two molecules of ATP are produced per glucose metabolized under anaerobic conditions, the cell must utilize additional glucose at a faster rate in order to maintain the pool of intracellular ATP. This step is accomplished through activation of the enzyme phosphofructokinase (Fig. 1.10), which, in turn, increases carbon flow through glycolysis. The increase in rate of glucose breakdown under anaerobic conditions is known as the Pasteur effect. This phenomenon is only observable when glucose concentrations are low, approximately 0.9 g/L (Walker, 1998).

If the concentration of glucose is high (>9 g/L), the Pasteur effect in *S. cerevisiae* is repressed. However, fermentative metabolism and ethanol formation will continue even with oxygen available (Ribéreau-Gayon et al., 2000), a phenomenon known as the Crabtree effect. Here, NADH generated from glycolysis is reoxidized by producing ethanol (fermentation) rather than the combined pathways of respiration (glycolysis, tricarboxylic acid cycle, and oxidative phosphorylation). From a winemaking perspective, this effect is very important because it allows fermentation in the presence of some oxygen. Although present in *Saccharomyces*, the Crabtree effect is not observed in some non-*Saccharomyces* yeasts (Walker, 1998).

1.5.1.2 Custers Effect

Dekkera/Brettanomyces are unique with regard to carbohydrate utilization. Like other yeasts, *Dekkera/Brettanomyces* possess the glycolytic pathway and so will produce ethanol from acetaldehyde along with NAD^+, the latter being reused for the conversion of glyceraldehyde-3-phosphate to 1,3-bisphosphoglycerate (Fig. 1.10). However, carbohydrates can also be oxidized to acetate (acetic acid) through acetaldehyde as follows:

$$C_6H_{12}O_6 + 2\ NAD^+ \rightarrow 2\ CH_3CHO + 2\ CO_2 + 2\ NADH + 2\ H^+$$

$$2\ CH_3CHO + 2\ NAD(P)^+ + 2\ H_2O \rightarrow 2\ CH_3COOH + 2\ NAD(P)H + 2\ H^+$$

Another source of acetic acid produced by *Dekkera/Brettanomyces* is the oxidation of ethanol, again with acetaldehyde as the intermediate:

$$CH_3CH_2OH + 2\ NAD(P)^+ + H_2O \rightarrow CH_3COOH + 2\ NAD(P)H + 2\ H^+$$

Due to the production of acetic acid, NADH accumulates within the cell and must be reoxidized to maintain the NAD^+/NADH balance. Under aerobic conditions, this is accomplished by oxygen accepting electrons from NADH to yield NAD^+ and H_2O (Walker, 1988). In fact, the availability of oxygen greatly affects the amount of acetic acid formed by *Dekkera/Brettanomyces* with far more acetic acid produced under aerobic conditions (Ciani and Ferraro, 1997; Aguilar Uscanga et al., 2003). Because O_2 helps to maintain the availability of NAD^+, the formation of ethanol during oxidation of carbohydrates is stimulated by oxygen, a phenomenon known as the Custers effect (Wijsman et al., 1984; Walker, 1998).

According to Ciani and Ferraro (1997), the lack of NAD^+ under anerobic conditions is not restored by glycerol production from dihydroxyacetone phosphate (Fig. 1.8). In fact, glycolytic pathways are temporarily inhibited in *Dekkera/Brettanomyces* when these yeasts are introduced into an anaerobic environment (Wijsman et al., 1984). Because of the lack of NAD^+, *Dekkera/Brettanomyces* will conduct a limited alcoholic fermentation with the production of primarily ethanol, not acetic acid (Ciani and Ferraro, 1997). Biochemically, another source of NAD^+ during growth of these yeasts in red wines may be the formation of volatile phenols (Section 11.2.2).

1.5.2 Sulfur

Microorganisms require a source of sulfur in order to synthesize sulfur-containing amino acids and other important metabolites (Fig. 1.12). Yeasts can assimilate either sulfur dioxide (SO_2) or sulfate (SO_4^{2-}), the latter requiring energy for transport into the cell (Breton and Surdin-Kerjan, 1977). These molecules will be sequentially reduced to sulfide (S^{2-}) through a series of reactions known as the sulfate reduction sequence (Rauhut, 1993). Sulfide can either be incorporated into organic molecules containing sulfur such as the amino acids methionine or cysteine or be exported as the sensorially offensive compound H_2S (Section 8.5.2). Most of these biochemical reactions can be reversed such that the yeast also release sulfite, the amount depending on conditions as well as yeast strain (Larsen et al., 2003).

1.5.3 Odor/Flavor Compounds

Besides ethanol and CO_2, minor concentrations of by-products such as glycerol, succinic acid, acetic and lactic acids, as well as acetaldehyde and a large number of other volatile and nonvolatile substances are formed by

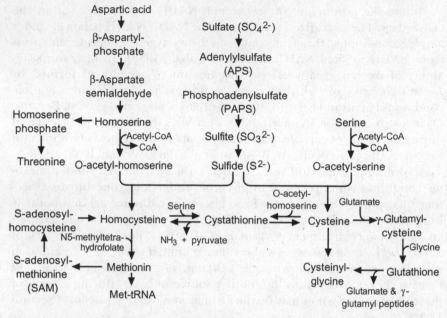

Figure 1.12.   Sulfur metabolism in *Saccharomyces*. Adapted from Wang et al. (2003) with the kind permission of Blackwell Publishing.

yeast during fermentation. Individually and collectively, these compounds play an important role in the sensory characteristics of wine (Rankine, 1967; Zeeman et al., 1982; Nykänen, 1986; Edwards et al., 1990; Herraiz et al., 1990; Lema et al., 1996).

Alcohols that contain more than two carbons are known as higher alcohols, or fusel oils. Isobutanol, *n*-propanol, isoamyl alcohol, and active amyl alcohol are produced in various proportions by various yeasts including *Candida, Hansenula, Pichia,* and *Saccharomyces* (Rankine, 1967; Edwards et al., 1990; Holloway and Subden, 1991; Webster et al., 1993; Lambrechts and Pretorius, 2000) and, potentially, play an important role in the sensory character of wine (Rankine, 1967). Sensory descriptors that are commonly used to describe higher alcohols include "fusel" (butanol), "alcoholic" (isobutyl alcohol), "marzipan" (active amyl alcohol and isoamyl alcohols), and "floral" or "rose" (phenethyl alcohol) (Lambrechts and Pretorius, 2000).

Higher alcohols are by-products of transamination reactions where yeasts will transfer amino groups between amino acids and α-keto-acids (Fig. 1.13). The α-keto-acids are decarboxylated into aldehydes, which are then converted into higher alcohols. Known as the Ehrlich pathway (Castor

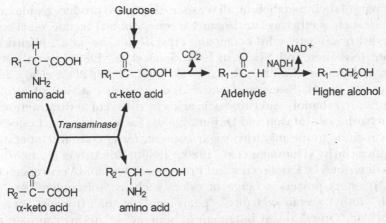

Figure 1.13.   Synthesis of higher alcohols by *Saccharomyces*.

and Guymon, 1952), valine, leucine, isoleucine, and threonine are deaminated to produce isobutanol (2-methyl-1-propanol), isoamyl alcohol (3-methyl-1-butanol), active amyl alcohol (2-methyl-1-butanol), and propyl alcohol, respectively (Chen, 1978). Quantitatively, the concentrations of fusel alcohols may range from 140 to 420 mg/L, with isoamyl alcohol generally accounting for more than 50% (Rankine, 1967; Muller et al., 1993).

By itself, the available pool of amino acids present in juice is not sufficient to account for concentrations of fusel oils found in wine. Early work by Castor and Guymon (1952) showed that the yields of fusel oils tend to be higher than the available corresponding amino acid. Chen (1978) determined that threonine, valine, isoleucine, and leucine only contributed 30%, 34%, 75%, and 80% of their corresponding higher alcohol, respectively. In fact, higher alcohols can also be synthesized by another pathway, specifically from pyruvate formed from glucose through glycolytic pathways (Lambrechts and Pretorius, 2000).

Formation of higher alcohols during fermentation is affected by many factors including total nitrogen and the amino acid content of grape musts (Bell et al., 1979; Ough and Bell, 1980; Webster et al., 1993). In general, elevated levels of must nitrogen will decrease the formation of high alcohols. Other factors include fermentation temperature (Crowell and Guymon, 1963), suspended or insoluble solids (Crowell and Guymon, 1963; Klingshirn et al., 1987; Edwards et al., 1990), and oxygen concentration of juice/must (Guymon et al., 1961).

As part of their metabolism, all yeasts are known to produce a wide range of esters such as ethyl acetate, isoamyl acetate, isobutyl acetate, ethyl butyrate, ethyl hexanoate, ethyl octanoate, ethyl decanoate, and 2-phenylethyl acetate (Nykänen and Nykänen, 1977; Soles et al., 1982; Edwards et al., 1990; Webster et al., 1993; Rojas et al., 2001; 2003; Plata et al., 2003; Lee et al., 2004). Esters are synthesized through reaction of alcohols (commonly, ethanol) and carboxylic acids by different acyltransferases or "ester synthases" (Mason and Dufour, 2000). Factors that affect ester synthesis include grape maturity, sugar content, fermentation temperature, and juice clarity (Houtman et al., 1980a; 1980b; Edwards et al., 1990).

As described by Lambrechts and Pretorius (2000) and Verstrepen et al. (2003), esters possess a range of odors such as "solvent-like" or "nail polish" (ethyl acetate), "fruity," "pear," or "banana" (isoamyl acetate), "floral" or "fruity" (ethyl butanoate), "sour apple" (ethyl caproate and ethyl caprylate), and "flowery," "roses," or "honey" (phenyl ethyl acetate). Given the number of different esters present in varying concentrations produced by yeasts, there is little doubt that wine aroma can be strongly influenced by these compounds (Schreier, 1979).

1.5.4 Glycosidases

Monoterpenes, important odor and flavor compounds in grapes (Rapp and Mandery, 1986), include geraniol, nerol, linalool, citronellol, and α-terpineol (Fig. 1.14). These compounds are distributed in various locations within the grape berry (Wilson et al., 1986). Although potentially important in imparting fragrant, "floral" aromas, many are naturally present in grapes as glycosidically bound forms that are not volatile (Fig. 1.15). To release more "grape flavors" into wine by removing the sugar moiety from the terpene, specific enzymes known as glycosidases can be used (van Rensburg and Pretorius, 2000). In fact, some have attempted to use specific cultures of yeasts or bacteria with high enzymatic activities (Günata et al., 1986; 1990; McMahon et al., 1999; Cabaroglu et al., 2003). For instance, Delcroix et al. (1994) investigated the use of strains of *Saccharomyces* with higher β-glucosidase activity but noted few differences in the concentrations of terpenes and no differences in sensory quality of the resultant wines. More recently, Mendes Ferreira et al. (2001) and Rodríguez et al. (2004) noted that selected species of non-*Saccharomyces* yeasts (*Candida, Kloeckera, Pichia,* and *Metschnikowia*) had high glycosidase activities that could potentially contribute different aromas to wines. Although *Brettanomyces* has glycosidase activity, Mansfield et al. (2002) determined that the enzyme(s) were not active against glycosides isolated from grapes.

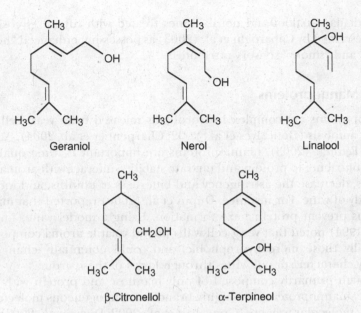

Figure 1.14.   Monoterpenes present in some cultivars of *Vitis vinifera*.

Enzymes can also be added to grape musts rather than using viable cultures as a means to increase quality. As an example, Yanai and Sato (1999) applied a β-glucosidase obtained from *Debaryomyces hansenii* and were able to produce wines with much higher concentration of terpenoids,

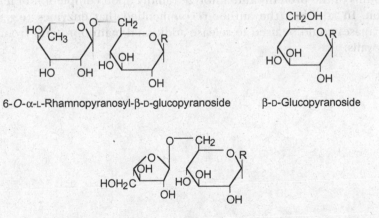

6-*O*-α-L-Rhamnopyranosyl-β-D-glucopyranoside   β-D-Glucopyranoside

6-*O*-α-L-Arabino-furanosyl-β-D-glucopyranoside

Figure 1.15.   Monoterpene glycosides present in some cultivars of *Vitis vinifera*. The R-groups represent geraniol, nerol, linalool, citronellol, or α-terpineol.

in particular linalool and nerol. Wines treated with fungal glycosidases were described by Cabaroglu et al. (2003) as possessing enhanced "honey," "lime," and "smoky" sensory attributes.

1.5.5 Mannoproteins

Mannoproteins are complex hydrocolloids released from yeast cell walls during autolysis (Goncalves et al., 2002; Charpentier et al., 2004). According to Feuillat (2003), mannoproteins are important to wine quality as these contribute to protein and tartrate stability, interact with aroma compounds, decrease the astringency and bitterness of tannins, and increase the body of wine. For instance, Dupin et al. (2000) reported that mannoproteins prevent protein haze formation. Using a model wine, Lubbers et al. (1994) noted that yeast cell walls bound volatile aroma compounds, especially those more hydrophobic, and could potentially change the sensory characteristics of wines through losses of these aromas.

Though primarily composed of only mannose and protein with some glucose, mannoproteins are highly branched heterogeneous molecules of varying molecular weights (Goncalves et al., 2002; Doco et al., 2003). The role of each fraction toward wine quality remains unknown, but certain processing practices can influence the amount present. For instance, *batonnage* (stirring) during aging of wines on lees results in an increase in these compounds, whereas microoxygenation had little effect (Doco et al., 2003). Because the release of mannoproteins can require long periods of time, Feuillat (2003) suggested using yeast strains that produce large amounts of the proteins and autolyze rapidly upon completion of fermentation. In addition, the author recommended that enzymes (e.g., β-1,3 glucanase) could be used to release additional mannoproteins from yeast cell walls.

CHAPTER 2

LACTIC ACID BACTERIA

2.1 INTRODUCTION

Lactic acid bacteria comprise an ecologically diverse group of microorganisms united by formation of lactic acid as the primary metabolite of sugar metabolism (Davis et al., 1985b; 1988; Lonvaud-Funel, 1999; Carr et al., 2002; Liu, 2002). These bacteria utilize sugars by either homo- or heterofermentative pathways (Section 2.4.1) as well as L-malic acid, a major acid present in grape must (Section 2.4.3). Whereas growth of some bacteria in certain wines is desirable (i.e., malolactic fermentation or MLF), growth of other species can lead to spoilage.

2.2 TAXONOMY

The lactic acid bacteria isolated from grape musts or wine belongs to two families representing three genera. The Lactobacillaceae are represented by the genus *Lactobacillus*, and the Streptococcaceae are represented by the genera *Oenococcus* and *Pediococcus*.

2.2.1 *Lactobacillus*

Lactobacillus represents a highly diverse group of Gram-positive, micro-aerophilic bacteria that microscopically appear as long to short rods or even coccobacilli (Fig. 2.1) (Kandler and Weiss, 1986). Species within this genus are generally catalase-negative, although a few strains decompose peroxide by a non-heme-containing pseudo-catalase (Johnston and Delwiche, 1962; Kono and Fridovich, 1983; Beyer and Fridovich, 1985). *Lactobacillus* spp. are either homo- or heterofermentative with regard to hexose metabolism (Section 2.4.1). Physiological characteristics used to identify some species of *Lactobacillus* found in grape musts or wines are presented in Table 2.1.

Various species of *Lactobacillus* that have been isolated from grapes and wines worldwide including *L. brevis, L. buchneri, L. casei, L. cellobiosus, L. curvatus, L. delbrueckii, L. diolivorans, L. fructivorans, L. heterohiochii, L. hilgardii, L. jensenii, L. kunkeei, L. leichmanni, L. nagelli, L. paracasei, L. plantarum, L. trichodes, L. vermiforme,* and *L. yamanashiensis* (Douglas and Cruess, 1936; Vaughn, 1955; Fornachon, 1957; Kitahara et al., 1957,

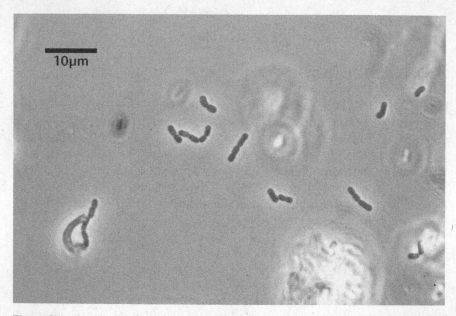

Figure 2.1. *Lactobacillus brevis* as viewed with phase-contrast microscopy at a magnification of 1000×. Photograph provided with the kind permission of WineBugs LLC.

Table 2.1. Characteristics of some *Lactobacillus* found in wines.

Characteristic	*L. brevis*	*L. hilgardii*	*L. kunkeei*	*L. plantarum*
Ammonia from arginine	+	+	–	–
Catalase	v	+	w	v
Gas from glucose	+	+	+	–
Hydrolysis of esculin	v	–	–	+
Lactic acid from glucose	DL	DL	L	DL
Mannitol from fructose	+	+	+	–
Fermentation of:				
Arabinose	+	–	–	v
Fructose	+	+	+	+
Lactose	v	v	–	+
Mannitol	–	–	+	+
Maltose	+	+	–	+
Melezitose	–	v	–	+
Ribose	+	+	–	+
Sucrose	v	v	+	+
Trehalose	–	–	–	+
Xylose	v	+	–	v

(+) 90% or more of the strains are positive; (–) 90% or more of the strains are negative; (v) variable response of strains; (w) weak reaction.
Data from Kandler and Weiss (1986), Dicks and van Vuuren (1988), Pilone et al. (1991), Hammes et al. (1992), and Edwards et al. (1993; 1998).

Du Plessis and van Zyl, 1963a; Pilone et al., 1966; Carr et al., 1977; Chalfan et al., 1977; Maret and Sozzi, 1977; 1979; Costello et al., 1983, Lafon-Lafourcade et al., 1983b; Nonomura, 1983; Davis ct al., 1986a; 1986b; Dicks and van Vuuren, 1988; Sieiro et al., 1990; Edwards et al. 1993; 1998a; 2000; Mills, 2001; Gorga et al., 2002; Beneduce et al., 2004; Du Plessis et al., 2004).

Recent evidence has resulted in changes in the taxonomy of the lactobacilli. Reflecting this, *L. cellobiosus* is currently regarded as a synonym of *L. fermentum*, and *L. leichmanni* is now referred to as *L. delbrueckii* subsp. *lactis* (Kandler and Weiss, 1986). *L. trichodes* and *L. heterohiochii* (Kitahara et al., 1957) are now considered synonyms of *L. fructivorans* (Weiss et al., 1983). Edwards et al. (1998a; 2000) isolated two novel *Lactobacillus* spp. from commercial grape wines undergoing sluggish/stuck alcoholic fermentations. Based on phenotypic and phylogenetic evidence, *L. kunkeei* and *L. nagelii* were proposed as new species. Few reports are available describing *L. vermiforme* (Sharpe et al., 1972; Garvie, 1976), and it is not clear whether the bacterium represents a separate species or is a synonym of a closely related species, *L. hilgardii*.

2.2.2 Oenococcus

Wine bacteria belonging to the genus *Oenococcus* have been previously classified as *Leuconostoc gracile*, *Leuconostoc citrovorum*, and *Leuconostoc oenos* (Pilone and Kunkee, 1965; Garvie, 1967a; Kunkee, 1967a). Later phylogenetic studies revealed that *L. oenos* represented a distinct subline separate from other *Leuconostoc* spp. (Martinez-Murcia et al., 1993), a finding that resulted in reassignment of this bacterium to a new genus, *Oenococcus* (Dicks et al., 1995). Given the diversity in physiological characteristics such as carbohydrate fermentation patterns, Tracey and Britz (1987) suggested that it is possible that *O. oeni* could represent more than one species.

Strains of *O. oeni* are described as Gram-positive, nonmotile, facultatively anaerobic, catalase-negative, ellipsoidal to spherical cells that usually occur in pairs or chains (Fig. 2.2) (Garvie, 1967a; 1986a; Holzapfel and Schillinger, 1992; Dicks et al., 1995). Cells can be difficult to differentiate microscopically from short rods of Lactobacillus (Fig. 2.1). The species is heterofermentative, converting glucose to equimolar amounts of primarily D-lactic acid, CO_2, and ethanol or acetate (Krieger et al., 1993; Cogan and Jordan, 1994; Cocaign-Bousquet et al., 1996). The bacterium produces gas from glucose, hydrolyzes esculin, forms D-lactic acid from glucose and

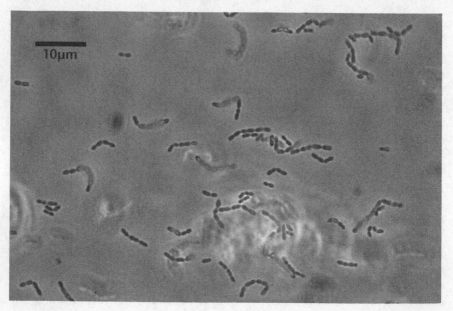

10μm

Figure 2.2.   *Oenococcus oeni* as viewed with phase-contrast microscopy at a magnification of 1000×. Photograph provided with the kind permission of WineBugs LLC.

mannitol from fructose, and may produce ammonia from arginine (Pilone et al., 1991; Holzapfel and Schillinger, 1992; Dicks et al., 1995).

Although only one species is assigned to this genus, *O. oeni* belongs to a heterogeneous group evidenced by wide variability in the fermentation of specific carbohydrates (Lafon-Lafourcade et al., 1983b; Tracey and Britz, 1987; Davis et al., 1988; Kelly et al., 1989; Edwards et al., 1991). Most strains of *O. oeni* utilize L-arabinose, fructose, and ribose but not galactose, lactose, maltose, melezitose, raffinose, or xylose. By comparison, Lafon-Lafourcade et al. (1983b) noted that only 11% of the strains evaluated in their study utilized both fructose and glucose, contrary to the findings of others (Pilone and Kunkee, 1972; Beelman et al., 1977; Izugabe et al., 1985; Edwards et al., 1991). Davis et al. (1988) determined that only 55% of the strains studied fermented ribose, 27% fermented D-arabinose, and 45% fermented sucrose. Strain A-9 described by Chalfan et al. (1977) fermented glucose but not fructose. Although discrepancies in carbohydrate fermentations are probably the result of strain characteristics, differences in techniques used to determine carbohydrate fermentability (Pardo et al., 1988; Jensen and Edwards, 1991) and the nutritional composition of media given the fastidious nature of *Oenococcus* (Garvie, 1967a; 1967b) may also cause variable results.

O. oeni has the ability to metabolize malic acid found in grapes to form lactic acid through MLF (Sections 2.4.3 and 6.4.2). Though other species of lactic acid bacteria have been investigated and used as commercial starters for MLF, strains of *O. oeni* appear to have the physiological properties to consistently tolerate the environmental challenges of wine while producing desirable results within an amount of time acceptable to the winemaker.

2.2.3 *Pediococcus*

Of the approved species of *Pediococcus* (Garvie, 1986b; Weiss, 1992), only four have been isolated from wines; *P. damnosus*, *P. parvulus*, *P. inopinatus*, and *P. pentosaceus* (Davis et al., 1986a; 1986b; Edwards and Jensen, 1992). Several researchers previously reported isolation of *P. cerevisiae* from wines (Maret and Sozzi, 1977; 1979; Costello et al., 1983; Lafon-Lafourcade et al., 1983b; Fleet et al., 1984). The species is now considered invalid because it represents at least two species including *P. damnosus* and *P. pentosaceus* (Garvie, 1974; Raccach, 1987). *P. damnosus* and *P. parvulus* appear to be more commonly found in wines than the other species.

Pediococci are characterized as being Gram-positive, nonmotile, catalase-negative, and aerobic to microaerophilic bacteria (Garvie, 1986b; Pilone et al., 1991; Weiss, 1992). Members of this genus are

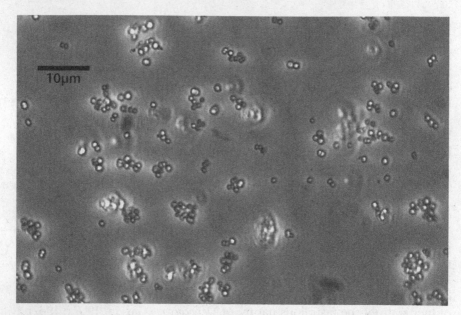

Figure 2.3.   *Pediococcus damnosus* as viewed with phase-contrast microscopy at a magnification of 1000×. Photograph provided with the kind permission of WineBugs LLC.

homofermentative (Section 2.4.1), with glucose converted to either L- or DL-lactate (Garvie, 1986b). Under glucose limitation, Pasteris and Strasser de Saad (2005) noted that a strain of *P. pentosaceus* degraded glycerol to pyruvate, the latter being further metabolized to either acétate or diacetyl or 2,3-butanediol through "active-acetaldehyde" (Section 2.4.5). Growing cultures commonly possess the ability to form L-lactate from L-malic acid (Raccach, 1987; Edwards and Jensen, 1992). Pediococci are chemoorgano-trophs and require complex growth factor and amino acid requirements. In addition, these are the only lactic acid bacteria that divide in two planes, which results in the formation of pairs, tetrads or large clumps of spherical cells as shown in Fig. 2.3 (Garvie, 1986b; Axelsson, 1998). Characteristics for three species of *Pediococcus* are listed in Table 2.2.

2.3 NUTRITIONAL REQUIREMENTS

Lactic acid bacteria have very limited biosynthetic capabilities and, reflecting this, are described as nutritionally fastidious. Early work by Du Plessis (1963) noted that all strains of wine lactic acid bacteria required nicotinic acid, riboflavin, pantothenic acid, and either thiamine or pyridoxine.

Table 2.2. Characteristics of *Pediococcus*.

Characteristic	*P. damnosus*	*P. parvulus*	*P. pentosaceus*
Ammonia from arginine	–	–	+
Catalase	–	–	v
Gas from glucose	–	–	–
Hydrolysis of esculin	+	+	+
Lactic acid from glucose	DL	DL	DL
Mannitol from fructose	–	–	–
Fermentation of:			
Arabinose	–	–	+
Fructose	+	+	+
Lactose	–	–	v
Mannitol	–	–	–
Maltose	v	+	+
Melezitose	v	–	–
Ribose	–	–	+
Sucrose	v	–	–
Trehalose	+	v	+
Xylose	–	–	v

(+) 90% or more of the strains are positive; (–) 90% or more of the strains are negative; (v) variable response of strains.
Data from Garvie (1986b), Pilone et al. (1991), and Weiss (1992).

More recently, Garvie (1986b) reported that all species of *Pediococcus* required nicotinic acid, pantothenic acid, and biotin, whereas none required thiamine, *p*-aminobenzoic acid, or cobalamine. Several amino acids (glutamic acid, valine, arginine, leucine, and isoleucine) appear to be essential for growth. Garvie (1967b) reported similar results but included cysteine, tyrosine, and others depending on the strain of *Oenococcus* (*Leuconostoc*). In addition, purines (guanine, adenine, xanthine, and uracil) and folic acid are also required by many species. Finally, it should be noted that lactic acid bacteria cannot utilize diammonium phosphate as a nitrogen source and so must rely on amino acids.

Another important nutrient is the so-called tomato juice factor (Garvie and Mabbitt, 1967). This nutrient was named for the fact that many lactic acid bacteria isolated from grape musts or wines seemed to grow better on media supplemented with either fruit or vegetable juices or serums such as tomato or apple (Section 13.6). This requirement varies with growth conditions and strains (Garvie, 1984). In fact, Tracey and Britz (1987) were able to grow a number of strains of *O. oeni* in the absence of the tomato juice factor, although growth was much slower. Amachi (1975) ascertained its structure to be a derivative of pantothenic acid, 4-O-(α-D-glucopyranosyl)-D-pantothenic acid.

2.4 METABOLISM

2.4.1 Glucose

After completion of alcoholic fermentation, low concentrations of hexose sugars may remain in the wine. These include glucose and fructose with lesser amounts of mannose and galactose. Among the five-carbon sugars (pentoses), arabinose, ribose, and xylose are the most common. Further, there may be sufficient quantities of sugar to support the growth of lactic acid bacteria in "dry" wines.

Lactic acid bacteria utilize sugars (e.g., glucose) to form lactic acid by either the homo- or heterofermentative pathway. The homofermentative pathway, illustrated in Fig. 2.4, results in the transformation of glucose to pyruvate through the Embden–Meyerhof–Parnas pathway (EMP, or glycolysis), eventually yielding lactic acid. NADH produced by the oxidation of glyceraldehyde-3-phosphate to 1,3-bisphosphoglycerate is reoxidized to NAD^+ in the formation of lactate from pyruvate through the action of lactate dehydrogenases (LDHs). The LDH enzymes vary in their stereospecificity and can yield D- or L-lactic acid or the racemic mixture (DL).

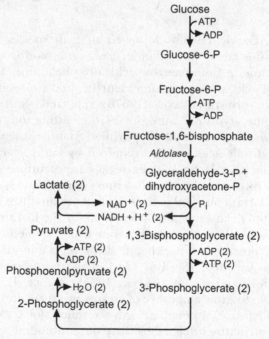

Figure 2.4.   Homofermentative pathway illustrating the production of lactic acid.

Though 1 mole of glucose should produce 2 moles of lactic acid, the actual yield is closer to 1.8 moles of lactic acid (Gottshalk, 1986). Energetically, glycolysis yields 2 moles ATP per mole glucose. The diagnostic enzyme present in those microorganisms that possess this pathway, aldolase, catalyzes the conversion of 1 mole of fructose-1,6-bisphosphate to 2 moles of glyceraldehyde-3-phosphate (Fig. 2.4). As was the case with *Saccharomyces*, these bacteria cannot metabolize pentoses. Examples of lactobacilli that are obligate homofermenters are *L. delbrueckii* and *L. jensenii*.

Obligate heterofermenters (e.g., *O. oeni*, *L. brevis*, *L. hilgardii*, *L. fructivorans*, and *L. kunkeei*) lack aldolase and must divert the flow of carbon through a different series of reactions, the pentose phosphate, or phosphoketolase, pathway (Fig. 2.5). From 1 mole of glucose, heterofermentative bacteria produce 1 mole each of lactate, CO_2, and either acetic acid or ethanol. In reality, these bacteria produce 0.8 mole lactate from glucose (Gottshalk, 1986). Unlike homofermentative microorganisms, these bacteria do not have aldolase but possess phosphoketolase, the

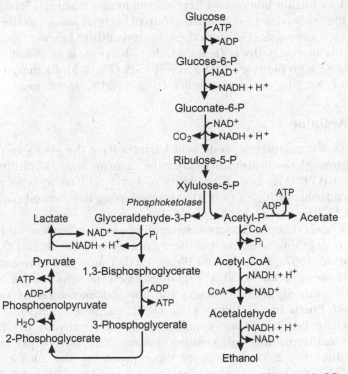

Figure 2.5. Heterofermentative pathway showing production of lactic acid, CO_2, and either ethanol or acetic acid.

enzyme responsible for the cleavage of xylulose-5-phosphate to form glyceraldehyde-3-phosphate and acetyl phosphate. Due to the biosynthesis of five-carbon sugars in this pathway (ribulose-5-phosphate and xylulose-5-phosphate), some strains can utilize the pentoses present in wine such as ribose, xylose, and arabinose. An important consequence of only half of the carbon from glucose going to glyceraldehyde-3-phosphate is formation of only 1 mole of ATP per mole glucose. However, heterofermentative bacteria can gain additional energy though conversion of acetyl-phosphate to acetate (Fig. 2.5).

From the winemaker's perspective, Fig. 2.5 highlights an important facet of successful management of these bacteria. Specifically, acetic acid production can result from conversion of both hexose and pentoses under even slight oxidative conditions. Under reductive conditions, cells experience a shortage of NAD^+ and so acetyl phosphate is converted to ethanol rather than acetate. Conversely, acetyl phosphate can be used to produce energy (ATP) under oxidative conditions in formation of acetate and increased volatile acidity (Section 11.3.1).

Besides obligate homo- and heterofermentative bacteria, Kandler and Weiss (1986) also described a third group of bacteria known as the facultative heterofermenters. Although these bacteria utilize hexoses through the homofermentative pathway (Fig. 2.4), they also possess an inducible phosphoketolase with pentoses acting as inducers (Fig. 2.5). Examples of wine bacteria belonging to this group are *L. casei* and *L. plantarum*.

2.4.2 Arginine

Many heterofermentative lactic acid bacteria have the ability to produce energy through the utilization of arginine in formation of ornithine, NH_3, CO_2, and ATP (Fig. 2.6). The ability of lactic acid bacteria to produce ammonia from arginine can be determined using the method outlined in Section 15.4.1.

It has been thought that most heterofermentative lactobacilli produce NH_3 from arginine, whereas homofermentative lactobacilli and *O. oeni* do not (Garvie, 1967a; Kandler and Weiss, 1986; Tonon and Lonvaud-Funel, 2002). However, Pilone et al. (1991) questioned the sensitivity of Nessler's reagent commonly used to detect the low concentrations of ammonia produced. Furthermore, Pilone et al. (1991) suggested that some heterofermentative lactobacilli are capable of carrying out only the first two biochemical steps, thus only yielding one molecule of NH_3 per molecule of arginine (Fig. 2.6). Because of these problems, the authors recommended that the concentration of the amino acid be increased from 0.3% to 0.6% w/v. Using an increased concentration of L-arginine, Pilone et al.

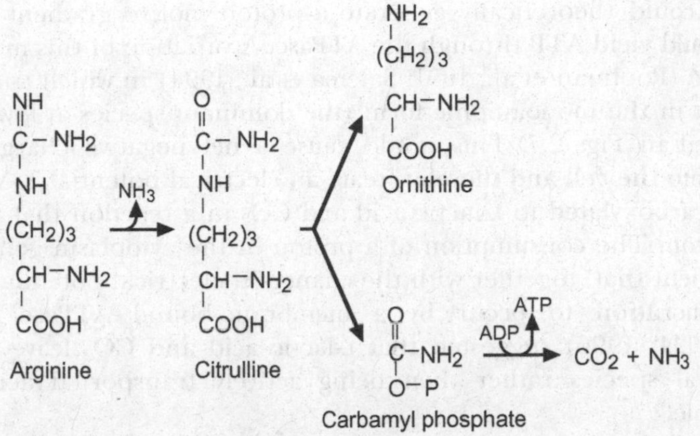

Figure 2.6. Formation of ornithine, ammonia, and carbon dioxide from arginine by some heterofermentative lactic acid bacteria.

(1991) found that some strains of *O. oeni* did, in fact, produce ammonia. In addition to problems related to concentration of arginine, Liu et al. (1995b) noted that fructose is inhibitory to arginine degradation by some bacterial strains.

2.4.3 Malate

Although malic acid stimulates the growth of *O. oeni* (Firme et al., 1994), the biochemical benefit of MLF to the bacterium has been a mystery because formation ATP or other direct energy could not be detected (Pilone and Kunkee, 1972). This prompted researchers to suggest that the reaction must serve a non-energy-yielding function (Kunkee, 1967b; Pilone and Kunkee, 1972). However, it became clear that MLF does, in fact, produce energy in an indirect means based on the chemiosmotic theory (Gottschalk, 1986), which holds that viable microorganisms maintain a pH gradient across cell membranes and it is this gradient that allows energy (ATP) to be produced. Under normal conditions, a higher concentration of H^+ exists outside compared with the interior of the cell. As a proton (H^+) travels through a membrane-associated enzyme complex (ATPase) following the concentration gradient from high to low concentration, this allows the bacterium to generate one molecule of ATP from ADP and inorganic phosphate (P_i). The model described requires that the membrane be impermeable to protons except at specific sites where the ATPase complex is located.

Cox and Henick-Kling (1989; 1995) were able to demonstrate that MLF yielded ATP and proposed that the ability of a cell to expel lactate and protons could theoretically generate a proton motive gradient that, in turn, would yield ATP through the ATPase. A variation of this model was proposed (Poolman et al., 1991; Salema et al., 1994) in which L-malate is taken up in the monoanionic form (the dominant species at low pH) as illustrated in Fig. 2.7. This would cause a net negative charge to be moved into the cell and thereby create an electrical potential. L-Malate is then decarboxylated to L-lactic acid and CO_2 in a reaction that requires one proton. The consumption of a proton in the cytoplasm generates a pH gradient that, together with the change in electrical potential, allows ATP generation to occur by a membrane-bound ATPase. Salema et al. (1994; 1996) suggested that L-lactic acid and CO_2 leave the cell as neutral species rather then being actively transported across the membrane.

2.4.4 Mannitol and Erythritol

As stated previously, many heterofermentative lactic acid bacteria gain additional energy by converting acetyl phosphate to acetate instead of ethanol. Although an additional ATP can be produced, the cell requires regeneration of NAD^+, a process achieved using an alternative electron acceptor, fructose (Wisselink et al., 2002). The reduction of fructose to mannitol by lactic acid bacteria catalyzed by mannitol dehydrogenase is shown in Fig. 2.8.

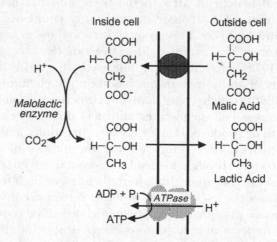

Figure 2.7.   Proposed model for energy generation (ATP) by *Oenococcus oeni* through conversion of malic acid to lactic acid and carbon dioxide. Modified from Poolman et al. (1991) with the kind permission of the *Journal of Bacteriology*.

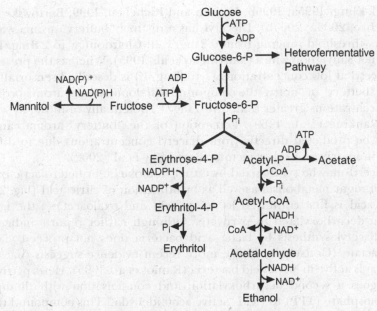

Figure 2.8. Formation of mannitol and erythritol. Adapted from Veiga-da-Cunha et al. (1993) with the kind permission of the *Journal of Bacteriology*.

Mannitol formation is used as a laboratory diagnostic test for the separation of heterofermentative from homofermentative bacteria (Section 15.4.10). Although primarily a property of heterofermentative lactic acid bacteria, a few homofermentative strains can also produce small amounts of the sugar alcohol (Wisselink et al., 2002).

Veiga-da-Cunha et al. (1993) observed that *O. oeni* produced another sugar alcohol, erythritol, anaerobically from glucose but not from fructose or ribose. In the presence of oxygen, synthesis of this sugar alcohol was absent. In agreement, Firme et al. (1994) reported erythritol production by this bacterium under N_2 or CO_2 environments. As with the formation of mannitol, synthesis of erythritol is probably related to the cell's need to reoxidize NADPH under anaerobic conditions.

2.4.5 Diacetyl and Other Odor/Flavor Compounds

One of the most important odor active compounds produced by lactic acid bacteria is 2,3-butandione, or diacetyl (Fornachon and Lloyd, 1965; Collins, 1972; El-Gendy et al., 1983; Rodriguez et al., 1990; Martineau and

Henick-Kling, 1995a; 1995b; Nielsen and Richelieu, 1999; Bartowsky and Henschke, 2004a; 2004b). Diacetyl has a distinct "buttery" aroma with a sensory threshold ranging from 0.2 mg/L in Chardonnay to 2.8 mg/L in Cabernet Sauvignon wines (Martineau et al., 1995). Whereas the presence of diacetyl at low concentrations (1 to 3 mg/L) is described sensorially as being "buttery" or "nutty," the compound will dominate the aroma profiles at concentrations greater than 5 to 7 mg/L, potentially resulting in spoilage (Rankine et al., 1969). Perception of the "buttery" aroma cannot always be predicted directly from diacetyl concentrations due to differences in matrix and other factors (Bartowsky et al., 2002).

Diacetyl may be synthesized by either homolactic or heterolactic pathways of sugar metabolism as well as by utilization of citric acid (Fig. 2.9). Citric acid is first converted to acetic acid and oxaloacetate; the latter is then decarboxylated to pyruvate. Although earlier reports indicated that diacetyl synthesis by lactic acid bacteria does not proceed via α-acetolactate (Gottschalk, 1986), more recent evidence suggests that this pathway is active in lactic acid bacteria (Ramos et al., 1995). Here, pyruvate undergoes a second decarboxylation and condensation with thiamine pyrophosphate (TPP) to yield "active acetaldehyde." This compound then reacts with another molecule of pyruvate to yield α-acetolactate, which, in

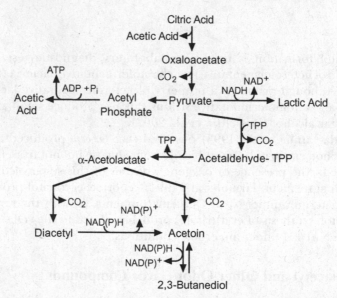

Figure 2.9.   Biochemical formation of diacetyl, acetoin, and 2,3-butanediol by lactic acid bacteria. Adapted from Ramos et al. (1995), Bartowsky and Henschke (2004b), and Ribereau-Gayon et al. (2000). TPP refers to thiamine pyrophosphate.

turn, undergoes oxidative decarboxylation to produce diacetyl (Ramos et al., 1995; Bartowsky and Henschke, 2004b). Diacetyl can be further transformed into acetoin as well as 2,3-butanediol. An alternate pathway involving the reaction of "active acetaldehyde" with acetyl CoA has been proposed, but the responsible enzyme, diacetyl synthetase, has never been isolated (Ribéreau-Gayon et al., 2000).

During growth, malic and citric acid utilization by lactic acid bacteria may occur concomitantly, although utilization of citric acid proceeds at a much slower rate (Pimentel et al., 1994). Thus, complete conversion of citric acid does not necessarily coincide with completion of MLF, and levels of citric acid remaining in the wine post-MLF may be sufficient to stimulate bacterial formation of diacetyl and acetic acid.

Microbial formation of diacetyl is a dynamic process, and concentrations in the wine depend on many factors including bacterial strain, wine type, and redox potential (Martineau and Henick-Kling, 1995a; 1995b; Nielsen and Richelieu, 1999). For instance, MLF during or just after alcoholic fermentation, when high populations of yeast are present, yields lower amounts of diacetyl due to rapid yeast reduction to acetoin and butylene glycol. By comparison, MLF occurring in low-density populations of viable yeast results in correspondingly higher concentrations of diacetyl. In general, diacetyl levels produced by *O. oeni* are relatively low compared with *Lactobacillus* or *Pediococcus*, which can synthesize objectionable concentrations. Prahl and Nielsen (1995) illustrated that the reversible reaction between diacetyl and SO_2 can result in rapid decreases in the concentration of diacetyl from 30% to 60%. Because the reaction is transitory, however, objectionable levels may return after several weeks of storage. Factors that impact the synthesis of diacetyl by lactic acid bacteria are summarized in Table 2.3.

Besides diacetyl, *O. oeni* can also synthesize higher alcohols and other compounds as by-products of their metabolism (Tracey and Britz, 1989; Edwards and Peterson, 1994; De Revel et al., 1999; Maicas et al., 1999; Delaquis et al., 2000). More recent evidence indicates that *O. oeni* possesses β-glucosidase activity (Section 1.5.4), an enzyme responsible for hydrolysis of monoglucosides, which can alter the sensory characteristics of a wine (Grimaldi et al., 2000; Boido et al., 2002; Mansfield et al., 2002; Ugliano et al., 2003; D'Incecco et al., 2004).

Osborne et al. (2000) reported that *O. oeni* can metabolize acetaldehyde producing ethanol and acetic acid. In some cases, this may be desirable because excess acetaldehyde may result in wine spoilage (Kotseridis and Baumes, 2000; Liu and Pilone, 2000). However, acetaldehyde also plays a role in the color development and stabilization of red wines (Timberlake and Bridle, 1976). More recently, Morneau and Mira de Orduna (2005)

Table 2.3.   Factors that affect diacetyl synthesis by lactic acid bacteria.

Factor	Impact
Bacterium	• Synthesis varies with genus, species and strain.
Inoculation rate	• Lower initial inoculums (10^4 vs. 10^6 CFU/mL) of bacteria favors synthesis.
Wine contact with yeast lees (*sur lies*)	• Conflicting studies where synthesis (Boulton et al., 1996) and degradation (Bartowsky and Henschke, 2004a) have been reported.
Wine contact with air	• Nonenzymatic reaction of α-acetolactate to diacetyl favored.
Addition of SO_2	• Binds diacetyl yielding sensory inactive adduct. • Inhibits bacteria
Addition of citric acid	• Favors synthesis (increases acetic acid too).
Temperature	• More diacetyl retained in wines undergoing MLF at 18°C than 25°C.
pH	• Lower pH retards growth of bacteria but favors synthesis.

CFU, colony-forming units; MLF, malolactic fermentation.
Adapted from Bartowsky and Henschke (2004a) and Boulton et al. (1996).

demonstrated that acetaldehyde degradation was strain as well as pH and SO_2 dependent.

Although *O. oeni* produces a variety of volatile compounds, the debate continues regarding the exact contribution of malolactic fermentation to the sensory properties of a wine. Early work by Kunkee et al. (1964) and Rankine (1972) indicated that except for its role in deacidification, MLF did not have a measurable influence on sensory properties of wine. On the other hand, more recent studies suggested differential changes in wine aroma and flavor (McDaniel et al., 1987; Laurent et al., 1994; Henick-Kling, 1995; Sauvageot and Vivier, 1997; Nielsen and Richelieu, 1999; Delaquis et al., 2000; Gambaro et al., 2001; Boido et al., 2002). For example, Sauvageot and Vivier (1997) noted that Chardonnay wines that completed MLF were perceived as being higher in "hazelnut," "fresh bread," and "dried fruit" aromas, whereas Pinot noir lost "strawberry" and "raspberry" sensory notes. Aside from influencing flavor and aroma, MLF may increase the body and mouthfeel, possibly due to the production of polyols such as glycerol and erythritol (Henick-Kling et al., 1994). Pripis-Nicolau et al. (2004) determined that lactic acid bacteria could metabolize methionine to produce 3-(methylsulphanyl)propionic acid. The authors described this compound as "chocolate" and "roasted" and theorized that this acid could contribute to the sensory complexity of wines post-MLF.

CHAPTER 3

ACETIC ACID BACTERIA

3.1 INTRODUCTION

Considered to be spoilage microorganisms in winemaking (Drysdale and Fleet, 1988; Du Toit and Pretorius, 2002), growth of acetic acid bacteria results in oxidation of ethanol to acetic acid (the process of acetification). In addition, other odor- and flavor-active metabolites as well as polysaccharides including dextrans and levans may be formed (Colvin et al., 1977; Tayama et al., 1986). The latter can create problems during post-fermentation clarification and stability.

3.2 TAXONOMY

Acetic acid bacteria are Gram-negative, aerobic, catalase-positive rods belonging to the family Acetobacteraceae (Holt et al., 1994; Ruiz et al., 2000; Du Toit and Pretorius, 2002). *Acetobacter* and *Gluconobacter* are described as being rod to ellipsoidal in shape, although there is often considerable microscopic variation among species as well as their respective strains (De·Ley et al., 1984). Even verified pure cultures may exhibit considerable morphological heterogeneity ranging from club-shaped, curved, to filamentous rods that may be observed as occurring singly, in pairs, or as short chains (Fig. 3.1).

According to Du Toit and Pretorius (2002), acetic acid bacteria are now divided into four genera: *Acetobacter, Acidomonas, Gluconobacter,* and *Gluconoacetobacter.* This differs from the last edition of *Bergey's Manual* where *Acetobacter* and *Gluconobacter* were the only recognized genera (De Ley and Swings, 1984; De Ley et al., 1984). *Gluconobacter oxydans, Acetobacter aceti, A. pasteurianus, Gluconacetobacter liquefaciens* (formerly *A. liquefaciens*), and *Gluconacetobacter hansenii* (formerly *A. hansenii*) have been isolated from grapes and wine (Vaughn, 1955; Joyeux et al., 1984a; Drysdale and Fleet, 1985; 1988; Du Toit and Lambrechts, 2002). Additional reorganization of this group with the inclusion of different species is probable given the application of phylogenetic studies (Cleenwerck et al., 2002). A taxonomic history of acetic acid bacteria can be found in Adams (1998).

Gluconobacter oxydans is generally found growing in sugar-rich environments where alcohol is either absent or present in low concentrations. Typical sites of isolation include flower parts and deteriorating fruit. By comparison, *Acetobacter* spp. are generally isolated from fermented sub-

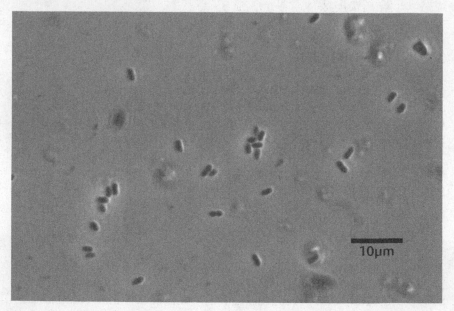

Figure 3.1. *Acetobacter aceti* as viewed with phase-contrast microscopy at a magnification of 1000×. Photograph provided with the kind permission of WineBugs LLC.

strates (wine, beer, etc.), although these microorganisms can also be found in decaying fruit undergoing fermentation.

3.3 NUTRITIONAL REQUIREMENTS

Nutritional requirements vary depending on the available carbon source and other variables. Some strains require *p*-aminobenzoic acid, niacin, thiamin, and/or pantothenic acid (De Ley and Swings, 1984; De Ley et al., 1984). Single amino acids cannot be used as sole sources of nitrogen and carbon. However, some strains of *Acetobacter* can grow without amino acids using NH_4^+ and ethanol as sole sources of nitrogen and carbon, respectively. De Ley and Swings (1984) and De Ley et al. (1984) noted that no "essential" amino acids are known for *Gluconobacter* or for *Acetobacter*.

Acetic acid bacteria normally require oxygen for growth (obligate aerobes). Because of this, limiting oxygen contact with a wine has served as a means to limit bacterial growth. However, there is evidence that acetic acid bacteria do not necessarily die with a lack of oxygen. In fact, *Acetobacter* may enter a "viable-but-not-culturable" state under oxygen deprivation (Section 6.5.2).

3.4 METABOLISM

3.4.1 Carbohydrates

Gluconobacter can utilize sugar alcohols (mannitol, sorbitol, or glycerol) or hexoses (glucose or fructose) as carbon sources. Acids are formed from propanol, butanol, glycerol, erythritol, mannitol, arabinose, ribose, fructose, galactose, mannose, and maltose by a majority of strains (De Ley and Swings, 1984).

Gluconobacter lacks phosphofructokinase, the enzyme responsible for catalyzing formation of fructose-1,6-diphosphate from fructose-6-phosphate in glycolysis (Fig. 1.7). Because this pathway is not operative, glucose is metabolized using the oxidative pentose phosphate pathway, also known as the hexose monophosphate shunt (Fig. 3.2). The conversion of ribulose-5-phosphate to fructose-6-phosphate and glyceraldehyde-3-phosphate involves a series of reactions catalyzed by transketolases and transaldolases that transfer carbonyl groups between molecules. The glyceraldehyde-3-phosphate formed is converted to acetate via pyruvate. Although the major pathway for sugar utilization, Olijve and Kok (1979) pointed out that this shunt is inhibited in *Gluconobacter* at pH 3.5 to 4.0.

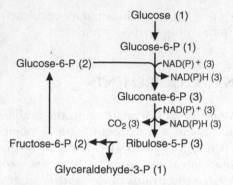

Figure 3.2.   Oxidative pentose phosphate pathway by which *Gluconobacter* utilizes glucose. NAD(P)$^+$ indicates that either NAD$^+$ or NADP$^+$ are used as coenzymes.

While these bacteria can grow within this pH range, their nutritional requirements are expanded.

Gluconobacter does not contain a functional tricarboxylic cycle (Fig. 1.9) and therefore cannot oxidize acetate or lactate to CO_2 and water (De Ley and Swings, 1984). However, these bacteria can directly oxidize glucose to form gluconate and, to a lesser degree, ketogluconates (Weenk et al., 1984; Seiskari et al., 1985). Depending on strain and environmental conditions, gluconic acid may accumulate, with concentrations approaching 30 g/L in grape musts inoculated with *G. oxydans* (Drysdale and Fleet, 1989a). Weenk et al. (1984) reported that utilization of gluconate may also be limited by environmental pH (>3.5) and glucose levels (<10 mM).

Sources of carbon for *Acetobacter* is species and strain dependent. Whereas all strains of *A. aceti* studied by De Ley et al. (1984) exhibited growth on ethanol, mannitol, acetate, and lactate, most, but not all (>50%), grew on propanol, fructose, and glucose. De Ley and Schell (1959) noted growth of *A. aceti* on a number of sugars, acids, and alcohols. By comparison, *A. pasteurianus* utilizes fewer carbon sources (ethanol, acetate, and lactate), whereas *A. liquefaciens* metabolizes more (ethanol, proposal, erythritol, ribitol, mannitol, sorbitol, galactose, fructose, glucose, gluconate, sucrose, maltose, acetate, glycerate, and lactate).

Like *Gluconobacter*, *Acetobacter* spp. metabolize sugars by the oxidative pentose phosphate pathway, but unlike *Gluconobacter*, these microorganisms do have an active Krebs cycle (Kitos et al., 1958). While glycolysis is either absent or very weak (De Ley et al., 1984), several species are capable of oxidizing polyols to corresponding ketoses. Both acetate and lactate can be oxidized to CO_2 and water and many strains form 2-keto-gluconic acid, 5-keto-gluconic acid, and/or 2,5-diketo-gluconic acid from glucose. The most preferred carbon sources for growth of *Acetobacter* are (in descending order) ethanol, glycerol, and lactate.

Growth of *Botrytis cinerea* on grapes, followed by dehydration and concentration of infected berries, can represent an important source of glycerol for acetic acid bacteria. Sponholz and Dittrich (1985) detected gluconic acid and keto-gluconates in *Botrytis*-infected grapes and believed the sources were acetic acid bacteria rather than mold. In the case of mold-infected fruit, glycerol concentrations have been reported to be over 15 g/L (Dittrich et al., 1974; 1975; Nieuwoudt et al., 2002). Glycerol is thought to contribute to the perception of sweetness and body (viscosity) at concentrations greater than the reported threshold of 4 to 5 g/L (Noble and Bursick, 1984), although this sensory impact has more recently been questioned (Nieuwoudt et al., 2002).

The conversion of glycerol to dihydroxyacetone is known as "ketogenesis." As ketogenesis requires oxygen and is inhibited at alcohol levels >5% v/v (Yamada et al., 1979; Aldercreutz, 1986), the presence of substantial amounts of dihydroxyacetone in wine likely reflects "carry-over" from infected musts rather than formation during alcoholic fermentation. Du Toit and Pretorius (2002) reviewed other research where very high concentrations, upwards of 2500 mg/L, were reported. Ketogenesis may have an important impact on sensory properties of the wine due to the loss of glycerol and the formation of dihydroxyacetone (Yamada et al., 1979; Drysdale and Fleet, 1989a; 1989b; Nieuwoudt et al., 2002). Containing a carbonyl functional group, dihydroxyacetone can also bind to SO_2 so that grapes infected with *Gluconobacter* have a higher requirement for sulfites (Swings, 1992). Finally, dihydroxyacetone can also react with the amino acid proline to yield a sensory-active compound described as being "crust-like" (Margalith, 1981).

3.4.2 Ethanol

Whereas *Acetobacter* produces limited amounts of acetic acid through carbohydrate metabolism, much more of the acid is synthesized through the oxidation of ethanol (Eschenbruch and Dittrich, 1986). Two membrane-bound enzymes, an alcohol dehydrogenase and an aldehyde dehydrogenase, are involved in this conversion (Saeki et al., 1997). Alcohol dehydrogenase oxidizes ethanol to acetaldehyde, which is further oxidized to acetic acid by the aldehyde dehydrogenase as follows:

$$CH_3CH_2OH + PQQ \rightarrow CH_3CHO + PQQH_2 \text{ (alcohol dehydrogenase)}$$

$$CH_3CHO + PQQ + H_2O \rightarrow CH_3COOH + PQQH_2 \text{ (aldehyde dehydrogenase)}$$

Pyrroloquinoline (PQQ) is part of the membrane-bound dehydrogenases. Unlike many other microorganisms that use NAD$^+$ as a coenzyme, *Acetobacter* uses PQQ as the preferred hydrogen-acceptor that transfers electrons generated from these reactions. Electrons are initially transferred to ubiquinone (Fig. 3.3), which will be re-oxidized by a membrane-associated oxidase. Eventually, oxygen is the final electron acceptor, resulting in formation of H_2O and a proton motive force necessary for energy production through a membrane-bound ATPase. As such, acetic acid bacteria are thought to have absolute requirements for oxygen and, hence, are described as obligate aerobes.

Although most strains of *Gluconobacter* can oxidize ethanol at low concentrations (<5% v/v), the microorganism cannot survive in the alcoholic environment of wine even when aerated (Drysdale and Fleet, 1989b). Conversely, *Acetobacter* can grow and spoil wine at much higher alcohol concentrations. Drysdale and Fleet (1989b) and others have demonstrated growth of *Acetobacter* in wines containing greater than 10% v/v ethanol.

Acetic acid bacteria can produce large quantities of acetaldehyde, approaching 250 mg/L (Du Toit and Pretorius, 2002). Acetaldehyde has a sensory threshold of 100 to 120 mg/L (Berg et al., 1955) and therefore can be a sensory defect in wines. Acetaldehyde has been described as being "nutty," "sherry-like," "bruised apple," "green," "grassy," or even "vegetative" (Zoecklein et al., 1995; Kotseridis and Baumes, 2000; Liu and Pilone, 2000).

Under conditions of low oxygen as would occur in barreled wine, *Acetobacter* will produce acetaldehyde rather than acetic acid (Drysdale and Fleet, 1989b). As an explanation, the authors speculated that while the alcohol dehydrogenase is active under low oxygen, the activity of acetaldehyde dehydrogenase could be impeded (Fig. 3.3). Besides the sensory

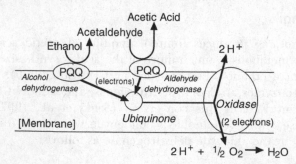

Figure 3.3. Oxidation of ethanol with the formation of acetic acid by acetic acid bacteria. Adapted from Adams (1998) with the kind permission of Springer Science and Business Media.

impact, acetaldehyde rapidly binds with SO_2 to form an addition compound of limited volatility and odor (Section 5.2.1). As such, it may be possible to mitigate objectionable properties of acetaldehyde through careful additions of sulfites.

Besides acetic acid and acetaldehyde, *Acetobacter* also produces ethyl acetate. This pungent ester contributes to the sensory interpretation of "volatile acidity" (Section 11.3.1). However, synthesis is affected by available oxygen resulting in less ethyl acetate being produced under low oxygen conditions (Drysdale and Fleet, 1989b).

CHAPTER 4

MOLDS AND OTHER MICROORGANISMS

4.1 INTRODUCTION

Molds are filamentous fungi that are classified based on the morphology of asexual or vegetative mycelial elements and their spore structures. A typical vegetative structure of molds consists of individual hyphal elements, collectively called the mycelium. Hyphae are of three types: (a) penetrative hyphae called rhizoids, which serve to enter the substrate to glean and transport nutrients; (b) stolons, which have a larger diameter than rhizoids and serve to link the mycelial mass, and (c) aerial asexual reproductive hyphae, also known as conidiophores or sporangiophores. Depending on the mold, asexual spores may be produced within an enclosed structure (the sporangium) or appear exposed at the tips of vesicles. Spores serve an important role in mold dispersal, being carried by air currents and foraging insects. Under appropriate conditions of humidity and temperature, spores germinate to yield the vegetative body (mycelium), which can continue to alternatively propagate and sporulate.

As an example, *Aspergillus* produces asexual spores (conidiospores or conidia) on a structure called an aspergillum, which consists of a swollen

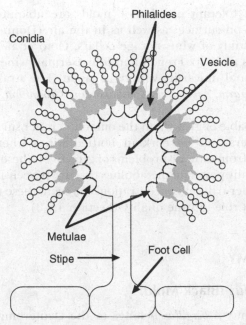

Figure 4.1. Basic structure of *Aspergillus*. Adapted from Chang et al. (2000) with the kind permission of Elsevier Ltd.

vesicle and either one or two layers of specialized cells (phialides with or without metulae). A typical structure is illustrated in Fig. 4.1. The aspergillum, stipe, and foot cell together are called the conidiophore.

4.2 ECOLOGY HABITATS

Molds are ubiquitous with various genera commonly found on grapes. Common examples include *Aspergillus*, *Botrytis*, and *Penicillium*, and, to a lesser extent, *Phythophthora*, *Moniliella*, *Alternaria*, and *Cladosporium* (Rosa et al., 2002). Mold growth plays an important role in the physical and chemical stability as well as the sensory properties of the future wine. For example, uncontrolled proliferation of mold on grapes just prior to harvest rapidly leads to growth of secondary contaminants (yeasts and bacteria), which, in turn, leads to a deteriorative state called "rot." Recognizing the importance of mold growth to wine quality, grape contracts generally include specification for the extent of infections. Depending on the extent to which mold and rot is present, the winemaking staff may find it necessary to modify their processing protocol (Section 7.5).

Although not tolerant of ethanol, molds are ubiquitous in wineries and are present on surfaces as well as in the air (Donnelly, 1977). In a comprehensive study of wine storage cellars, Goto et al. (1989) isolated 108 mold strains from six French and four German wineries in addition to an experimental university winery in Japan. The strains belonged to six genera, *Alternaria*, *Aspergillus*, *Cladosporium*, *Geosmithia*, *Penicillium*, and *Verticillium*.

Molds are capable of growth on the outer and inner surfaces of wooden storage containers and on cork in bottled wines where seepage has occurred. Aside from esthetic problems of growth on these surfaces, molds produce sensorially powerful metabolites that are perceivable at parts-per-billion or parts-per-trillion concentrations. As such, these compounds can play a significant role in wine quality (Section 4.5.3).

4.3 TAXONOMY

4.3.1 *Aspergillus* (Black Mold)

The asexual genus *Aspergillus* belongs to the Deuteromycetes (formerly *Fungi imperfecti*). Initial growth of *Aspergillus* appears as a white mycelium similar to *Penicillium*, but as conidia mature, the colony develops a black coloration. One species, *A. niger*, plays an important role in bunch rot, particularly in warm climates, which led to its description as the "hot weather mold." *Aspergillus* along with a second mold, *Rhizopus*, and acetic acid bacteria are also associated with bunch or sour rot.

4.3.2 *Botrytis* (Gray Mold)

Gray mold is caused by *B. cinerea* and can affect a wide range of plants as well as stored fruits and vegetables. The disease is encouraged by cool, damp conditions and is characterized by a gray, "fluffy," surface mold overlaying a soft, commonly brown rot (Alur, 2000). Early in the growth cycle, colonies appear white, changing to gray as conidia form and mature. *Botrytis cinerea* growing on a grape is illustrated in Fig. 4.2. Although in most cases, growth of the mold decreases grape quality, infections are encouraged, in some instances, to produce high-quality dessert wines (so-called noble rot).

Upon germination, hyphal elements of *Botrytis* penetrate berries through microscopic cracks and other lesions (Donèche, 1993). Growth of the fungus cracks and loosens the grape skin such that when the berry is gently squeezed, the skin separates easily from pulp. This observation

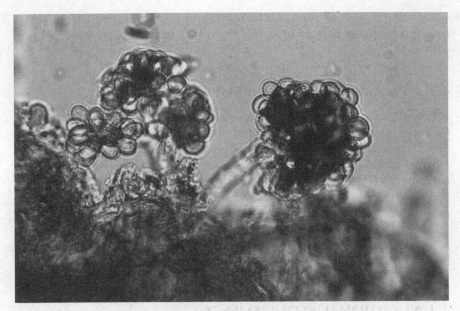

Figure 4.2. *Botrytis cinerea* as viewed with brightfield microscopy at a magnification of 400×. Photograph provided with the kind permission of WineBugs LLC.

resulted in *Botrytis* earning the nickname of "slip-skin" mold. Another visual manifestation of an early stage of *Botrytis* development is browning of the berries, a "chocolate" hue known as *pourri plein* (Donèche, 1993).

Where conditions within and surrounding the cluster remain wet and the relative humidity high (>90%) after infection, the cycle from germination to sporulation may take less than 3 days during and after which the disease may follow two (often overlapping) courses. Optimal temperature and relative humidity for infection lies between 15°C/59°F and 20°C/68°F and >90% RH (Bulit and Lafon, 1970). Under these conditions, growth of *Botrytis* as well as native fungi (*Penicillium, Aspergillus, Cladosporium,* and *Rhizopus*) and acetic acid bacteria rapidly ensues, yielding "bunch rot" (Jackson, 2000). Although originally believed to arise only from early fall rains, other factors such as fog, dew, and irrigation practices coupled to microclimates that maintain relatively a high humidity during the growing season may also promote the infection.

Where warm, sunny, and windy weather follows the primary infection by *Botrytis,* free water in the cluster evaporates, and the fruit begins a process of dehydration, yielding the so-called noble rot (Donèche, 1993). Although mold and bacterial growth continues to consume a portion of the sugars and acids, this is countered by a concentration of flavors and sugars due to dehydration. Fermentation of musts prepared from these infected grapes yields what many regard as high-quality sweet (white) wines.

4.3.3 *Penicillium* (Blue-Green Mold)

Penicillium is a diverse genus, with more than 200 recognized species. Colonies normally grow well and sporulate on a number of laboratory media, producing gray-green or gray-blue colors (Pitt, 2000). *P. expansum* (formerly, *P. glaucum*) is frequently isolated and is also known as the "apple rot fungus" (Pitt, 2000). This species spoils apples, pears, tomatoes, avocados, mangoes, and grapes and is the major source of the mycotoxin patulin in fruit juices (Section 4.5.2).

Most *Penicillium* species grow well over lower temperature ranges, with nearly all capable of growth below 5°C/41°F. Its growth at low temperature has led to *Penicillium* being referred to as the "cold weather mold."

4.4 NUTRITIONAL REQUIREMENTS

Like other microorganisms, molds have requirements for various nutrients including sulfur, phosphorus, magnesium, and potassium as well as trace amounts of iron, copper, calcium, manganese, zinc, and molybdenum (Carlile et al., 2001). Most can grow over a wide pH range (2 to 8.5) and produce various hydrolytic enzymes such as amylases, pectinases, proteinases, and lipases. Because of this, most species can utilize a wide range of carbon and nitrogen sources. In fact, *A. niger* has the ability to grow only on glucose as a sole organic source (Carlile et al., 2001). Though many can grow under conditions of low water activity, *Botrytis* need a water activity (a_w) of 0.93 for spore germination (Alur, 2000).

Molds generally require high concentrations of oxygen. Reflecting their oxidative requirements, mold growth in stored juice and concentrate is generally restricted to surface contamination. However, *P. expansum* and *P. roqueforti* are able to grow at oxygen levels as low as 2% (Pitt, 2000).

4.5 METABOLISM

4.5.1 Glucose

Aspergillus and *Penicillium* utilize glucose through the Embden–Meyerhof–Parnas pathway (Fig. 1.7), although some species possess a limited hexose monophosphate shunt or oxidative pentose phosphate pathway (Fig. 3.2). As pointed out by Donèche (1993), young mycelia of *B. cinerea* will also exhibit oxidation of glucose by the Entner–Doudoroff pathway (Fig. 4.3). Glyceraldehyde-3-phosphate formed from this reaction can be oxidized to pyruvate by enzymes of the Embden–Meyerhof–Parnas pathway. As molds are obligate aerobes, the tricarboxylic acid, or Krebs, cycle (Fig. 1.9) is active with NADH being reoxidized to NAD^+ through oxidative phosphorylation. Although several organic acids are synthesized through the Krebs cycle, namely citric, itaconic, and malic, *Aspergillus* also produces significant amounts of gluconic acid from the oxidation of glucose by glucose oxidase (Chang et al., 2000). In oxygen-poor environments, glycerol accumulates due to the activity of glycerol dehydrogenase, an enzyme that allows for regeneration of NAD^+ (Fig. 1.8).

4.5.2 Mycotoxins

A major concern with the growth of certain molds on grapes is the production of mycotoxins such as ochratoxin A and patulin (Scott et al., 1977; Battilani and Pietri, 2002; Cabañes et al., 2002; Delage et al., 2003). Patulin causes gastrointestinal problems and skin rashes, and ochratoxin A is a potent nephrotoxin and carcinogen. Molds known to produce mycotoxins when grown on grapes include *Aspergillus* and *Penicillium*. Recently, Rosa et al. (2002) isolated 101 strains of *A. niger*, of which 24 could produce ochratoxin A. Similarly, Moeller et al. (1997) and Abrunhosa et al. (2001) found that *Penicillium* spp. could produce one or more mycotoxins (isofumigaclavine A, isofumigaclavine B, festuclavine, patulin, roquefortine C, and PR toxins) when inoculated into grape must. Finally, Esteban et al. (2004) noted that strains of *Aspergillus* could produce ochratoxin A over a wide range of temperatures.

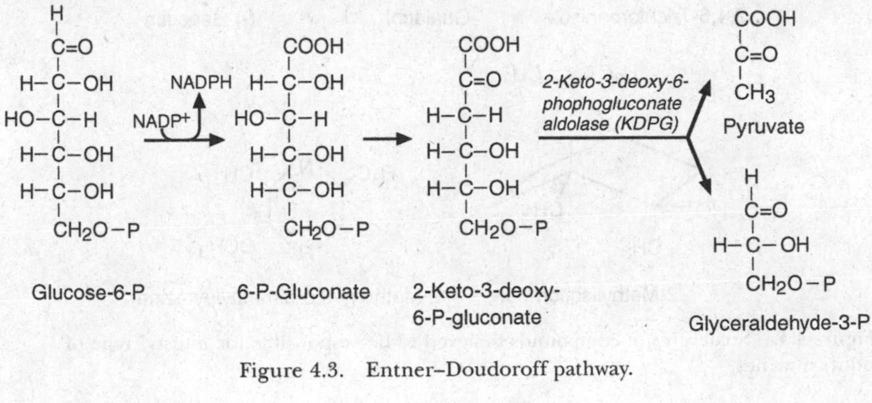

Figure 4.3.   Entner–Doudoroff pathway.

Ochratoxin A has been found in grapes and wines from Argentina and Brazil (Rosa et al., 2002), Canada (Ng et al., 2004), France (Sage et al., 2004), Italy (Battilani et al., 2004), Portugal (Serra et al., 2003), South Africa (Stander and Steyn, 2002), Spain (Bellí et al., 2004; Blesa et al., 2004), and the United States (Siantar et al., 2003). Stander and Steyn (2002) noted higher concentrations in a few late harvest wines compared with other wines analyzed, probably due to some *Aspergillus* and/or *Penicillium* growth prior to harvest. Beginning January 1, 2005, the European Union established a maximum concentration of 2.0 µg/L on all domestic and imported wines.

4.5.3 Odor/Flavor Compounds

Molds synthesize numerous compounds that impart odors and flavors to wines. Kaminski et al. (1974) reported that *Aspergillus* and *Penicillium* produced 3-methylbutanol, 3-octanone, 3-octanol, 1-octen-3-ol, 1-octanol, and 2-octen-1-ol along with a number of higher alcohols (isobutyl alcohol and phenethyl alcohol), aldehydes (including benzaldehyde), and ketones. Among these odors, the authors determined that 1-octen-3-ol, described as being "musty" or "mushroom," was the most predominant and represented from 36.6% to 93.1% of the total volatiles produced. Other compounds possessing "musty" or "moldy" sensory attributes are also synthesized (Fig. 4.4).

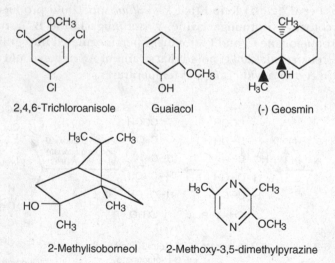

2,4,6-Trichloroanisole   Guaiacol   (-) Geosmin

2-Methylisoborneol   2-Methoxy-3,5-dimethylpyrazine

Figure 4.4.   Structures of compounds believed to be responsible for "musty" type of odors in wines.

B. cinerea metabolizes aromatic terpenes including linalool, geraniol, and nerol in the formation less volatile compounds such as β-pinene, α-terpineol, and other oxides (Nigam, 2000). In addition, the mold produces esterases that degrade the esters, compounds that give many white wines their fruity character, and can also synthesize sotolon ("honey-like") and 1-octeno-3-ol (Nigam, 2000).

Molds also produce large amounts of glycerol, a compound that can be utilized as a carbon source for acetic acid bacteria (Section 3.4.1). Ravji et al. (1988) examined glycerol production by four common vineyard molds involved in rot, *Botrytis cinerea* (gray rot), *Aspergillus niger* (black-mold rot), *Penicillium italicum* (blue-mold rot), and *Rhizopus nigricans* (Rhizopus rot) and concluded that glycerol production per unit of mycelial weight varied not only with the individual mold but also with grape variety and method of juice extraction.

4.6 OTHER MICROORGANISMS

4.6.1 *Bacillus*

The genus *Bacillus* is composed of Gram-positive (sometimes Gram-variable), aerobic, catalase-positive, endospore-producing species. Although these bacteria are common soil-borne, the occurrence of *Bacillus* sp. growing in alcoholic beverages has been observed infrequently. Early research by Gini and Vaughn (1962) reportedly found *B. subtilis, B. circulans,* and *B. coagulans* in spoiled wines, and Murrell and Rankine (1979) isolated a strain of *B. megaterium* from brandy. More recent reports of isolation from wines of eastern European origin include that of Bisson and Kunkee (1991). In these cases, Kunkee (1996) noted that growth of *Bacillus* in bottled wine did not affect palate or olfactory properties and, thus, the defect was limited to sediment or haze formation. Besides wines, *B. subtilis* has also been found on cork (Álvarez-Rodríguez et al., 2003).

B. thuringiensis, a bacterium used as a biological insecticide on grapes, was subsequently isolated from fermenting grape juice in a commercial winery (Bae et al., 2004). Although the bacterium could apparently remain viable in wine, growth and multiplication was inhibited.

4.6.2 *Clostridium*

Due to the normally acidic nature of grape wines (pH < 4.0), the presence of *Clostridium* has rarely been reported. According to Sponholz (1993), growth is restricted to low-acid, high-pH (>4.0) grape and apple juices and wines. Growth of these bacteria imparts a "rancidness" in wines due to synthesis of *n*-butyric acid.

4.6.3 *Streptomyces*

Streptomyces has been implicated in the formation of "musty" compounds in corks as well as in the winery (Silva Pereira et al., 2000). One causative molecule is 2,4,6-trichloroanisole (Buser et al., 1982). Recently, Mara and Bisson (2005) characterized more than 150 isolates of *Streptomyces* from various winery environments. Of these isolates, 12 were able to grow well on trichlorophenol, with synthesis of trichloroanisole. The authors concluded that differential occurrence of species or strains may account for environmental trichloroanisole in one winery but not another.

Besides trichloroanisole, other metabolites such as guaiacol, geosmin, 2-methylisoborneol, and 2-methoxy-3,5-dimethylpyrazine (Fig. 4.4) can also cause "musty" or "moldy" aromas (Silva Pereira et al., 2000; Simpson et al., 2004). Moreover, Silva Pereira et al. (2000) pointed out that cork taints can originate from other microorganisms. As an example, a putrid odor is produced by *Armillaria mellea*, a microorganism that infects oak trees and causes "yellow spot" on corks.

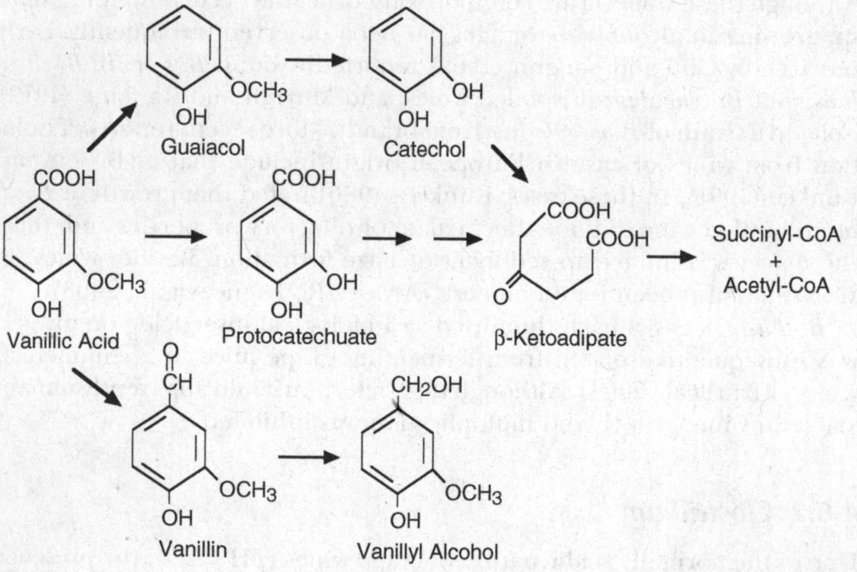

Figure 4.5. Formation of guaiacol, catechol, and other compounds from vanillic acid by *Streptomyces* isolated from corks. Adapted from Álvarez-Rodríguez et al. (2003).

Guaiacol is another compound found in wines produced by some molds or bacteria when grown on cork. Here, *Streptomyces* is capable of degrading vanillic acid to produce guaiacol as well as other compounds (Fig. 4.5). Earlier work by Simpson et al. (1986) suggested that tainted wines contained 0.07 to 2.63 mg/L guaiacol, whereas unspoiled wines from the same bottling had much lower concentrations, 0.003 to 0.006 mg/L.

Other sensory-active metabolites produced by *Streptomyces* and some Cyanobacteria and molds are geosmin (*trans*-1,10-dimethyl-*trans*-9-decalol) and 2-methylisoborneol. Geosmin has an odor reminiscent of "cooked beets," "earthy," or "freshly tilled soil" and 2-methylisoborneol has been similarly described as being "earthy" or "musty." Darriet et al. (2001) detected geosmin present in a "musty" Cabernet Sauvignon wine, specifically the (–) isomer, which has a much lower sensory threshold than (+) geosmin. The authors further determined that both *Streptomyces* and *Penicillium* spp. synthesize (–) geosmin. As geosmin has been found in freshly crushed grapes (Darriet et al., 2000), microorganisms that develop on grapes could therefore potentially yield "musty" wines.

SECTION II

VINIFICATION AND WINERY PROCESSING

CHAPTER 5

MANAGING MICROBIAL GROWTH

5.1 INTRODUCTION

Controlling the growth of microorganisms at critical junctures during the winemaking process is vital to success. Important concerns include not only maximizing the fermentative performance of *Saccharomyces* but also managing the growth of undesirable yeasts and bacteria. For example, whereas some winemakers may value limited activity of *Brettanomyces* in their red wines, massive infections should be prevented. Control of microbiological activity involves the use of chemical additives (preservatives and sterilants) as well as physical removal by filtration or other methods and control of cellar/tank temperatures.

5.2 PRESERVATIVES AND STERILANTS

Whereas chemical preservatives may inhibit the growth of microorganisms without necessarily killing (e.g., they are bacteriostatic or fungistatic), use of sterilants, when properly administered, kills the entire (100%) target population (e.g., they are bacteriocidal or fungicidal). Preservatives such as SO_2 should not be expected to be a *carte blanche* remedy for controlling problematic microorganisms during vinification. Though preservatives are used individually, combinations can often result in an enhanced effect in that some pairings can act synergistically (i.e., the combined effect of two preservatives greatly surpasses the individual effect of each).

Presented here are general descriptions of preservatives and sterilants as well as filtration options that can be used at different stages during the winemaking process. A variety of additives may be used to control the growth of microorganisms. These include sulfur dioxide, lysozyme, dimethyl dicarbonate, and sorbic acid. The use of other additives such as fumaric acid, nisin, carbon monoxide, and particular plant extracts has also been explored. Additional antimicrobial chemicals that are used in a winery as sanitizers are discussed in Chapter 9.

5.2.1 Sulfur Dioxide

Sulfur dioxide is widely recognized in both the wine and food industries for its antioxidative and antimicrobial properties. The current legal limit for SO_2 in wines in the United States is 350 mg/L, a concentration well above levels normally used by winemakers. Nevertheless, wines that contain greater than 10 mg/L must disclose this information on the label. Because SO_2 is a metabolite of yeasts during fermentation (Section 1.5.2), wines will usually contain some sulfite even though additions were not made during processing.

5.2.1.1 Forms of Sulfur Dioxide

Once dissolved in water, sulfur dioxide exists in equilibrium between molecular SO_2 ($SO_2 \bullet H_2O$), bisulfite (HSO_3^-), and sulfite (SO_3^{2-}) species as illustrated below:

$$SO_2 + H_2O \leftrightarrows SO_2 \bullet H_2O$$

$$SO_2 \bullet H_2O \leftrightarrows HSO_3^- + H^+$$

$$HSO_3^- \leftrightarrows SO_3^{2-} + H^+$$

This equilibrium is dependent on pH, with the dominant species at wine pH (3 to 4) being the bisulfite anion (Fig. 5.1). Besides being in equilibrium with the molecular and sulfite species, bisulfite also exists in "free" and "bound" forms. Here, the molecule will react with carbonyl compounds (e.g., acetaldehyde), forming addition products or adducts such as hydroxysulfonic acids.

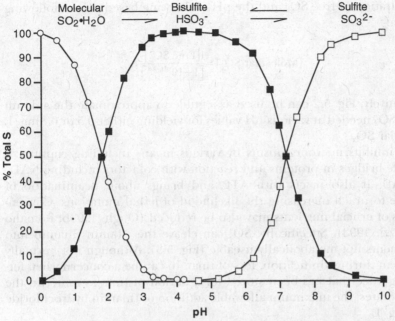

Figure 5.1. Relative abundance of molecular SO_2, bisulfite, and sulfite at different pH values.

In addition to acetaldehyde, molecules present in grape juice that react with bisulfite are pyruvic acid, α-keto-glutaric acid, dihydroxyacetone, diacetyl, anthocyanin pigments, and others (Ough, 1993b).

$$HSO_3^- + R - \overset{\overset{\displaystyle O}{\|}}{C} - H \rightleftharpoons R - \overset{\overset{\displaystyle OH}{|}}{\underset{\underset{\displaystyle H}{|}}{C}} - SO_3^-$$

5.2.1.2 *Microbial Inhibition*

It is generally believed that the molecular sulfur species is the antimicrobial form of sulfur dioxide. Because $SO_2 \cdot H_2O$ does not have a charge, the molecule enters the cell and undergoes rapid pH-driven dissociation at cytoplasmic pH (generally near 6.5) to yield bisulfite and sulfite. As the intracellular concentration of molecular SO_2 decreases due the internal equilibrium, more molecular SO_2 enters the cell, further increasing intracellular concentrations.

The amount of molecular SO_2 present in any wine is not normally measured directly. Rather, the concentration is calculated knowing the concentration of free SO_2 and the pH of the wine based on the following formula:

$$[\text{Molecular SO}_2] = \frac{[\text{Free SO}_2]}{\left[1 + 10^{\text{pH}-1.8}\right]}$$

Alternatively, Fig. 5.2 can be used as a guide to approximate the amount of free SO_2 needed at various pH values for yielding either 0.5 or 0.8 mg/L molecular SO_2.

SO_2 inhibits microorganisms by various means including rupture of disulfide bridges in proteins and reaction with cofactors including NAD^+ and FAD. It also reacts with ATP, and brings about deamination of cytosine to uracil increasing the likelihood of lethal mutations. Concentrations of crucial nutrients may also be reduced (Ough, 1993b; Romano and Suzzi, 1993). Specifically, SO_2 can cleave the vitamin thiamin into components not metabolically useable (Fig. 5.3). Although not normally a problem during vinification, loss of thiamin can be a concern when fermenting juice that has been stored for a period of time as muté. In the United States, the maximum allowable addition of thiamin hydrochloride is 0.6 mg/L.

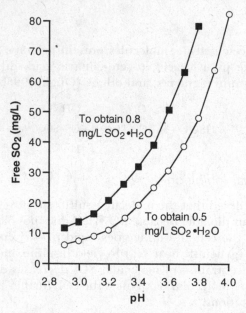

Figure 5.2.   Amounts of free SO_2 at given pH to yield either 0.5 or 0.8 mg/L molecular SO_2.

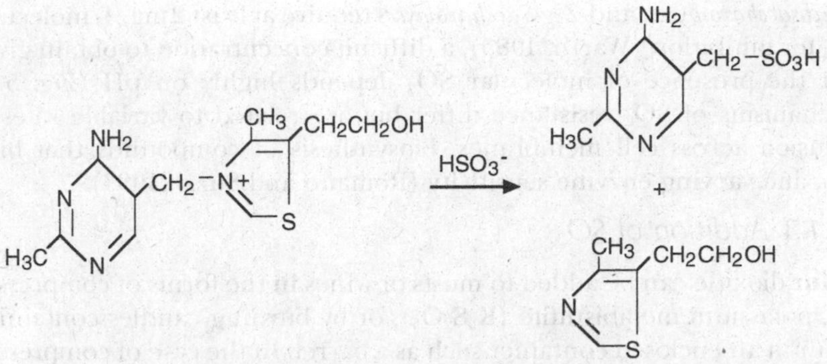

Figure 5.3. Breakdown of thiamin by SO_2.

The concentration of molecular SO_2 needed to prevent growth of microorganisms varies with wine/juice pH, temperature, population density and diversity, stage of growth, alcohol level, and other factors. The frequently cited addition level of 0.8 mg/L molecular SO_2 was suggested by Beech et al. (1979) as the amount needed in white table wines to bring about a 10^4 CFU/mL reduction in 24 h in populations of several spoilage microorganisms. Differences in sensitivity to SO_2 between genera of yeasts and bacteria found in wines are known to exist (Warth, 1977; 1985; Du Toit et al., 2005). For example, work by Davis et al. (1988) with lactic acid bacteria isolated from Australia red wines indicated that strains of *L. oenos* (*O. oeni*) were less tolerant to sulfur dioxide than strains of *P. parvulus*. Davis et al. (1988) further suggested that wines with high total SO_2 concentration may be more likely to support the growth of *Pediococcus* than *L. oenos*. In contrast, Hood (1983) reported that pediococci were less tolerant to bound SO_2 than lactobacilli or leuconostocs. In practice, many winemakers attempt to maintain 0.4 to 0.6 mg/L molecular SO_2 to control *Brettanomyces* and other spoilage microorganisms during wine aging. Although Du Toit et al. (2005) suggested that a concentration of 0.8 mg/L molecular SO_2 was needed to prevent the growth of *Acetobacter pasteurianus*, the authors further noted that this concentration did not completely eliminate the bacterium.

Whereas some winemakers encourage non-*Saccharomyces* yeasts (Section 8.4), others desire to limit their growth due to synthesis of undesirable odors and flavors. Although SO_2 can suppress the non-*Saccharomyces* populations prior to alcoholic fermentation (Constantí et al., 1998; Egli et al., 1998; Henick-Kling et al., 1998; Ciani and Pepe, 2002; Cocolin and Mills, 2003), its use at the crusher may not inhibit yeast species that have greater resistance to the additive. In fact, some species of *Pichia*, *Saccharomycodes*,

Schizosaccharomyces, and *Zygosaccharomyces* require at least 2 mg/L molecular SO_2 for inhibition (Warth, 1985), a difficult concentration to obtain given that the presence of molecular SO_2 depends highly on pH (Fig. 5.1). Mechanisms of SO_2 resistance differ but are related to variable rates of diffusion across cell membranes, biosynthesis of compounds that bind SO_2, and varying enzyme sensitivity (Romano and Suzzi, 1993).

5.2.1.3 Addition of SO_2

Sulfur dioxide can be added to musts or wines in the forms of compressed gas, potassium metabisulfite ($K_2S_2O_5$), or by burning candles containing sulfur in an enclosed container such as a barrel. In the case of compressed SO_2, the amount required for a 30 mg/L addition using a gas of 90% purity can be calculated:

$$\frac{30 \text{ mg SO}_2}{L} \times \frac{1 \text{ g}}{1000 \text{ mg}} \times \frac{1 \text{ lb}}{454 \text{ g}} \times \frac{3.8 \text{ L}}{1 \text{ gallon}} \times \frac{1}{0.90} \times 1000 \text{ gallons}$$

$$= 0.279 \text{ lb SO}_2 \text{ per } 1000 \text{ gallons}$$

Most commonly, sulfur dioxide is incorporated into grape must or wine as potassium metabisulfite. Theoretically, 2 moles of SO_2 can be derived from each mole of $K_2S_2O_5$. Therefore, the theoretical yield of SO_2 from potassium metabisulfite would be a ratio of molecular weights for each species:

$$\frac{2 \text{ SO}_2}{K_2S_2O_5} = \frac{S_2O_4}{K_2S_2O_5} = \frac{128.12}{222.32} = 0.576$$

From the above equation, 57.6% of $K_2S_2O_5$ is theoretically available to form SO_2. In practice, this figure is high given that $K_2S_2O_5$ decomposes upon prolonged storage.

Knowing the relative proportion of SO_2 in $K_2S_2O_5$, the amount required for any addition can be calculated in either g/L or lb/1000 gallons. Thus, the weight of $K_2S_2O_5$ required to yield a 30 mg/L SO_2 addition can be calculated:

$$\frac{30 \text{ mg SO}_2}{L} \times \frac{1 \text{ g}}{1000 \text{ mg}} \times \frac{1 \text{ lb}}{454 \text{ g}} \times \frac{3.8 \text{ L}}{1 \text{ gallon}} \times \frac{1}{0.576} \times 1000 \text{ gallons}$$

$$= 0.436 \text{ lb K}_2S_2O_5 \text{ per } 1000 \text{ gallons}$$

5.2.2 Dimethyl Dicarbonate

Sold under the trade name Velcorin™, the sterilant dimethyl dicarbonate (DMDC) is approved for use in the United States for table as well as low-alcohol and dealcoholized wines at a maximum concentration of 200 mg/L over the life of the wine (Ough, 1993a). The United States Food and Drug Administration also permits its use in juice drinks and certain non-carbonated juice beverages (Anonymous, 1987). The additive does not possess any residual activity because DMDC will undergo hydrolysis to yield carbon dioxide and methanol.

$$H_3CO - \overset{O}{\overset{\|}{C}} - O - \overset{O}{\overset{\|}{C}} - OCH_3 \rightarrow 2\ CH_3OH + 2\ CO_2$$

DMDC has been examined for general control of many wine microorganisms (Daudt and Ough, 1980; Porter and Ough, 1982; Ough et al., 1988a). In a wine containing 10% v/v ethanol, a concentration of 25 mg/L was found effective against *Saccharomyces*, *Brettanomyces*, and *Schizosaccharomyces* (Ough, 1993a). The additive is also inhibitory against acetic and lactic acid bacteria (Table 5.1). Although a synergy between DMDC and SO_2 against *Saccharomyces* has been reported (Ough et al., 1988a), this effect was not observed by Renee Terrell et al. (1993). Rather, these authors noted that DMDC was more effective than combinations of SO_2 and/or sorbic acid and suppressed fermentation of grape juice more effectively at higher temperatures.

The antimicrobial effect of DMDC results from inactivation of microbial enzymes (Ough, 1993a). Porter and Ough (1982) reported the

Table 5.1. **Concentrations of dimethyl dicarbonate (Velcorin™) required for control of various yeasts and bacteria inoculated at 500 CFU/mL.**

Microorganism	Dimethyl dicarbonate (mg/L)
Yeasts	
Candida krusei	100–200
Hansenula anomala	25–50
Kloeckera apiculata	25–50
Rhodotorula spp.	30–200
Saccharomyces spp.	40–200
Torulopsis spp.	75–100
Zygosaccharomyces	50–150
Bacteria	
Acetobacter pasteurianus[a]	190–250
Lactobacillus brevis	200
Lactobacillus buchneri	30
Pediococcus	300

[a]Inoculation at 250 CFU/mL.
Adapted from Ough (1993a) and J. Just (personnal communication, 2005).

mechanism of action is denaturation of the fermentative pathway enzymes glyceraldehyde-3-phosphate dehydrogenase and alcohol dehydrogenase.

Although effective, DMDC suffers from both formulation and dosing problems. First, the chemical is only sparingly soluble in water and requires thorough mixing to ensure uniform distribution in wines to be bottled. Additionally, the compound has a melting point of 15.2°C/59.4°F (Ough, 1993a) and, therefore, must be slightly warmed prior to addition. Given these limitations, Velcorin™ is added to wine using special equipment that delivers the optimal dose in each application. Although the dosing equipment is expensive, mobile systems are available in some areas.

Another concern regarding the use of Velcorin™ is safety. This chemical is toxic by ingestion and inhalation, is a skin and eye irritant, and is combustible when exposed to an open flame. As such, safety precautions must be taken when using Velcorin™ in a winery. The dosing equipment does reduce safety risks given the toxicity of the chemical.

5.2.3 Lysozyme

Lysozyme is a low molecular weight protein (14,500 Da) derived from egg white that brings about lysis of the cell wall of Gram-positive bacteria (*Oenococcus*, *Lactobacillus*, and *Pediococcus*). Activity toward Gram-negative bacteria (*Acetobacter* and *Gluconobacter*) is limited because of the protective outer layers in this group (Conner, 1993). The enzyme has no effect on yeasts or molds.

Given its specificity, lysozyme finds applications among white, rose, and blush wine producers wanting to prevent malolactic fermentation as well as wineries wanting to reduce initial populations of lactic acid bacteria before fermentation (Nygaard et al., 2002; Delfini et al., 2004). Gerbaux et al. (1997) reported that pre-fermentation additions of 500 mg/ L inhibited MLF but, when added post–alcoholic fermentation, the concentrations required for microbial stability could be reduced to 125– 250 mg/L. Delfini et al. (2004) reported differential sensitivity to lysozyme among species of *Lactobacillus* or *Pediococcus* studied where some strains survived up to 500 mg/L, generally higher concentrations than for *Oenococcus*. Lysozyme is now approved for use in the United States at concentrations up to 500 mg/L.

Because lysozyme is a protein, the presence of phenolics as well as the degree of clarification will affect activity. Plant phenolics are well-known to react with enzymes, thereby decreasing activity (Rohn et al., 2002). Reflecting this, lysozyme is more active in white wines than reds, most likely due to the differences in polyphenolic content (Daeschel et al., 2002;

Delfini et al., 2004). In support, Daeschel et al. (2002) reported that activity of lysozyme was greatly reduced in a Cabernet Sauvignon wine after 28 days but not in Riesling. Delfini et al. (2004) reported that lysozyme binds to suspended solids thus decreasing activity after clarification.

Lysozyme may also influence protein stability in white wines. Thus, fining trials should be conducted prior to bottling wines treated with the enzyme. Although evidence is lacking, utilization of lysozyme may have an indirect sensory impact on palate structure similar to that of protein-aceous fining agents used in red wines.

5.2.4 Sorbic Acid

Sorbic acid (2,4-hexandienoic acid) is a short-chain fatty acid that is used in grape juices and in sweetened, bottled wines to prevent re-fermentation by *Saccharomyces* (De Rosa et al., 1983; Renee Terrell et al., 1993). The maximum concentration allowed in the United States is 300 mg/L, whereas the Office International de la Vigne et du Vin (O.I.V.) places the limit at 200 mg/L. In practice, concentrations of 100 to 200 mg/L are typically used. At recommended levels, sorbic acid is generally effective in control-ling *Saccharomyces*, but other yeasts exhibit differential resistance (Warth, 1977; 1985). For example, *Kloeckera apiculata* and *Pichia anomala* (formerly *Hansenula anomala*) are inhibited at 156 to 168 mg/L, respectively, whereas *Schizosacccharomyces pombe* and *Zygosaccharomyces bailii* require at least 672 mg/L (Warth, 1985). Mechanisms of inhibition are not fully under-stood but probably due to morphological differences in cell structure, changes in genetic material, alteration in cell membranes, as well as inhibi-tion of enzymes or transport functions (Sofos and Busta, 1993).

Bacteria are not affected by sorbic acid, and, in fact, several species can metabolize the acid to eventually yield 2-ethoxyhexa-3,5-diene, a compound that imparts a distinctive "geranium" odor/tone to wines (Section 11.3.5). Other odor/flavor–active compounds detected in spoiled wines treated with sorbic acid include l-ethoxyhexa-2,4-diene and ethyl sorbate (Chisholm and Samuels, 1992), the latter of which has been associated with off-flavors in sparkling wines (De Rosa et al., 1983). Whereas Chisholm and Samuels (1992) described ethyl sorbate as possess-ing a "honey" or "apple" aroma, De Rosa et al. (1983) thought the com-pound imparted a very unpleasant "pineapple–celery" odor upon short-term (6 month) storage. Based on this observation, De Rosa et al. (1983) recommended that sorbates should not be used in sparkling wine production.

Because sorbic acid is relatively insoluble in water (1.5 g/L at room temperature), the additive is usually sold as the salt, potassium sorbate, which is readily soluble (58.2 g/L). When calculating the amount of potassium sorbate to add, it is necessary to consider the differences in molecular weights between the salt and acid forms:

$$\frac{\text{Concentration}}{\text{to add (mg/L)}} = \frac{150.22 \text{ g/mole(salt)}}{112.13 \text{ g/mole(acid)}} \times \text{Desired final concentration (mg/L)}$$

The required amount of potassium sorbate should be first hydrated in wine or water prior to addition to wine.

Sulfur dioxide is thought to work synergistically with sorbic acid, lowering the concentration of the acid needed for the control of fermentative yeasts. This property was demonstrated by Ough and Ingraham (1960) who reported that SO_2 and sorbic acid added to sweetened table wine at 80 mg/L each had a greater inhibitory action than either SO_2 at 130 mg/L or sorbic acid alone at 480 mg/L. In contrast, Parish and Carroll (1988) reported an antagonistic interaction between SO_2 and sorbate against *S. cerevisiae* where the inhibition exerted by each compound individually was greater than the combination. This antagonism may be explained by sorbate reacting with SO_2 to yield an adduct product of the following structure (Heintze, 1976):

$$CH_3 - CH - CH_2CH = CH - COOH$$
$$|$$
$$HSO_3$$

If used to stabilize a sweet wine, sorbic acid should be added just prior to bottling. Additions should be carried out in stainless steel or other containers that can be cleaned and sanitized. Use of wood tanks for prebottling mixing and storage should be avoided. Here, residual sorbic acid trapped in the wood may be utilized by resident lactic acid bacteria in the production of "geranium-tone," which can continually leech into wines subsequently processed through that tank.

The amount of ethanol present will also dictate the appropriate amounts of sorbic acid to add. Although Ough and Ingraham (1960) recommended adding 150 mg/L to a wine with 10% to 11% v/v ethanol, that concentration decreased at alcohol concentrations of 12% (100 mg/L) or 14% v/v (50 mg/L).

5.2.5 Other Preservatives and Sterilants

A number of other antimicrobials have been examined for control of undesirable yeast and bacterial activity in juice and wine. Generally, these have met with relatively limited commercial success or have not gone beyond laboratory-scale research.

5.2.5.1 Fumaric Acid

Fumaric acid is approved by the United States Alcohol and Tobacco Tax and Trade Bureau for both controlling growth of lactic acid bacteria and as an acidulant at maximum concentration of 3.0 g/L. Being a relatively strong organic acid, fumaric has received attention as an acidulating agent in wine, rather than the more expensive tartaric acid (Cofran and Meyer, 1970). In this regard, fumaric acid additions of 1 g/L are equivalent to tartaric acid additions of 1.2 g/L.

The most important function of fumaric acid is its ability to inhibit malolactic fermentation (Ough and Kunkee, 1974; Pilone et al., 1974). In this regard, Ough and Kunkee (1974) reported that none of the wines in their study containing 1.5 g/L fumaric acid underwent MLF, even after 12 months of storage. However, fumaric acid is degraded during alcoholic fermentation by *Saccharomyces* forming L-malic acid (Pilone et al., 1973). Fumaric acid might be useful to reduce initial bacterial populations in musts, such as some species of *Lactobacillus* (Section 6.6.2). Although it is not known whether these spoilage bacteria can metabolize fumaric acid, Pilone et al. (1973) noted that the acid could be degraded to L-lactic acid by wine leuconostocs (*Oenococcus*), possibly by the same mechanism of yeast.

One concern regarding the use of fumaric acid is its limited solubility in wine. In fact, fumaric acid is sparingly soluble in water (6.3 g/L at 25°C/77°F) compared with tartaric acid, which is soluble at 1,390 g/L at 20°C/68°F (Anonymous, 1983). However, the acid is more soluble at higher temperatures (10.7 g/L at 40°C/104°F) or in 95% ethanol (57.6 g/L at 30°C/86°F). Margalit (2004) suggested solubilizing 50 to 80 g/L in hot water and addition to must or wine while the solution is hot.

. Like other antimicrobial acids, the efficacy of fumaric acid depends on pH where less activity is noted with increasing pH (Doores, 1993). Further, fumaric acid can impart a "harsh" taste (Margalit, 2004). Because of this and solubility concerns, caution should be exercised when used in wine.

5.2.5.2 Nisin

Nisin is an antimicrobial bacteriocin (Section 6.6.3) produced by some strains of *Lactococcus lactis* subsp. *lactis* (Hurst and Hoover, 1993). The site of primary action is the bacterial membrane where the antibiotic disrupts cytoplasmic membranes (Hurst and Hoover, 1993).

Like lysozyme, nisin is an effective inhibitor of Gram-positive bacteria. Radler (1990a; 1990b) determined that most lactic acid bacteria are inhibited by nisin, even in low concentrations, while alcoholic fermentation was not affected. However, the author reported that species varied in their response, with *L. casei* being the least sensitive. Daeschel et al. (1991) successfully used nisin-resistant strains of *O. oeni* to conduct MLF in a wine previously treated with nisin to control other spoilage bacteria. Others have reported that nisin killed 100% of *O. oeni* present as a biofilm on stainless steel (Nel et al., 2002). Despite its potential, nisin is not currently approved for use in wine in the United States.

5.2.5.3 Carbon Monoxide

For several years, research conducted at California State University (Fresno) has focused on the utilization of carbon monoxide as an alternative to sulfur dioxide and other preservatives and sterilants (Muller et al., 1996). In laboratory studies using model systems for juice and wine, control of several spoilage yeasts has been examined. At inocula levels of 1×10^5 to 2×10^5 CFU/mL, *Brettanomyces* and *Hansenula* were controlled by exposure to 90 mg/L CO compared with *Kloeckera*, which required 240 mg/L. *Zygosaccharomyces bailii* in the juice system was controlled at treatment levels of 480 mg/L, whereas in wine at 7.5% and 10% v/v ethanol, activity was delayed for 14 days. By comparison with the spoilage yeast, *Saccharomyces* was not affected at CO levels of 1000 mg/L.

Despite these results, there has been reluctance to consider carbon monoxide for microbiological control. This may result from fears regarding its inherent toxicity as well as from the public perception of the use of toxic compounds to treat wine. Indeed, carbon monoxide is a toxic, colorless, and odorless gas whose detection requires specialized equipment. However, in terms of potential risks to human health, SO_2 is considerably more toxic than carbon monoxide.

5.2.5.4 Plant Extracts

One plant extract that has been proposed as an antimicrobial agent in wine is paprika seed. The active ingredient(s) remains unknown but was named "paprika seed antimicrobial substance," or PSAS, by Yokotsuka et al. (2003). These authors were able to demonstrate that extracts of paprika seeds exhibited strong antimicrobial activity against *S. cerevisiae*, even when added at only 16 mg/L. PSAS appears to be very effective (100% kill) when added during the course of alcoholic fermentation.

5.3 FILTRATION

In the cellar, removal of particles can be achieved using either macro- or microfiltration. Macrofiltration has traditionally been accomplished using diatomaceous earth filters or by pad filtration, whereas sterile filtration can be achieved only by the use of a integrity-tested membrane (0.45 μm). Other types of filtration, namely ultrafiltration and reverse osmosis, are also occasionally used but for different purposes.

With development of cross-flow technology during the past 20 years, the ability to remove not only particulates but a range of soluble compounds ranging from colloidal macromolecules to ionic species has become common practice. Microfiltration is the separation of submicrometer-sized particles (>0.2 μm) and is the most common system used. Ultrafiltration (UF) allows the removal of smaller particles and colloids (0.001 to 0.2 μm), and reverse osmosis (RO) separates low molecular weight components (i.e., ethanol and acetic acid) as well as ions.

Classically, filtration has been achieved using perpendicular-flow ("dead-end") systems to achieve wine clarity during cellaring or sterility at bottling. In these cases, juice/wine impacts the filter perpendicular to the matrix ("head-on") and particulates are either retained or pass through depending on the size of the pores (Fig. 5.4). Depending on the stage in processing (cellaring vs. bottling), the two types of filters used with perpendicular-flow systems are depth ("nominal") and membrane ("absolute" or "sterile").

In contrast with perpendicular-flow, micropore, UF, and RO systems use cross-flow (tangential) designs. Here, juice/wine is circulated such that the membrane surface is continuously swept of particulates while liquid migrates through the pores (Fig. 5.4).

5.3.1 Perpendicular-Flow Filtration ("Nominal" or "Depth")

Depth filtration systems use either paper pads, diatomaceous earth ("DE" or "powder"), or a combination of the two media to remove particulates. In the case of pad filtration, particles accumulate at the filter surface and are entrapped within a matrix of narrowing channels and/or retained by charge interactions with the pad. Diatomaceous earth filtration utilizes the same filtration principles as pads (i.e., particle exclusion and

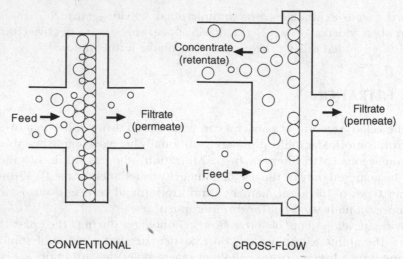

CONVENTIONAL   CROSS-FLOW

Figure 5.4.. Comparison between conventional ("dead-end") and cross-flow filtration systems.

entrapment), but the medium is precoated onto supports known as septa. Both pads and DE cannot be reused and are discarded after use.

As these media have a range of pore diameters and other slight imperfections, it is not possible to guarantee either their porosity or integrity. Even the so-called sterile pads may contain perforations/channels larger than those required to remove wine bacteria. Pads are therefore assigned nominal ratings that refer to the average particle retention based on batch laboratory testing at defined pressure differentials (ΔP) across the pad. Again, it must be noted that filtration through a nominal pad or system will not remove the microorganisms to yield a sterile wine. If sterility is desired, then filtration through a membrane with a maximum pore size of 0.45 μm is required.

5.3.2 Perpendicular-Flow Filtration ("Absolute" or "Sterile")

Sterile filtration requires the use of membranes of sufficient porosity that microorganisms are physically excluded. To remove bacteria, membranes with maximum pore sizes of 0.45 μm are recommended. Although attempts have been made to use membranes with pore sizes of <0.45 μm (e.g., 0.2 μm), flow rates are commonly too slow to be economically feasible. Where spoilage yeasts are believed to represent a threat, some winemakers have relied on membranes with larger pores (0.65–1.0 μm). However, caution should be exerted when using membranes with pore sizes >0.45 μm because of variation in sizes of microorganisms to be removed. For instance, older and larger yeast cells may be removed by a 1 μm filter whereas younger and smaller cells may not be, thereby maintaining the risk of re-fermentation in the bottled wine. Furthermore, use of >0.45 μm membranes will not remove bacteria. Confounding the decision of pore size selection, Millet and Lonvaud-Funel (2000) noted that the size of *Acetobacter* cells decrease during lengthy residency in wine, allowing the bacterium to pass through 0.45 μm membranes.

Membrane filtration utilizes cartridges that when properly cleaned and stored can be reused at a later date. For those cartridges that can be stored, it is common to thoroughly back-flush the membranes twice, first with cold water and then with warm water because the latter can "bake-on" macro-molecules such as proteins. Membranes can be stored wet, most commonly in food-grade 95% v/v ethanol or a solution of citric (1200 mg/L) and SO_2 (600 mg/L). In all cases, the membrane manufacturer should be consulted for specific storage and back-flushing recommendations.

5.3.2.1 Bubble Point

Although the pore sizes for absolute membranes are guaranteed, membrane integrity should be verified before use. This is accomplished by membrane integrity testing, a process referred to as "bubble pointing." The bubble point is based on the fact that the pressure needed to force a gas bubble through a pore is inversely proportional to the diameter or size of the pore. In other words, a higher pressure is required to force gas through a small pore than through a large pore.

To perform a bubble test, the membrane must be kept constantly wet. Gas, normally N_2, is then connected to the inlet side of the filter and the outlet is connected to tubing that is placed into a vessel containing water. As gas is allowed to flow into the filter housing, the pressure is slowly increased using the gas tank regulator. The bubble point is that pressure where a stream of bubbles first appears on the outlet or filtrate side of the membrane. If the measured bubble point pressure is less than the manufacturer's specified value, the membrane has been compromised and should not be used. The bubble point should be recorded prior to start-up and at the end of bottling as well as after any interruptions (i.e., upon return from worker break periods).

5.3.2.2 Diffusion Test

Like bubble pointing, the diffusion test allows evaluation of the integrity of filter membranes. The membrane is first wetted and gas pressure on the upstream side of the filter is increased to that recommended by the manufacturer, commonly 80% of the bubble point pressure. Using an inverted graduated cylinder, gas is collected on the downstream side of the filter for 1 min. The amount of gas collected is then compared with the maximum diffusional flow specification assigned to the membrane.

5.3.2.3 Potential Limitations

Membranes have very limited capacity in terms of particle removal ("dirt-handling" capacity). Wines of low clarity or that have obvious suspended solids will quickly clog or plug membranes, greatly lowering flow rates. As a consequence, application of 0.45 μm membranes should be restricted to final pre-bottling filtration of brilliantly clear wine. For smaller lots of wine, membrane cartridges can be placed in tandem so that the wine first passes through a 1 μm filter prefilter prior to the final filtration through an 0.45 μm sterile membrane.

As with any sterile bottling, it is essential that the filtrate not be contaminated enroute to the bottle after filtration. Thus, the bottling line must be designed and constructed to be readily sterilized. Sterilization may be achieved by exposure to either steam or hot water (>82°C/180°F). In either case, contact time, as measured at the furthest point from the hot water/steam source, should be 20 min minimally. Because of costs associated with generation of large amounts of hot water and increasing problems with *Zygosaccharomyces*, some bottling facilities are utilizing chemical sterilants followed by sterile water rinses (Chapter 9).

5.3.3 Cross-Flow/Tangential Filtration

Cross-flow or tangential filtration is finding increased interest and application in winemaking and juice/concentrate production. Unlike perpendicular-flow filtration where the juice/wine impacts the membrane perpendicularly, cross-flow filtration relies on the product flowing across the surface of the membrane tangentially, producing the filtrate (permeate), which passes through the membrane, and the retentate (concentrate), which is retained in the system. Given their small pore size, ultrafilter membranes are rated for a molecular weight cutoff (i.e., an approximate molecular weight of colloidal compounds that will be retained and not pass through the membrane).

By design, cross-flow systems are partially self-cleaning arising from the turbulent flow along the membrane surface, which continuously dislodges particles. Because permeate is being continuously removed and new juice/wine is being added, the retentate will eventually become so laden with particulates that the system must be completely cleaned before continuing operation. To lengthen the time between cleanings, cross-flow systems commonly utilize a series of back-flush cycles, which force permeate backwards through the membrane to remove strongly entrapped particulates.

Although small particulates can be removed, tangential filtration should not be considered a substitute for sterile filtration. Currently, it is not possible to determine when pores or tears large enough to allow microorganisms to pass through occur because an equivalent verification method as the bubble point is not available. In fact, a population of $>10^4$ CFU/mL of *Zygosaccharomyces* was found in a permeate obtained from a commercial cross-flow filtration system, indicating the importance of using sterile perpendicular-flow membranes at bottling (K.C. Fugelsang, personal observation).

CHAPTER 6

MICROBIAL ECOLOGY
DURING VINIFICATION

6.1 INTRODUCTION

As outlined in Chapters 1 to 4, several genera and species of microorganisms can be found in grape musts and wines at various times during the winemaking process. For instance, *Saccharomyces*, *Brettanomyces*, and *Pediococcus* can be found together in wine (Fig. 6.1). Even with this wide diversity of microorganisms, vinification commonly involves a sequential development of microorganisms (Fig. 6.2). In general, non-*Saccharomyces* yeasts will be the first group to dominate during vinification (Fig. 6.2A), followed by *Saccharomyces* that normally completes alcoholic fermentation (Fig. 6.2B). After the primary fermentation is finished, malolactic fermentation may be induced by *Oenococcus* or other lactic acid bacteria (Fig. 6.2C). During the aging of wines, several different yeasts and bacteria may grow, many of which bring about spoilage (Fig. 6.2D).

Dominance by a specific species or by a group of microorganisms at any given stage during vinification depends on many factors including the microorganisms present and grape conditions prior to harvest (humidity, physical damage due to birds or harvesters, use of fungicides, and degree

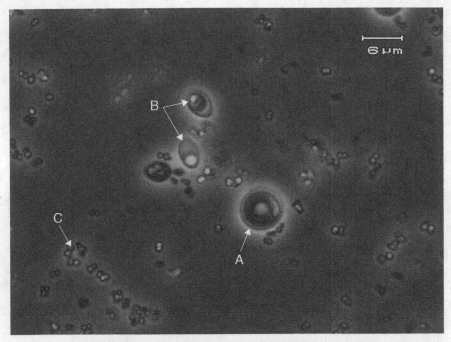

Figure 6.1. Mixture of (A) *Saccharomcyes*, (B) *Brettanomyces,* and (C) *Pediococcus* in a wine as viewed with phase-contrast microscopy at a magnification of 1000×. Photograph provided through the courtesy of R. Thornton and E. Akaboshi.

of maturity). Moreover, the chemical and physical environments of juice/ must/wine (Chapters 7 and 8), winery cleaning and sanitation programs (Chapter 9), as well as metabolic interactions between microorganisms (Section 6.6) also play key roles.

Understanding microbial ecology during vinification is further compli- cated by mounting evidence that microorganisms can also exist in a state known as "viable-but-non-culturable" (VBNC). Although no growth in an appropriate medium can indicate microbial death, this is not necessarily the case. By definition, microorganisms in the VBNC state fail to grow on microbiological media yet display low levels of metabolic activity (Oliver, 2005). Typically, the VBNC state is induced in response to stress such as osmotic pressure, temperature, oxygen concentration, and others. Cells may be resuscitated under favorable environmental conditions. One con- sequence of VBNC is the potential failure to detect microorganisms during vinification, leading to false conclusions regarding population dynamics.

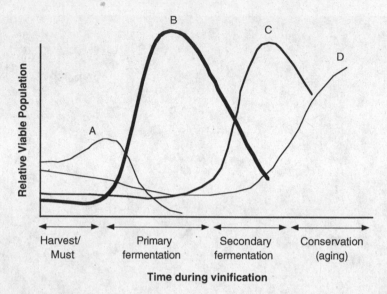

Figure 6.2.   Generalized growth of (A) non-*Saccharomyces* yeasts, (B) *Saccharomyces*, (C) *Oenococcus oeni*, and (D) spoilage yeasts and/or bacteria during vinification of a wine.

Microorganisms found in wine and believed to be able to enter a VBNC state are *Acetobacter aceti, Brettanomyces bruxellensis, Candida stellata, Lactobacillus plantarum, Saccharomyces cerevisiae,* and *Zygosaccharomyces bailii* (Millet and Lonvaud-Funel, 2000; Divol and Lonvaud-Funel, 2005; Du Toit et al., 2005; Oliver, 2005).

6.2 NON-*SACCHAROMYCES* AND *SACCHAROMYCES* YEASTS

6.2.1 Grapes and Musts

It is generally accepted that in the case of sound, undamaged grapes, the viable population of yeasts ranges from 10^3 to 10^5 CFU/mL (Parish and Carroll, 1985; Fleet and Heard, 1993). The most frequently isolated native species is *Kloeckera apiculata*, which may account for more than 50% of the total yeast flora recovered from fruit. Lesser numbers of other yeasts, such as species of *Candida, Cryptococcus, Debaryomyces, Hansenula, Issatchenkia, Kluyveromyces, Metschnikowia, Pichia,* and *Rhodotorula,* have also been reported (Heard and Fleet, 1986; Holloway et al., 1990; Longo et al., 1991; Fleet and Heard, 1993; Sabate et al., 2002). If grape juice or concentrate is added to a grape must to increase alcohol yield or as a sweetening agent in wine, other yeasts may be present (e.g., *Zygosaccharomyces*).

Saccharomyces can be found in grape musts, but the populations are often less than 50 CFU/mL (Fleet and Heard, 1993). This is true even in vineyards where fermented pomace is returned as a soil amendment. Failure to routinely isolate *Saccharomyces* from the vineyard could reflect the preference of this yeast for the high-sugar environments of grape juice and fermentation.

To date, only a few studies are available describing the growth of *Issatchenkia* or *Metschnikowia* spp. in grapes or wines. Whereas Guerzoni and Marchetti (1987) and Clemente-Jimenez et al. (2004) found *I. orientalis* in grape musts, other studies did not report isolating this species or its anamorph, *Candida krusei* (Romano et al., 1997; Constantí et al., 1998; Mills et al., 2002; Sabate et al., 2002; van Keulen et al., 2003). Mora and Mulet (1991) and Clemente-Jimenez et al. (2004) found another species of *Issatchenkia*, *I. terricola*, formerly known as *Pichia terricola* (Deak and Beuchat, 1996). Along with *C. krusei*, Guerzoni and Marchetti (1987) isolated *K. apiculata*, *I. occidentalis*, and *M. pulcherrima* from rotting grapes.

6.2.2 Alcoholic Fermentation

Once grapes are crushed, non-*Saccharomyces* yeasts multiply and reach peak populations sometime during the early stages of alcoholic fermentation (Fig. 6.2A). Heard and Fleet (1988) noted that *K. apiculata* and *C. stellata* tend to dominate early to mid-stages of fermentation but other genera of non-*Saccharomyces* may also be found. Peak populations can be as high as 10^6 to 10^8 CFU/mL depending on conditions (Fleet et al., 1984; Heard and Fleet, 1988).

Although *Hanseniaspora/Kloeckera* normally represent the dominant native genera present on grapes at harvest, their activity was thought to be restricted to pre-fermentation and early stages of alcoholic fermentation. In fact, many non-*Saccharomyces* yeasts are believed to possess lower ethanol tolerances compared with *Saccharomyces* (Deak and Beuchat, 1996). This factor probably contributes to the frequently observed die-off of these yeasts shortly after the start of alcoholic fermentation when the ethanol concentration reaches 5% to 6% v/v. In agreement, Heard and Fleet (1988) and Clemente-Jimenez et al. (2004) reported that inoculation of various species of *Kloeckera*, *Candida*, *Issatchenkia*, *Metschnikowia*, and *Pichia* in grape juice yielded maximum ethanol concentrations of less than 6%. In contrast, Heard and Fleet (1988) reported that in mixed culture fermentation at 10°C/50°F, *K. apiculata* or *C. stellata* were able to achieve populations of 10^7 CFU/mL and completed fermentation. Similarly, Erten

(2002) observed that *K. apiculata* survived longer during fermentations at 10°C/50°F and 15°C/59°F than at vinifications conducted above 20°C/68°F, in agreement with the findings of Mora and Rossello (1992). It therefore appears that low fermentation temperatures (10°C/50°F to 15°C/59°F) could extend the alcohol tolerance of these yeasts thereby extending their growth cycle (Gao and Fleet, 1988). In fact, Heard and Fleet (1988) suggested that the contributions of non-*Saccharomyces* yeasts to wine quality could increase with a decrease in fermentation temperature. Egli et al. (1998) reported that wines fermented at cooler temperatures and not inoculated with *Saccharomyces* were "more aroma intense" than those inoculated with *Saccharomyces*.

Although ethanol tolerance is one factor affecting the growth of non-*Saccharomyces* yeasts, there are clearly others as well. In studying the ecology of *Torulaspora delbrueckii* during alcoholic fermentations, Mauricio et al. (1991) concluded that a lack of oxygen and subsequent lack of sterol and phospholipid biosynthesis resulted in the die-off of this species. In agreement, Hansen et al. (2001) concluded that the death of *T. delbrueckii* and *Kluyveromyces thermotolerans* inoculated in mixed culture with *Saccharomyces* was not due to the formation of toxic by-products but rather the lack of oxygen. More recently, Nissen and Arneborg (2003) noted that early deaths of *Kluyveromyces* and *Torulaspora* could not be explained by nutrient depletion or the presence of toxic compounds and appeared to be affected by cell-to-cell contact mechanism with high cell populations of *Saccharomyces*. In a comprehensive study, Pina et al. (2004) suggested many factors affect the ethanol tolerance of non-*Saccharomyces* yeasts including species, presence of oxygen, and the addition of survival factors.

As populations of non-*Saccharomyces* yeasts decline, *Saccharomyces* will dominate and complete alcoholic fermentation (Fig. 6.2B). Recognizing this, many winemakers will inoculate musts with commercial cultures of *Saccharomyces* in order to control the fermentation (Section 8.3). When added to must/juice at recommended levels, the population of actively growing *Saccharomyces* should exceed 1×10^6 to 3×10^6 CFU/mL. During the height of alcoholic fermentation, populations normally reach at least 10^7 CFU/mL, if not higher. Normally, by the time peak populations are reached and the culture is in stationary phase, at least half the fermentable sugar has been utilized. *Saccharomyces* will continue to utilize the remaining sugar, most notably glucose and fructose, until dryness is reached (≤0.2% residual fermentable sugar).

It should be noted that *Saccharomyces* is glucophilic; that is, the yeast prefers glucose over fructose. As such, glucose present in a grape must will be exhausted before complete utilization of fructose. In that fructose is sensorially sweeter than glucose, this may be of importance in the case of stuck fermentations (Section 8.5.1).

6.2.3 Post-fermentation

The distribution of yeast species in cellar-aging wine includes *Dekkera/ Brettanomyces* (Section 11.2.2), film yeasts (Section 11.2.3), *Saccharomycodes* (Section 11.2.4), and *Zygosaccharomyces* (Section 11.2.5), all of which can result in serious wine spoilage.

6.3 *DEKKERA/BRETTANOMYCES*

Brettanomyces spp. have been identified in spoilage from all wine-producing areas of the world (van der Walt and van Kerken, 1959; 1961; Heresztyn, 1986; Ibeas et al., 1996; Mitrakul et al., 1999). *Brettanomyces* is generally observed in barrel-aging red wines, although infections have also been noted in Chardonnay and Sauvignon Blanc. Ciani and Ferraro (1997) report isolation of *Brettanomyces* from sparkling wines.

6.3.1 Grapes and Musts

Some researchers have reported isolating these yeasts from grapes, but confirmed reports are rare. Parish and Carroll (1985) found *Brettanomyces* in grapes undergoing fermentation but could not confirm their origin because the grapes were hand-harvested and the fermentations were not performed aseptically. While Romano and Suzzi (1993) indicated that *Brettanomyces* could be isolated from grapes and during fermentation with or without SO_2, Guerzoni and Marchetti (1987) reported low populations of *Brettanomyces* on healthy and sour-rot infected grapes. As Romano and Suzzi (1993) and Guerzoni and Marchetti (1987) did not identify these yeasts to species level, it is not known if these isolates were capable of growing in wine. It is possible though that *Brettanomyces* could be brought into a winery on deteriorated fruit because Dias et al. (2003b) noted that the yeast can survive alcoholic fermentation. With newer detection technologies such as genetic probes, it is possible that researchers will eventually confirm the presence of *Brettanomyces* in grape samples.

6.3.2 Alcoholic and Post-fermentation

As *Brettanomyces* is not thought to be found on grapes, the microorganism is believed to spread through a winery by importation of infected wine, poor sanitation of hoses, tanks, or other equipment, or even by the common fruit fly (Fugelsang et al., 1993; Licker et al., 1999). Because these yeasts are unable to survive passage through the gastrointestinal tract of the adult fruit fly (Shihata and Mrak, 1951), another mechanism of dispersal is passive adherence to the body surfaces of adult fruit flies. Although previous surveys of air samples taken from wineries failed to detect *Brettanomyces* (Donnelly, 1977), Beech (1993) pointed out that some spoilage problems in ciders tentatively attributed to *Brettanomyces* were enhanced through exposure to air. The observations of Beech (1993) were later confirmed by Connell et al. (2002) who found *Brettanomyces* in air samples drawn from different areas within a commercial winery.

The most frequently cited locale for *Brettanomyces* within the winery is wood cooperage. *Brettanomyces* is capable of utilizing the disaccharide cellobiose (Blondin et al., 1982), a carbohydrate resulting from the toasting process required to bend staves in the production of barrels. Because new barrels have higher concentrations of cellobiose, these appear to be "better" sites for growth than previously used cooperage. Substantial populations can also build up in difficult to clean sites such as processing equipment and transfer lines and valves where organic deposits may accumulate over the course of a season. Other important reservoirs for the microorganisms include drains and isolated pockets of juice and wine.

In contrast to other yeasts found in wines, *Brettanomyces* tends to grow very slowly. Growth of *Brettanomyces* in barrel-aging wines follows a bell-shaped pattern that typically reaches maximum population density 5 to 7 months after vinification (Fugelsang and Zoecklein, 2003). The lengthy time for development of maximum cell number depends on many factors but includes available nutrients and fermentable sugars. Chatonnet et al. (1995) suggested that very low concentrations of residual sugar in a wine (0.275 g/L as glucose, fructose, galactose, and trehalose) was enough to support growth of this spoilage yeast and adversely affect wine aroma. Thus, wine considered to be dry by generally accepted standards may, in fact, support growth of *Brettanomyces*, even in the bottle after months of storage.

6.4 LACTIC ACID BACTERIA

6.4.1 Grapes and Musts

Like yeasts, lactic acid bacteria are also present in vineyards. However, given their nutritional requirements, species diversity and population density is limited. Sound, undamaged fruit contains $<10^3$ CFU/g and so populations in grape musts during the early stages of processing tend to be low (Lafon-Lafourcade et al., 1983b). Species that have been isolated from grape musts include *Lactobacillus hilgardii, L plantarum, L. casei, O. oeni, Leuconostoc mesenteroides,* and *P. damnosus* (Lonvaud-Funel et al., 1991). Where fruit deterioration has occurred, substantial populations may develop. It is not known whether increases in native bacterial populations are in response to increased availability of nutrients (resulting from fruit degradation) or some other factor(s) coincidental to the infection. Besides grapes, indigenous (native) populations are frequently isolated from cooperage, poorly sanitized pumps, valves, and transfer lines as well as from fillers and drains in bottling rooms (Donnelly, 1977).

6.4.2 Alcoholic and Malolactic Fermentations

Once the fruit is crushed, the ecology of lactobacilli and other lactic acid bacteria is complex with different species dominating at different times during vinification. In a comprehensive study by Costello et al. (1983), these authors isolated *L. jensenii, L. buchneri, L. hilgardii, L. brevis, L. cellobiosis, L. plantarum, Leuconostoc oenos* (*O. oeni*), and *Pediococcus* spp. from musts and wines at different times during vinification, in agreement with others (Lafon-Lafourcade et al., 1983b; Fleet et al., 1984; Davis et al. 1986a; 1986b; Sierro et al., 1990). Similarly, several species of *Lactobacillus* were isolated from commercial Washington State wines from which strains of *O. oeni* and *P. parvulus* were also isolated (Edwards et al., 1991; 1993; Edwards and Jensen, 1992). In those studies, each wine lot was sampled once during vinification so sampling the same lots at different times probably would have yielded different species.

Growth and decline of any particular species is influenced by a variety of conditions, including nutritional status, pH, and alcohol, as well as cellar temperature and the interactive impact of yeast and other bacteria. For instance, wines from warmer regions typically have pH values in excess of 3.5, conditions favorable to the growth of lactobacilli and other bacteria (Davis et al., 1986a). Winemaking decisions, such as yeast selection, fermentation and storage temperature, timing of SO_2 additions, racking, fining, and filtration play important roles in this regard.

Although species diversity in a must can increase shortly after crush, viable cell number normally remains relatively low (<10^3 to 10^4 CFU/mL) for some period of time (Lonvaud-Funel et al., 1991). Even if inoculated prior to alcoholic fermentation, most lactic acid bacteria experience a rapid die-off during alcoholic fermentation, commonly to populations below 100 CFU/mL. As an example, Edwards et al. (1990) observed the population of *O. oeni* decrease from 10^6 CFU/mL to less than 30 CFU/mL, resulting in a delayed malolactic fermentation. Sometime after completion of alcoholic fermentation, the population of *Oenococcus oeni* may increase to conduct malolactic fermentation (Section 8.6). An exception to this observation has been seen with infections of certain *Lactobacillus* spp. (Section 6.6.2) where rapid growth during alcoholic fermentation has been noted.

In general, growth of *Pediococcus* in wines is undesirable due to formation of excessive amounts of diacetyl and other adverse odors or flavors (Du Plessis and van Zyl, 1963b; Radler and Gerwarth, 1971) as well as biogenic amines (Section 11.3.6). Some species of *Pediococcus* are also capable of degrading glycerol to acrolein (Section 11.3.7), a compound that reacts with anthocyanins producing a bitter taint in wine (Davis et al., 1988; Sponholz, 1993; Du Toit and Pretorius, 2000).

Although growth of some species of *Pediococcus* in wines is undesirable, Edwards and Jensen (1992) reported that several wines from which pediococci had been isolated were not spoiled. In agreement, Edwards et al. (1994) reported that *P. parvulus* altered the bouquet of a Cabernet Sauvignon wine that had not undergone MLF but that the wine was not considered to be flawed. Furthermore, Silver and Leighton (1981) reported that strain B44-40, thought initially to be *Leuconostoc oenos* (*Oenococcus oeni*) but now believed to be a species of *Pediococcus* (Kelly et al., 1989), brought about malolactic fermentation without formation of off-odors or flavors. It is therefore possible that the growth of pediococci in wine may not adversely affect quality and may actually add desirable flavors and aromas under certain circumstances. Further research elucidating the impact of these microorganisms on the chemical composition, bouquet, and flavor of wines is warranted as *Pediococcus* spp. have been isolated from wines worldwide (Costello et al., 1983; Fleet et al., 1984; Edwards and Jensen, 1992; Manca de Nadra and Strasser de Saad, 1995).

The occurrence and survival of *Lactobacillus* in wine depends, in large part, on pH and ethanol (Davis et al., 1986a) In high-pH wines (>3.5), *Lactobacillus* spp. often dominates, whereas *O. oeni* will be present in higher relative populations at lower pH (Davis et al., 1986b; Henick-Kling, 1993). Regarding ethanol, growth of *L. plantarum* ceases at concentrations of 5% to 6% v/v, while the more ethanol tolerant *L. casei* and *L. brevis* have been successfully used to induce MLF (Wibowo et al., 1985; Kosseva et al., 1998). However, it is clear that differences between strains exist. In support, G-Alegría et al. (2004) noted that other strains of *L. plantarum* tolerated much higher concentrations of ethanol, up to 13%.

Like the pediococci, most species of *Lactobacillus* are considered to be spoilage microorganisms (Lafon-Lafourcade et al., 1983b; Wibówo et al., 1985). Growth of these bacteria in bottled wines can result in haze formation, sediment, gassiness, off-odors, and/or excessive volatile acidity and/ or lactic acid (Splittstoesser and Stoyla, 1987). *L. fructivorans* has been implicated in spoilage of fortified wines in California, spoilage that is characterized by mycelial/fiber-like growth in wines (Section 11.3.4). Other *Lactobacillus* spp. have been implicated in causing stuck or sluggish alcoholic fermentations (Section 6.6.2).

6.4.3 Post-fermentation

Once alcoholic and malolactic fermentations are completed, lactic acid bacteria may still grow in the wine resulting in spoilage. Wines commonly contain small amounts of arabinose, glucose, fructose, and trehalose (Liu and Davis, 1994), sugars that can be metabolized by microorganisms including lactic acid bacteria. Even in post-MLF wines, lactobacilli and/or pediococci may grow (Fig. 6.2D), leading to problems of haziness, gassiness, excessive acetic acid and volatile acidity (Section 11.3.1), acrolein (Section 11.3.7), biogenic amines (Section 11.3.6), excessive diacetyl (Section 2.4.5), ethyl carbamate (Section 11.3.2), "geranium" odor/tone due to metabolism of sorbic acid (Section 11.3.5), formation of mannitol (Section 11.3.8), mousiness (Section 11.3.3), ropiness (Section 11.3.9), and utilization of tartaric acid (Section 11.3.10).

6.5 ACETIC ACID BACTERIA

6.5.1 Grapes and Musts

Acetic acid bacteria are commonly associated with grapes and are normally present in musts. Sound, unspoiled grapes are reported to have 10^2 to 10^3 cells/g, whereas deteriorated fruit can yield up to 10^6 cells/g (Sponholz, 1993). Splittstoesser and Churney (1992) report recovery of *G. oxydans* from New York State grapes at levels ranging upward to 10^4 CFU/g fruit. In a later study, Du Toit and Lambrechts (2002) noted a wide range in populations in South African Cabernet Sauvignon musts, 10^4 to 10^7 CFU/mL, over two vintages. Where mold growth, particularly *Botrytis cinerea*, and fruit deterioration develop, populations of acetic acid bacteria increase and diversify to include not only *G. oxydans* but *A. aceti* or *A. pasteurianus* as well. In agreement, González et al. (2005) reported a major presence of both *G. oxydans* and *A. aceti* in spoiled grapes. Acetic acid bacteria can also proliferate during cold soaking prior to fermentation (Section 7.4.2).

6.5.2 Alcoholic and Post-fermentation

Where fruit deterioration has not occurred and alcoholic fermentation begins quickly, populations of acetic acid bacteria decline to <100 CFU/mL

(Joyeux et al., 1984b). In agreement, González et al. (2005) reported a reduction in strain diversity and viable populations of *Acetobacter* during alcoholic fermentation. *Gluconobacter* cannot survive the alcoholic environment of wine even when aerated (Drysdale and Fleet, 1989b).

There is no doubt that the presence (or absence) of oxygen greatly impacts growth of acetic acid bacteria. Drysdale and Fleet (1989b) noted that increases in the oxygen content of a wine resulted in stronger growth of *Acetobacter*. *A. pasteurianus* exhibited some growth in wine held at 50% dissolved oxygen whereas *A. aceti* did not. However, Joyeux et al. (1984b) were able to isolate acetic acid bacteria from wine at the bottom of barrels suggesting that these bacteria can, in fact, survive in a more anaerobic environment than previously thought. In support, Drysdale and Fleet (1989b) reported that both *A. aceti* and *A. pasteurianus* remained viable at low concentrations of oxygen. Furthermore, Ribéreau-Gayon (1985) noted viable populations ($>10^2$/mL) of *A. aceti* in properly topped-off and stored barrel-aging wines. More recently, Du Toit et al. (2005) found that *Acetobacter pasteurianus* could survive for a prolonged period of time in a wine maintained relatively anaerobic.

Millet and Lonvaud-Funel (2000) have suggested that *Acetobacter* might exist in wines in a VBNC state. Although *Acetobacter* in the VBNC state were difficult to culture (grow) using standard media, the authors noted that even a brief exposure to air can be enough to encourage their growth. Transitory exposure to oxygen may occur during pumping, transferring, and/or fining. Even when wines are subsequently stored properly, such exposure may have been sufficient to stimulate rapid and continued growth (Joyeux et al., 1984b). In low oxygen environments, acetic acid bacteria are likely to utilize electron acceptors other than oxygen, quite possibly quinones as discussed by Du Toit and Lambrechts (2002) and as illustrated in Fig. 6.3. Du Toit et al. (2005) noted that addition of activated charcoal used to remove phenolic compounds from a wine also affected growth of *A. pasteurianus*.

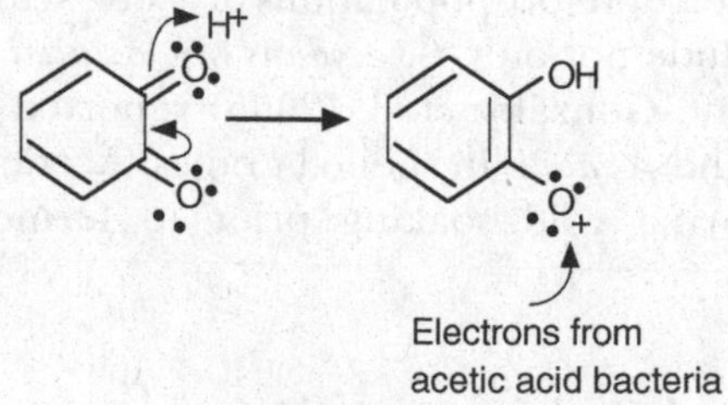

Figure 6.3.   Utilization of simple phenolics as electron acceptors under low oxygen conditions (C.J. Muller, personal communication, 1995.)

In slowly fermenting or stuck alcoholic fermentations where CO_2 evolution is not sufficient to prevent oxygen incursion, growth of acetic acid bacteria may be stimulated. Under these conditions, growth may result in the formation of gluconate by direct oxidation of glucose, as well as acetic acid and acetaldehyde from oxidation of ethanol (Section 3.4.2). Fructose is either not directly utilized or only to a limited extent by some of these bacteria (De Ley et al., 1984; Joyeux et al., 1984b). As a result of the disproportionate utilization of glucose relative to fructose by yeast and acetic acid bacteria, and the higher relative sweetness of fructose, the resultant wine may exhibit a "sweet-sour" character (Sections 6.2.2, 7.3.1, and 8.5).

6.6 MICROBIAL INTERACTIONS

Another factor that influences microbial ecology during vinification is the various interactions that occur between microorganisms. As recently reviewed by Alexandre et al. (2004). Many of these interactions result in the suppression and potential death of one or more species of the population.

6.6.1 *Saccharomyces* and *Oenococcus*

The interactions between *Saccharomyces* and *O. oeni* during vinification may be either stimulatory (Lüthi and Vetsch, 1959; Beelman et al., 1982; Feullat et al., 1985; Guilloux-Benatier et al., 1985) or inhibitory (Beelman et al., 1982; King and Beelman, 1986; Cannon and Pilone, 1993; Henick-Kling and Park, 1994; Larsen et al., 2003) to the bacterium. Inhibitory interactions have been reported where the viability of *O. oeni* declined from 10^5 to $10^7 CFU/mL$ to undetectable populations soon after inoculation into wine (Fornachon, 1968; Beelman et al., 1982; Liu and Gallander, 1983; Ribéreau-Gayon, 1985; King and Beelman, 1986; Lemaresquier, 1987; Wibowo et al., 1988; Semon et al., 2001). This rapid decline in bacterial viability has been frequently reported, even when *Saccharomyces* and *O. oeni* are co-inoculated at similar population densities (Fig. 6.4).

Two theories have been proposed to explain yeast antagonism of malolactic bacteria. First, the faster growing *Saccharomyces* may remove nutrients from a grape must (Amerine and Kunkee, 1968; Fornachon, 1968; Beelman et al., 1982) that are important to the nutritionally fastidious malolactic bacteria (Du Plessis, 1963; Garvie, 1967b). In support, Beelman et al. (1982) demonstrated that during growth in synthetic media, yeast depleted

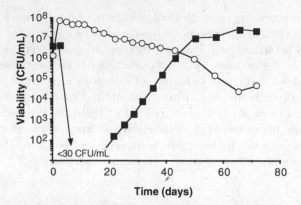

Figure 6.4.   Viability of *S. cerevisiae* (open circles) and *O. oeni* (closed squares) with bacteria inoculated on the same day as the yeast. Adapted from Semon et al. (2001) with the kind permission of the *Australian Journal of Grape and Wine Research*.

certain amino acids to concentrations that may not be sufficient to support bacterial growth. These authors suggested that as yeast enter stationary/ death phases and lyse, nutrients released into the wine facilitate the recovery of *O. oeni*. It has been demonstrated that the autolytic activity of wine yeasts during aging on lees can affect the concentrations of amino acids, peptides, and proteins in wine (Charpentier and Feuillat, 1993; Alexandre et al., 2001; Martinez-Rodriguez et al., 2001).

. More recent studies have demonstrated that the removal of nutrients by yeast does not always explain the observed inhibition of *O. oeni*. For instance, Larsen et al. (2003) reported that the addition of supplemental nutrients to a wine fermented by *S. cerevisiae* strain V1116 did not relieve the observed bacterial inhibition. Studying the impact of yeast autolysis on MLF, Patynowski et al. (2002) also concluded that nutrient depletion by *S. cerevisiae* was not responsible for the observed bacterial inhibition. Furthermore, this research showed that the yeast produced an unidentified inhibitory factor(s) that was progressively lost during aging. These results suggested that the second proposed theory, the production of toxic metabolites by yeast, may be responsible for the inhibition of *O. oeni*.

Wine yeasts are known to produce compounds during alcoholic fermentation that are inhibitory to malolactic bacteria including ethanol (Costello et al., 1983; Davis et al., 1986a; Britz and Tracey, 1990), SO_2 (Eschenbruch, 1974; Dott et al., 1976; Eschenbruch and Bonish, 1976; Suzzi et al., 1985; Romano and Suzzi, 1993; Henick-Kling and Park, 1994; Larsen et al., 2003), medium-chain fatty acids (Edwards and Beelman, 1987; Lonvaud-Funel et al., 1988; Edwards et al., 1990; Capucho and San Ramao, 1994), and antibacterial proteins/peptides (Dick et al., 1992). Of these compounds, SO_2 is most commonly implicated in causing bacterial inhibition (Fornachon, 1963; Henick-Kling and Park, 1994; Larsen et al., 2003). SO_2 is an effective antimicrobial against wine lactic acid bacteria (Carr et al., 1976; Liu and Gallander, 1983; Ough and Crowell, 1987; Britz and Tracey, 1990). Of the various forms of sulfur dioxide, molecular SO_2 ($SO_2 \bullet H_2O$) is thought to be the most antibacterial (Macris and Markakis, 1974; King et al., 1981; Edinger, 1986). Larsen et al. (2003) concluded that in some cases high SO_2-producing strains cause inhibition of MLF, but other yeast strains inhibit MLF by means other than production of SO_2. In agreement, additional studies have also cast doubt on whether SO_2 produced by yeast is the sole mechanism for bacterial inhibition (King and Beelman, 1986; Lemaresquier, 1987; Wibowo et al., 1988; Eglinton and Henschke, 1996; Caridi and Corte, 1997). For instance, Wibowo et al. (1988) found that *S. cerevisiae* inhibited the growth of *O. oeni* in wine, but that inhibition was not due to the production of SO_2. Additionally, Eglinton and Henschke (1996) reported that production of SO_2 by *S. cerevisiae* strain AWRI 838 and related strains also did not account for wines that resisted malolactic fermentation.

Larsen et al. (2003) and others have demonstrated that the majority of SO_2 formed by yeast is bound rather than free and molecular forms. This finding suggests that bound SO_2 may be more inhibitory to lactic acid bacteria than previously thought. However, there is conflicting information as to whether some forms of bound SO_2, in particular acetaldehyde-bound SO_2, are inhibitory to wine bacteria. Early work by Fornachon (1963) reported that both *L. hilgardii* and *L. mesenteroides* were inhibited in a medium in which sulfurous acid and an excess of acetaldehyde had been added. The author determined that these bacteria could metabolize acetaldehyde-bound SO_2, an observation later confirmed for *O. oeni* (Osborne et al., 2000). Fornachon (1963) further noted that MLF could be prevented by the presence of bound SO_2 even when the amounts of free SO_2 are negligible, possibly due to any SO_2 liberated from metabolizing the acetaldehyde-bound SO_2 adduct. Hood (1983) suggested an alternative mechanism whereby the effect of bound SO_2 may be due to small amounts of free (and therefore molecular) SO_2 in equilibrium with the bound form. However, these results were contrary to those of Carr et al. (1976) who reported that acetaldehyde-bound SO_2 had no influence on the bacterium studied (*L. plantarum*).

Another theory for inhibition of malolactic bacteria was suggested by Wibowo et al. (1988) who proposed that *S. cerevisiae* may inhibit *O. oeni* through the yeast production of antibacterial proteins/peptides. In support, Dick et al. (1992) isolated two proteins produced by *S. cerevisiae* that showed activity against *O. oeni*. Osborne (2005) was able to isolate a low-molecular-weight protein (3 to 6kDa) that was inhibitory to *O. oeni*. Similar findings were reported by Comitini et al. (2005) using a different strain of *Saccharomyces*.

Medium-chain fatty acids produced by yeast during alcoholic fermentation have also been implicated in the inhibition of malolactic bacteria (Lonvaud-Funel et al., 1985; Edwards and Beelman, 1987; Carrete et al., 2002). Inhibition of *Saccharomyces* species and some lactic acid bacteria by medium-chain fatty acids has been reported in grape juice and silage (Pederson et al., 1961; Woolford, 1975). Although this hypothesis has not been conclusively shown, Lonvaud-Funel et al. (1985) and Edwards and Beelman (1987) reported decanoic acid to be inhibitory to the growth of malolactic bacteria. As an example, Edwards and Beelman (1987) noted that decanoic acid suppressed the growth of *O. oeni* PSU-1 at a concentration of 10 mg/L, a concentration reported to be present in some wines (Houtman et al., 1980b). In addition, Carrete et al. (2002) reported that decanoic acid acted synergistically with either low pH or ethanol to inhibit *O. oeni*. However, Edwards et al. (1990) found that MLF occurred more rapidly in wines containing 5 mg/L decanoic acid and other medium-chain fatty acids than in wines with lower levels.

6.6.2 *Saccharomyces* and *Lactobacillus*

Some winemakers have observed sluggish fermentations characterized by rapid growth of microorganisms dubbed the "ferocious" lactobacilli (Boulton et al., 1996). This spoilage has been characterized as being very swift with abundant bacterial growth during the early stages of vinification resulting in premature arrest of alcoholic fermentation. In these cases, acetic acid levels were extraordinarily high; generally ranging from 0.8 to 1.5g/L and, on occasion, 2 to 3g/L. However, a "cause and effect" relationship between growth of these bacteria and cessation of alcoholic fermentation was not previously known. Clearly, not all species were involved in this problem as illustrated by Edwards et al. (1993) who could not induce a sluggish alcoholic fermentation even with inoculation of different species of lactobacilli.

Inhibition of yeast by *Lactobacillus* had been previously observed in other foods (Barbour and Priest, 1988; Essia Ngang et al., 1990; Leroi and Pidoux, 1993), but the first confirmed report that *Lactobacillus* can inhibit *Saccharomyces* was that of Huang et al. (1996). In their study, early inoculation of strains YH-15, YH-24, and YH-37 resulted in slowed laboratory-scale Chardonnay fermentations conducted by *Saccharomyces cerevisiae* Epernay 2. As shown in Fig. 6.5, strain YH-15 was capable of reaching populations in excess of 10^9 CFU/mL within 48h. Furthermore, wine composition was

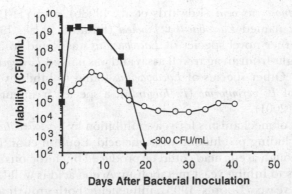

Figure 6.5. Growth of *Saccharomyces cerevisiae* Epernay (open circles) and *Lactobacillus kunkeei* strain YH-15 (closed squares) in a Chardonnay juice. Adapted from Huang et al. (1996) with the kind permission of the *American Journal of Enology and Viticulture*.

greatly affected (Table 6.1). Growth of YH-15 in these Chardonnay musts resulted in wines of much lower pH (3.30 and 3.75) and higher titratable acidity (1.52 and 0.51 g/100 mL) and volatile acidity (0.31 and 0.025 g/ 100 mL) compared with wines without bacteria. Ethanol yields were also significantly lower with YH-15.

Bacterial strains, YH-24, YH-37, and YH-15, were originally isolated from commercial wines. The first two have since been identified as wild

Table 6.1. Composition of Chardonnay wines fermented by either Prise de Mousse or Epernay in the absence or presence of *O. oeni* strains YH-24 and YH-37 or *L. kunkeei* strain YH-15.*

Yeast strain/composition	Control	YH-24	YH-37	YH-15
Prise de Mousse				
pH	3.75[c]	3.98[a]	3.95[b]	3.30[d]
Titratable acidity (g/100 mL)	0.51[b]	0.35[d]	0.36[c]	1.52[a]
Volatile acidity (g/100 mL)	0.025[c]	0.037[b]	0.038[b]	0.31[a]
Ethanol (% v/v)	13.2[b]	13.6[a]	13.5[a]	10.4[c]
Epernay				
pH	3.81[c]	4.07[a]	4.04[b]	3.38[d]
Titratable acidity (g/100 mL)	0.48[b]	0.37[c]	0.37[c]	1.39[a]
Volatile acidity (g/100 mL)	0.037[c]	0.054[b]	0.055[b]	0.29[a]
Ethanol (% v/v)	13.4[a]	13.2[a]	13.3[a]	10.7[b]

*Bacteria were inoculated two days before yeast inoculation.
Means in rows with different letters for a given yeast strain are significantly different (p < 0.05).
Adapted from Huang et al. (1996) and with the kind permission of the *American Journal of Enology and Viticulture*.

strains of *Oenococcus oeni* (Edwards et al., 1998b), and YH-15 is a novel species, later named *Lactobacillus kunkeei* (Edwards et al., 1998a). More recently, another novel species of *Lactobacillus* was found to inhibit yeast based on results from an agar well assay and was named *L. nagelii* (Edwards et al., 2000). Other species of *Lactobacillus* found to inhibit yeast include two strains of *L. vermiforme* (*L. hilgardii*), a species frequently found in wines (Mills, 2001).

A number of mechanisms for yeast inhibition by *Lactobacillus* have been proposed including production of acetic acid. Boulton et al. (1996) indicated that enough acetic acid could be produced by "ferocious" lactobacilli in 2 or 3 days to inhibit yeast metabolism. Acetic acid is well-known to be inhibitory to yeasts (Doores, 1993), influencing both growth and fermentative abilities (Pampulha and Loureiro, 1989; Ramos and Madeira-Lopes, 1990; Kalathenos et al., 1995). In agreement, Rasmussen et al. (1995) reported that addition of 4 g/L midway through grape juice fermentations slowed the fermentations. Later research by Edwards et al. (1999a) noted that the acetic acid concentration in musts inoculated with *L. kunkeei* or with both *S. cerevisiae* and *L. kunkeei* approached 5 g/L, an amount that can slow fermentations. The mechanism of yeast inhibition may partially be due to production of acetic acid (Edwards et al., 1999a) but other mechanisms are probable (Mills, 2001).

Processing strategies to control infections of lactobacilli may include the use of SO_2, low-temperature storage, and adjustment of must pH. Of these, *L. kunkeei* was determined to be highly sensitive to SO_2 (Edwards et al., 1999b). In addition, Gram-positive bacteria like *Lactobacillus* are sensitive to lysozyme (Section 5.2.3).

6.6.3 *Oenococcus, Pediococcus,* and *Lactobacillus*

Because of the increase in pH, MLF can be favorable to the growth of other lactic acid bacteria like *Pediococcus* (Davis et al., 1986a). However, Edwards et al. (1994) observed a definite antagonism by *O. oeni* against *Pediococcus* in Cabernet Sauvignon and Merlot wines that had undergone malolactic fermentation (Fig. 6.6). Here, the viability of all strains declined from an initial population of 10^5 to between <300 and 10^4 CFU/mL shortly after inoculation. In addition, the extent of this decline depended on the strain of pediococci. For example, the viability of WS-29A only decreased *ca.* 1 log in both wines, whereas growth of C5 in the MLF (+) Cabernet Sauvignon wine was not observed (<300 CFU/mL). The fact that viability of *P. parvulus* decreased in MLF (+) wines after inoculation supports the contention that MLF can impart microbiological stability to wines. However, this "stability" was not necessarily permanent in that most

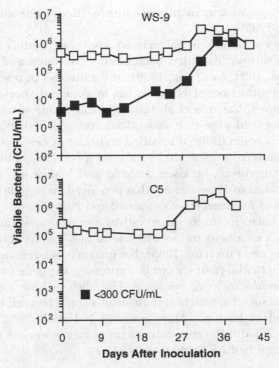

Figure 6.6.   Growth of *P. parvulus* strains WS-9 and C5 in Merlot wines that did not have MLF (open squares) and wines that completed the secondary fermentation (closed squares). Adapted from Edwards et al. (1994) with the kind permission of the *American Journal of Enology and Viticulture.*

strains overcame the initial inhibition and eventually grew to populations approaching or exceeding 10^6 CFU/mL. More recently, Walling et al. (2005) noted that wines that had undergone MLF were more resistant to ropiness caused by pediococci.

Other reports detailing the complicated interactions between wine lactic acid bacteria have been published. Although *O. oeni* can inhibit *Pediococcus* (Edwards et al., 1994), Davis et al. (1986a) noted that growth of *P. parvulus* was antagonistic to the survival of *O. oeni* after MLF in some red wines from Australia. In addition, Lonvaud-Funel and Joyeux (1993) observed strong inhibition of *O. oeni* by *L. plantarum* and *P. pentosaceus.* The authors theorized that the effect was due to accumulation of small (less than 1 kDa) compounds, quite possibly peptides or proteins. In other studies, synthesis of H_2O_2 by a strain of *L. hilgardii* was enough to inhibit

O. oeni and *P. pentosaceus* in mixed culture (Rodriguez and Manca de Nadra, 1995a; 1995b).

Many species of lactic acid bacteria are known to produce antibacterial proteinaceous substances called bacteriocins (De Vuyst and Vandamme, 1994; Jack et al., 1994; Nes et al., 1996), and a number of researchers have reported the production of bacteriocins by bacterial species present in wine. As examples, Navarro et al. (2000) isolated nine strains of *L. plantarum* from Rioja red wine that showed antibacterial activity. In addition, Yurdugul and Bozoglu (2002) identified an isolate of *Leuconostoc mesenteroides* subsp. *cremoris* from wine that produced a bacteriocin-like inhibitory substance. Furthermore, Strasser de Saad and Manca de Nadra (1993) isolated two strains of *P. pentosaceus* that produced an inhibitory substance against strains of *Lactobacillus, Oenococcus,* and *Pediococcus.*

The use of bacteriocins as preservatives has generated interest among researchers as a means to reduce the use of SO_2 due to potential health concerns (Yang and Purchase, 1985). For instance, Schoeman et al. (1999) developed bactericidal yeast strains by expressing the gene for *pedA* pediocin from *P. acidilactici* in *S. cerevisiae.* The authors have suggested that development of such bactericidal yeast strains could be used to inhibit the growth of spoilage bacteria. More recently, Nel et al. (2002) and Bauer et al. (2003) applied different bacteriocins as means to control malolactic fermentation and biofilm formation by *O. oeni.*

6.6.4 Other Interactions

Yeasts are well-known to produce and secrete so-called killer factors that are inhibitory to other yeasts (Jacobs and van Vuuren, 1991; van Vuuren and Jacobs, 1992; Shimizu, 1993). These factors are either proteins or glycoproteins and are lethal to sensitive yeasts. The ability to produce killer factors is widespread among yeasts including *Saccharomyces, Hansenula, Pichia, Kluyveromyces, Candida, Kloeckera, Hanseniaspora, Rhodotorula, Trichosporon, Debaryomyces,* and *Cryptococcus* (Shimizu, 1993). Radler et al. (1985) identified strains of *H. uvarum* that produced killer toxins that had activity toward sensitive strains of *S. cerevisiae.* More recently, Comitini et al. (2004) found two non-*Saccharomyces* yeasts capable of inhibiting a number of other yeasts including *Brettamomyces/Dekkera.* Here, *Pichia anomala* and *Kluyveromyces wickerhamii* were able to produce different molecular weight killer toxins that were stable in wines for 10 days. Given their stability in wine, these toxins have the potential to be used as fungicidal agents during the aging. Finally, Gerbaux et al. (2002) suggested that malolactic fermentation may also be inhibitory to the growth of *Brettanomyces* but data were not provided and additional research is required.

In an early report, Gilliland and Lacey (1964) noted that *Acetobacter* is capable of not only inhibiting *Saccharomyces* but also *Pichia, Schizosaccharomyces, Zygosaccharomyces, Candida,* and even *Brettanomyces.* Although the mechanism of inhibition was not specifically studied, these authors did not observe any rise in acidity and therefore proposed that *Acetobacter* produces some type of antibiotic under aerobic conditions. Although Drysdale and Fleet (1989a) noted that yeast growth was only slightly impaired by the presence of acetic acid bacteria, activity of *Acetobacter* resulted in stuck alcoholic fermentations depending on the juice and species present.

CHAPTER 7

HARVEST AND PRE-FERMENTATION PROCESSING

7.1 INTRODUCTION

The influence of microorganisms on winemaking begins in the vineyard and develops through fermentation and storage or aging of the wine. Grapes arriving at the winery reflect not only the condition of fruit at harvest but also the method of harvest and transport time. As visually defect-free fruit harbors yeasts (Section 6.2.1), molds (Section 4.2), and bacteria (Sections 6.4.1 and 6.5.1), it is important to use processing protocols that minimize the potential for further growth of undesirable native microflora.

7.2 HARVEST AND TRANSPORT

In that broken berries and juice will encourage uncontrolled enzymatic activity and microbial proliferation, mature fruit should be harvested in a manner that produces the least mechanical damage. As both factors are influenced by temperature, harvesting during the evening and early morning may improve quality by taking advantage of lower ambient temperatures. Ough and Amerine (1966) noted that the temperature of harvested fruit held in metal gondolas can rise appreciably, resulting in excessive enzymatic browning and microbial growth.

Transport time to the winery should be minimized to the extent possible. In this regard, widespread use of mechanical harvesters, although logistically and economically advantageous to growers, may create a winemaking challenge if there are significant delays between harvest and processing. In particular, infrequent sanitation of harvesting equipment may promote problems, especially as travel time to the winery increases. Further deterioration of mechanically harvested fruit during transit can be minimized by the addition of SO_2 (Section 5.2.1).

7.3 FRUIT QUALITY ASSESSMENT

Grower and winery contracts establish the maturity standards that are used to determine the optimal quality of incoming fruit during harvest. Such contracts may specify acceptable levels of sugar (soluble solids), pII, titratable acidity, and microbial spoilage. Other parameters can also be incorporated including levels of materials-other-than-grape (MOG) and the extent to which fruit exhibits physiological stress such as shriveling or ambering. Above all, the criteria and methodology used to establish adherence to standards should be agreed upon between the grower and winery prior to harvest.

7.3.1 Soluble Solids

Unlike most fruit where sucrose represents the major sugar, grapes consist primarily of glucose and fructose, with much smaller amounts of sucrose. Although sucrose is present in grapes at only 0.2% to 1% w/w (Hawker et al., 1976), this sugar is fermentable because *Saccharomyces* produces an extracellular invertase, an enzyme that catalyzes the hydrolysis of sucrose to yield glucose and fructose (Goldstein and Lampen, 1975).

Margalit (2004) proposed that optimal harvest sugar concentrations be 19°Brix to 23°Brix (white cultivars) and 21°Brix to 24°Brix (red cultivars). However, winemakers in the United States will frequently require more mature fruit optimal to maximize flavor development associated with physiological maturity (i.e., desired tannin structure). In the case of late harvest or *Botrytis*-infected fruit, the concentration of fermentable sugars can be much higher, ranging upward to 40% w/w.

Even though the concentrations of glucose and fructose vary with grape maturity, the ratio of the two sugars is generally close to 1:1 (Amerine et al., 1972). However, the rate of glucose uptake is approximately five times that of fructose during alcoholic fermentation by *Saccharomyces* (Dittrich, 1977). This results in a change in the glucose/fructose ratio from near 0.95 at the start of fermentation to 0.25 toward completion (Peynaud, 1984). Berthels et al. (2004) reported a similar pattern of sugar utilization among 17 strains of *Saccharomyces cerevisiae*. These reports show that *Saccharomyces* is glucophilic, preferentially utilizing glucose over fructose, possibly due to the inability of the cell to transport fructose later in fermentation (Schütz and Gafner, 1995). Concerns about glucose/fructose utilization may be of little practical importance to the winemaker except that fructose is sensorially sweeter than glucose (Godshall, 1997). While this can be used to the winemaker's advantage in production of light sweet wines, stuck fermentations result in a disproportionately sweet wine.

In addition to six-carbon monosaccharides (glucose and fructose) and 12-carbon disaccharides (sucrose), five-carbon pentoses are also present in grape musts. Among these, arabinose is the most abundant although rhamnose, xylose, and ribose have also been detected (Esau, 1967; Franta et al., 1986). According to Amerine et al. (1972), concentrations range from 0.8 to 2 g/L in grape musts. Although *Saccharomyces* cannot ferment pentoses, other yeasts (e.g., *Brettanomyces*), molds, and bacteria (e.g., some lactic acid bacteria) can metabolize some of these sugars. Thus, pentoses can serve as a source of carbon after alcoholic fermentation for some spoilage microorganisms.

7.3.2 pH and Titratable Acidity

Juice and wine pH play a major role in the development of yeast and bacterial populations. Species respond differently in terms of growth rate as well as production of sensorially important metabolites. In the case of *Saccharomyces*, growth and fermentation rates slow as pH decreases to near 3.0, concomitantly increasing the risk of sluggish or stuck fermentations. Kudo et al. (1998) found that the relative concentration of K^+ and H^+ in a grape must plays an important role in successful completion of alcoholic fermentation, with a minimum K^+/H^+ of 25:1 being necessary. Whereas molds can grow at very low pH (2.0), many lactic acid bacteria do not grow well at pH less than 3.5 (Section 6.4.2).

Determining optimal pH and titratable acidity for harvest is imprecise and difficult. If acidulation of a must is required, tartaric acid is normally the acid of choice. The amount of acid needed to correct acidity depends on existing titratable acidity, pH, and the buffer capacity of the must. When using tartaric acid within a pH range of 3.6 ± 0.3, an addition of 1 g/L will lower the pH by about 0.1 ± 0.03 units depending on buffering capacity (Margalit, 2004). Malic or citric acids can also be used, but addition of 1 g/L will decrease pH by a slightly lesser amount, 0.08 ± 0.02. Care should be taken when adding citric due to potential bacterial degradation forming diacetyl once alcoholic fermentation is complete (Section 2.4.5). Must/wine pH also plays a major role in the activity of various antimicrobial agents used in winemaking such as SO_2 (Section 5.2.1) as well as sorbic (Section 5.2.4) and fumaric (Section 5.2.5.1) acids.

7.3.3 Microbial Spoilage

Because mold spores are ubiquitous, controlling germination and growth in the vineyard by use of fungicides may only be part of the solution. For instance, inherent properties of the fruit as well as cultural practices play a significant role toward encouraging or discouraging mold infections. With tight-clustered varieties, the mechanical pressure of adjacent berries may disrupt cuticular waxes (Rosenquist and Morrison, 1989), thus predisposing these sites to greater incidences of infection. Thinner-skinned varieties are likewise more prone to *Botrytis* infection than other thicker-skinned cultivars. Finally, dense canopies, which minimize air circulation and light incursion, favor a relatively humid environment, which increases the potential for mold growth. Open canopies not only favor air circulation but spray penetration as well (English et al., 1990). Infections due to different species of molds can be found in Sections 4.1 to 4.5.

One of the most common methods to determine the extent of microbial damage on fruit arriving at the winery involves visual separation of sound and defective grapes followed by weighing each component and expressing the results on a weight percentage basis. Microbially compromised fruit can also be detected by "off-character" odors, sometimes even when rot is not easily seen. Alternative methods have been developed that provide for a more objective evaluation of microbial deterioration of fruit. Such methods revolve around quantification of key indicator metabolites of mold (glycerol and/or laccase), bacteria (acetic acid), and yeast (ethanol) activity using high-performance liquid chromatography (Kupina, 1984) or by other means.

Because chromatographic techniques are difficult to apply for routine analysis of musts, other methods have been explored in an attempt to better quantify mold damage in grapes. More recently, Marois et al. (1993) and Dewey et al. (2000; 2005) developed immunoassays for quantitation of *Botrytis*. Using their immunoassay, Marois et al. (1993) estimated that two samples per each gondola for hand-harvested fruit and one per each gondola for machine-harvested would be needed. Because many fungi produce laccases (Mayer and Harel, 1979), some have attempted to use activity of these enzymes to quantify mold or rot damage in grapes (Dubourdieu et al., 1984). The colorimetric method of Dubourdieu et al. (1984) relies on laccase-catalyzed oxidation of syringaldazine to its corresponding colored quinone and is now available as a field assay.

7.4 MUST PROCESSING

Management of the fruit upon arrival at the winery represents a critical quality control point in the production of wine. Decisions made at this stage may select for growth of some microbial species at the expense of others and, thereby, creating or diminishing the potential for problem fermentations. As noted previously (Chapter 6), incoming fruit will have a diverse microflora present but these can sometimes be controlled by changing intrinsic (e.g., suspended solids and use of SO_2) and extrinsic (e.g., temperature and limiting oxygen) conditions during must processing.

7.4.1 Enzymes

Pre-fermentation processing enzymes have been used for a long time by the wine and juice industries (van Rensburg and Pretorius, 2000). The first commercial enzyme preparations used were pectinases, enzymes responsible for the breakdown of pectin. Conventional application of pectinases pre-fermentation can enhance juice yield by degrading polysaccharides (e.g., pectin) that interfere with juice extraction. These preparations also improve the extraction of color (red wines) and flavor molecules trapped in grape skins as well as reduce the potential for post-fermentation instabilities. Newer commercial preparations may also contain secondary activities such as the ability to degrade cellulose to further enhance cell wall breakdown.

Besides pre-fermentation clarification, post-fermentation applications have been increasingly identified and include enhanced aromatic profiles, prevention of protein instability (especially white wines), and degradation of β-glucans. Aroma and flavor enhancing enzymes, most commonly β-glucosidases, can enhance the varietal character of Riesling and Gewürztraminer wines (Section 1.5.4). Proteolytic enzymes, responsible for hydrolysis of peptide linkages within amino acids of proteins, could potentially reduce hazes in finished wines, but enzymes not inhibited under wine conditions have yet to be found (van Rensburg and Pretorius, 2000). Polysaccharide (β-glucan) instability can be associated with mold growth on grapes and, on occasion, impact wine clarification and stability. To reduce this problem, β-glucanases are available either directly or as part of macerating enzyme preparations added either pre- or post-fermentation.

7.4.2 Suspended Solids

Modern white wine production techniques generally reduce the levels of suspended (insoluble) solids prior to fermentation. Conventional thought in this regard equates development and retention of fruit and varietal character while minimizing formation of those volatiles that reduce wine quality. For instance, Crowell and Guymon (1963) determined that wines produced from turbid grape musts contained higher concentrations of fusel alcohols, especially isobutyl and isoamyl alcohols, than those made from clarified juice. The authors further noted that addition of other inert solids such as cellulose and starch to fermenting musts leads to an increase in the production of higher alcohols. Houtman et al. (1980a) reported that addition of 1% to 2% juice lees increased the production of esters during fermentation but that larger additions (>5%) decreased ester concentrations. Delfini and Costa (1993) noted that various insoluble materials affected *Saccharomyces* in different ways with regard to production of acetic acid and acetaldehyde. Finally, lower amounts of isoamyl alcohol were present in wines where the musts were clarified by vacuum filtration, in comparison with those clarified by centrifugation or settling (Ferrando et al., 1998).

Conversely, overclarification of a must prior to alcoholic fermentation is also a concern. In general, yeasts do not grow well in highly clarified musts, leading to increased likelihood of slow or stuck fermentations (Williams et al., 1978; Edwards et al., 1990; Varela et al., 1999). In agreement, Groat and Ough (1978) found that different types of insoluble solids (e.g., grape solids, bentonite, talc, and diatomaceous earth) added to clarified juice resulted in faster and more complete alcoholic fermentations. According to these authors, the critical level of grape insoluble solids is between 0.1% and 0.5% v/v, below which fermentations were slower.

Clarification may also alter the nutritional status of the must. For instance, Ayestarán et al. (1995) observed that total nitrogen levels in unclarified musts were higher than those clarified by either gravity or vacuum filtration. By comparison, amino nitrogen levels were higher in clarified musts. Ferrando et al. (1998) found no effect of clarification by different methods (vacuum filtration, centrifugation, and settling) on ammonium nitrogen and total free amino acids in musts. Guitart et al. (1998) evaluated several commonly used pre-fermentation fining agents with regard to amino acid reduction and reported that silica gel additions removed the highest concentration of amino acids followed by enzyme treatment, cold-clarification, bentonite, and centrifugation. Pectic enzymes added after crush have also been reported to have an effect on the amino acid composition (Hernandez-Orte et al., 1998).

Besides potential changes in available nitrogen, clarification may also reduce other nutrients such as sterols (Delfini et al., 1993) essential for yeast cell membrane integrity. Changes in the fatty acid composition of musts have also been reported (Varela et al., 1999). However, it is thought that a major influence of clarification is the removal of trace amounts of oxygen trapped within the suspended solids. Crowell and Guymon (1963) reported that fusel oil production was reduced in fermentations where the suspended solids were deaerated before alcoholic fermentation, indicating trot insoluble solids can adsorb O_2.

Clarification can be used to reduce populations of non-*Saccharomyces* yeasts (Mora and Mulet, 1991). Although populations are initially reduced, growth of these yeasts may continue during cold-clarification and actually lead to denser populations than expected. Many of these yeasts may also be active during the course of fermentations conducted at lower temperatures (Section 6.2.2). The presence of insoluble solids in musts also influences subsequent malolactic fermentations (Liu and Gallander, 1982). Here, the authors reported that MLF was most rapid in wines fermented with the highest amount of solids.

Pre-fermentation clarification may be accomplished by cold-settling with or without fining agents, centrifugation, and/or filtration. In terms of removing unwanted native yeast and bacteria, Fleet and Heard (1993) noted that centrifugation was generally more effective than cold-clarification. At least one report noted differences in volatile acidity depending on clarification procedure (Aragon et al., 1998). The combination of cold-clarification and fining agents such as bentonite and proprietary formulations continues to be the method of choice among many producers.

7.4.3 Pre-fermentation Maceration (Cold-Soak)

Cold-soaking of musts prior to fermentation is a technique in red wine processing that encourages extraction of desirable grape flavor compounds and pigments from skins. Normally, crushed grape musts are held at cool temperatures (15°C/59°F to 20°C/68°F) for 12 to 24h but also up to 1 to

2 weeks. During this time, non-*Saccharomyces* yeasts can proliferate, due to their ability to grow better than *Saccharomyces* at lower temperatures, resulting in considerable changes to quality (Section 6.2.2).

Effective cold-soak requires that the must temperature be lowered rapidly. In this regard, the use of must chillers, dry ice, or liquid CO_2 can prove useful. Once the target temperature is achieved, it must be maintained and not allowed to slowly increase. Not infrequently, cold-soak temperatures can be warmer than expected, therefore encouraging proliferation of spoilage microorganisms such as *Lactobacillus*. Fortunately for the winemaker, the causative bacteria are relatively sensitive to the effects of both SO_2 and pH (Edwards et al., 1999b). This observation has led to the recommendation either to acidulate (pH <3.5) or, when this is not feasible, addition of 50 to 75 mg/L total SO_2. Lysozyme added prior to alcoholic fermentation (250 mg/L) has also been used as a means to reduce initial populations of Gram-positive bacteria (Section 5.2.3).

Even if the musts are maintained at cool temperatures, other undesirable microorganisms such as acetic acid bacteria or some non-*Saccharomyces* yeasts like *Kloeckera apiculata* can still grow. As an example, Du Toit and Lambrechts (2002) observed acetic acid bacteria to increase from 10^3 up to 10^5 CFU/mL in one Cabernet Sauvignon must held at 15°C/59°F to 18°C/64°F for 3 days with 40 to 50 mg/kg SO_2. As the other must already contained a high population of bacteria (10^5 CFU/mL), the population did not increase further. However, the bacteria quickly died off as alcoholic fermentation began (Du Toit and Lambrechts, 2002).

7.4.4 Thermovinification

Another technique to extract color from red grape cultivars is thermovinification. Here, a portion of the must is heated to 50°C/122°F to 60°C/140°F and mixed with the skins for a short period of time (Boulton et al., 1996). The must is then pressed prior to inducing alcoholic fermentation. In an interesting study by Malletroit et al. (1991), the effect of pasteurization on the microbiological and sensory quality of white grape juice and wines was investigated. The authors concluded that a pasteurization treatment equivalent to that used in the brewing industry is sufficient to reduce non-*Saccharomyces* populations prior to alcoholic fermentation without detrimental effects on wine quality.

7.4.5 Inert Gassing

Depending on cultivar and issues of winemaking style and philosophy, the practice of routine prefermentation purging of lines and tanks with nitrogen or carbon dioxide gas may not be necessary. In this regard, purging with inert gases lowers the concentration of physiologically useful oxygen available to *Saccharomyces* for cell membrane synthesis (Section 1.4.2). In the case of aromatic grape varieties (e.g., Muscat and Riesling), intentional exposure to oxygen may lead to irreversible oxidation and undesirable loss of aroma and flavor-active compounds (Section 1.5.4).

Grape processing and starter preparation can influence survival of yeasts during long-term (>2 weeks) fermentations. Starter aeration using sterile compressed air has been found to be beneficial (Wahlstrom and Fugelsang, 1988). Similarly, Valero et al. (2002) noted that incorporation of oxygen in grape musts affected synthesis of both higher alcohols and esters during fermentation. The winemaker can encourage a longer period of aerobic development by pouring rehydrated yeast starters over the top juice/must and not mixing into the tank.

7.5 PROCESSING MICROBIALLY DETERIORATED FRUIT

Microbially deteriorated fruit represents a challenge not only in terms of extraction of mold-produced odors and flavors into the juice and wine but also to the success of fermentation and clarification. *Botrytis* activity in grape musts results in (a) browning due to laccase, (b) problematic fermentations, and (c) post-fermentation clarity and stability problems. In the case of red grape cultivars, *Botrytis* also brings about significant deterioration/reduction in total anthocyanins (Ewart et al., 1989).

7.5.1 Enzymatic Browning

Enzymatic browning is especially a problem in white grape musts. Once berry integrity has been compromised at the crusher, oxidative enzymes (oxidases) and their respective substrates (phenols) are exposed to air, resulting in rapid formation of brown pigments (Traverso-Rueda and Singleton, 1973). Oxidases such as polyphenol oxidases (tyrosinases) originate from plant tissue (Mayer and Harel, 1979), whereas laccases come from molds.

Laccases are enzymes involved in the reduction of oxygen to produce water using phenolics as hydrogen donors (Fig. 7.1). One consequence of this activity is that juice will brown due to the reactivity of the oxidized phenolics (quinones) in formation of brown pigments, melanins. Although *Botrytis*-derived laccase is active against *o*-diphenols (Dubernet et al., 1977), Mayer and Staples (2002) pointed out that some fungal laccases can also oxidize monophenols. Unlike grape oxidases, which are inhibited by SO_2 even at low concentrations, laccases tend to be resistant. These enzymes

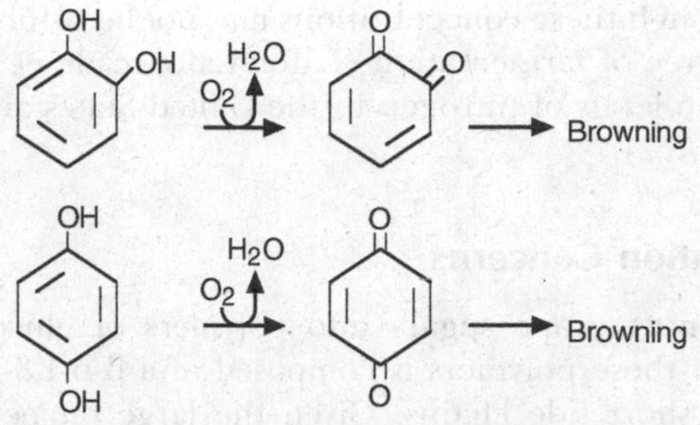

Figure 7.1. Reactions catalyzed by laccase.

also tolerate ethanol as illustrated by Somers (1984) who reported detecting activity in wines after 12 months of storage. Although one strategy to limit the effects of laccasses would be to minimize oxygen exposure during the processing of infected musts, application of heat is also used. However, the use of chelators, molecules that bind and therefore remove metals, has not been studied as a means to limit laccase activity. Like other oxidases, laccases require copper for activity, and chelators can be used to slow enzymatic browning in other foods (McEvily et al., 1992). Bentonite fining of juice can help reduce laccase activity, although this procedure may not completely remove the enzyme.

7.5.2 Fermentation Difficulties

Alcoholic fermentations are affected by *Botrytis* by several mechanisms. First, *Botrytis*-infected fruit tends to be higher in sugar due to the loss of water during the post-infection dehydration phase, resulting in greater osmotic stress upon *Saccharomyces* during fermentation. Second, *Botrytis* synthesizes a group of heteropolysaccharides, collectively referred to as "botryticine" (Donèche, 1993). Fermentation in the presence of botryticine leads to yeast inhibition coupled with higher than expected concentrations of acetic acid and glycerol, the relative amount depending on the stage of fermentation.

The nutritional status of *Botrytis*-infected fruit is another concern. For instance, musts from mold-damaged fruit are generally low in assimilable nitrogen (Dittrich, 1977). Sponholz (1991) reported decreases of 7% to 61% in the amino acid content of *Botrytis*-infected musts, and Henick-Kling (1994) noted that the losses of amino acids in botrytized grapes may be 1000 to 3500 mg/L. Nutrient deficiencies can be further increased by continued growth of non-*Saccharomyces* yeasts. In these cases, Peynaud (1984) recommends the addition of nitrogen supplements at 100 to

200 mg/L, although these concentrations may not be sufficient to achieve the desired degree of fermentation (Lafon-Lafourcade et al., 1979). Permissible addition levels of nitrogen in the United States are considerably higher (Section 8.2).

7.5.3 Clarification Concerns

Botrytis can convert grape sugars into polymers of glucose, known as glucans. One of these polymers is composed of a β-D-1,3-linked glucose backbone with short side chains. Given the large molecular weight of approximately 80 kD (Dubourdieu et al., 1981), these glucans greatly impede clarification and filtration even at concentrations as low as 10 mg/L (Villettaz et al., 1984). Although fining is generally ineffective in removing these molecules, some commercial glucanases have proved successful (Villettaz et al., 1984). Adding glucanases either to the juice at clarification or post-fermentation brings about hydrolysis of polymers (Dubourdieu et al., 1981). Because pre-fermentation additions to chilled juice require longer contact times and higher dosage rates, it is generally recommended that the enzyme preparation be added to the wine at first racking. Glucans can be measured using the method outlined in Section 17.5.3.

7.5.4 Management Strategies

Where rot cannot be avoided on incoming fruit, significant modification to traditional must and juice processing protocol is advised. Physical removal (grading) of fruit may be feasible in the case of wineries working with premium varietals. Processing of mold-damaged fruit for both red and white wines requires that both the extent of tissue maceration and time of juice and skin contact be minimized.

For white grape varieties, whole-cluster pressing is preferable to crushing. In this case, it may be necessary to separate the first 10+ gallons (per ton of grapes) collected from the press because this fraction is relatively rich in metabolites from mold growth. Subsequent press fractions should be individually evaluated for browning, phenolic extraction, and other markers of quality impairment. Once the must is prepared, some winemakers choose to sulfite at high concentrations (0.8 mg/L molecular SO_2) and immediately cold-clarify with bentonite in the hope of minimizing both microbiological and enzymatic deterioration. Polyvinyl polypyrolidone (PVPP) is sometimes added at a rate of 2 to 4 lb/1000 gal as well as gelatin–silica gel to reduce the concentrations of browning compounds and/or their precursors. Other winemakers prefer to allow any oxidized phenols to polymerize or precipitate at low temperatures for 24 h prior to the addition of SO_2 and PVPP. In any case, acidulation to pH <3.5 will aid in controlling spoilage bacteria.

Processing red musts infected with *Botrytis* is not as easily handled as white musts. Aside from SO_2 additions and minimizing pomace contact ("short-vating"), decreasing the pH through acidulation can minimize browning as well as limit the growth of spoilage bacteria. Heat processing of must and juice has also been used to prevent further growth of spoilage microorganisms as well as to denature browning enzymes including laccase. However, this treatment requires a heat exchanger(s) and may result in difficulties in post-fermentation clarification as well as lower wine quality.

7.6 JUICE STORAGE (MUTÉ)

Storage of unfermented juice (muté) for later blending or other purposes presents important and potentially challenging problems in terms of the growth of psychrotolerant ("cold-tolerant") and psychrophilic ("cold-loving") microorganisms. Although molds are capable of growth at the surface of refrigerated juice, yeasts represent a more serious problem. In an early report, Pederson et al. (1959) reported *Saccharomyces*, *Hanseniaspora*, *Torulopsis*, and *Candida* in juice stored at 0°C/32°F. Furthermore, Lawrence et al. (1959) recovered a number of different yeasts from commercial grape juice stored at −5°C/23°F to −2°C/28°F. Subsequently, Splittstoesser (1978) determined that psychrophilic species of *Candida* often comprise 95% of yeast contaminants in stored juice.

Filtration, in combination with SO_2 additions of 30 to 50 mg/L and subsequent cold storage, represent the most frequently used techniques for minimizing oxidation and microbiological activity during storage. However, SO_2 alone is not enough to completely halt yeast growth because many non-*Saccharomyces* are tolerant to high concentrations (Section 5.2.1). Since yeasts are inhibited at high pressures (Table 7.1), some winemakers

Table 7.1. Effect of CO_2 pressure on yeast viability.

CO_2 pressure (lb/in²)	Yeast population (cells/L)
0	104
29.4	15
44.1	11
58.8	6
73.5	3
88.2	0

Adapted from Schmitthenner (1950).

have combined low temperature ($<2°C/36°F$) and carbon dioxide ($51.3 lb/in.^2$ or $352 kPa$) to store juice. Others have reported successful storage of juice in Charmat tanks maintained at pressures of 70 to $90 lb/in.^2$ (480 to $620 kPa$). Delfini et al. (1995) studied the use of very high pressures to kill yeasts and bacteria in both grape musts and in wines. Where sorbic acid is used to stabilize stored juice against yeast activity, SO_2 should also be considered to prevent growth of lactic acid bacteria because these microorganisms can metabolize the acid, forming precursors that react with ethanol to impart a "geranium" odor to wine (Section 11.3.5).

CHAPTER 8

FERMENTATION AND POST-FERMENTATION PROCESSING

8.1 INTRODUCTION

During alcoholic fermentation, *Saccharomyces* converts grape sugars to alcohol and carbon dioxide as per the Gay–Lussac equation where 1 mole sugar yields 2 moles each of ethanol and carbon dioxide (Fig. 8.1). The rate of ethanol production by *Saccharomyces* varies with many factors but can reach 8×10^7 to 9×10^7 molecules of ethanol per yeast cell per second (Foy, 1994b).

The winemaker's challenge is to optimize those conditions that favor growth of fermentative yeasts (*Saccharomyces*) resulting in complete alcoholic fermentation (final reducing sugar concentrations of <0.2% w/v) while avoiding formation of undesirable odors or flavors. Ideally, completion of alcoholic and malolactic fermentations should result in a wine that is nutritionally insufficient to support further microbiological activity (i.e., the wine has become a "nutrient desert").

8.2 MUST SUPPLEMENTATION

The concentration and composition of nitrogen-containing compounds in grape must plays a crucial role in the nutrition of microorganisms involved

$$C_6H_{12}O_6 \longrightarrow 2\,CH_3CH_2OH \;+\; 2\,CO_2$$

Glucose or Fructose   Ethanol   Carbon Dioxide

Figure 8.1.   Utilization of sugars to yield ethanol and carbon dioxide as per the Gay–Lussac equation.

in fermentation as well as the potential for spoilage after fermentation. *Saccharomyces* will utilize most amino acids in must, with the exception of proline, which is not metabolized under anaerobic conditions (Bisson, 1991).

The nitrogenous components of grapes and must that are metabolically available to yeast are present as ammonium salts (NH_4^+) and amino acids primary, collectively known as "yeast assimilable nitrogen" (YAN). Therefore, a complete evaluation of the nutritional status of juice or must requires measurement of both fractions. The total nitrogen content of juice not only contains YAN but peptides and proteins as well. However, the latter two fractions are not thought to play a significant role in the nutritional needs of *Saccharomyces* during fermentation.

The concentration of yeast assimilable nitrogen present in must is vineyard-specific, varying with climate and soil type, grape variety, rootstock, fertilization and irrigation practices, harvest maturity (degree of ripening), as well as the extent of microbiological deterioration that may have occurred prior to harvest (Ough and Bell, 1980; Huang and Ough, 1989; 1991; Spayd et al., 1994). In general, many grape musts are considered to be deficient in nitrogen based on the estimated minimal requirement of 140 to 150 mg N/L. In a survey of grape musts from California, Oregon, and Washington, YAN ranged from 40 to 559 mg N/L (Table 8.1). Of the 1523 samples examined, Butzke (1998) noted that 13.5% were below 140 mg N/L. Spayd and Andersen-Bagge (1996) found that 25% of samples from central Washington State were below this concentration.

Table 8.1.   Yeast assimilable nitrogen (YAN) in 1523 grape musts obtained from Callifornia, Oregon, and Washington.

	Ammonium (mg N/L)	Primary amino acids (mg N/L)	YAN (mg N/L)
Average ±SD	79 ± 35	135 ± 51	213 ± 70
Range	5–325	29–370	40–559

Adapted from Butzke (1998) with the kind permission of the *American Journal of Enology and Viticulture.*

Whereas urea has historically been used as a nitrogen supplement during fermentation, this nitrogen source is no longer approved in many countries owing to its demonstrated involvement in the formation of ethyl carbamate (Section 11.3.2). Rather, nitrogen deficiencies are commonly corrected by the addition of diammonium phosphate (27% NH_4^+ and 73% PO_4^{3-}). In the United States, the maximum level of $(NH_4)_2HPO_4$ legally permitted to correct nutritional deficiencies is 960 mg/L (8 lb/1000 gal), a concentration that provides 203 mg N/L assimilable nitrogen. In Europe, the Office International de la Vigne et du Vin (O.I.V.) only allows a maximum addition of 300 mg/L $(NH_4)_2HPO_4$, and in Australia, additions are limited by maximum phosphate levels in the wine. There, a maximum concentration of 400 mg/L inorganic phosphate/L is permitted (Henschke and Jiranek, 1993).

In addition to $(NH_4)_2HPO_4$, some winemakers advocate the use of balanced nutritional formulations that also contain amino acids, minerals, vitamins, and/or other ingredients important for yeast growth. Because some of the ingredients in these formulations may not be currently approved for use in the United States, winemakers must petition the U.S. Alcohol and Tobacco Tax and Trade Bureau under Section 27 CFR 24.250 to use material not specifically authorized (R. Gahagan, personal communication, 2005).

Although adding inorganic nitrogen at yeast inoculation is convenient, this practice can encourage the growth of non-*Saccharomyces* yeasts also present. An alternative strategy is to delay nitrogen addition for 48 h (red fermentations) and 72 h (white fermentations) post-inoculation. Sablayrolles et al. (1996) reported that a single addition midway through fermentation was as effective as a single addition at the start. Addition of ammonium late in fermentation should be avoided because this compound is not consumed by *Saccharomyces* at this stage (Beltran et al., 2005). Furthermore, excessive ammonium in a wine may be enough to support the growth of spoilage microorganisms after fermentation has completed.

Some winemakers add yeast hulls (ghosts) to fermentations. Produced as by-products of commercial laboratory media manufacture and of brewing industries, hulls represent the remnants of yeast cell walls. These products provide some assimilable nitrogen and other nutrients and are able to absorb medium-chain fatty acid (C_8 and C_{10}), molecules potentially toxic to wine microorganisms (Munoz and Ingledew, 1989a; 1989b; Edwards et al., 1990). Because yeast hull preparations contain lipids (fats) that oxidize upon exposure to oxygen, they may degrade and develop a "rancid" character upon extended storage. At this point, the hulls should be discarded due to the potential for imparting these off-odors to the wine.

Besides nitrogen, winemakers may wish to adjust must acidity (Section 7.3.2), add antimicrobials like SO_2 (Section 5.2.1) and/or lysozyme (Section 5.2.3) to reduce populations of spoilage microorganisms, or incorporate fining agents (Puig-Deu et al., 1999).

8.3 ALCOHOLIC FERMENTATION

Although musts contain low populations of *Saccharomyces*, all will carry out fermentation under appropriate conditions. However, initiation of these fermentations may require more time than most winemakers are willing to accept, and the outcome is not always what was anticipated. For this reason, yeast starter cultures of *Saccharomyces* are used. Overall, the goal of using a starter culture is to initiate fermentation as quickly as possible while limiting the potential for spoilage by establishment of numerical dominancy over native species. Although commercial cultures of *Saccharomyces* can outgrow and inhibit indigenous populations of non-*Saccharomyces* (Henick-Kling et al., 1998), this is not always the case (Heard and Fleet, 1985; 1988). As evidence, Heard and Fleet (1985) observed growth of both *Kloeckera apiculata* and *Candida* in musts inoculated with *Saccharomyces*.

8.3.1 Historical Perspective of Starter Cultures

Before the development of commercial active dry yeast, winemakers wanting to use starter inocula were forced to propagate these from stock cultures maintained in the winery. This process involves transferring the pure culture to sterile juice and, over a period of several weeks, expanding the volume to a level of 1% to 5% (v/v) of the expected fermentation volume through a series of aseptic transfers. When the appropriate volume of starter was ready and was not microbiologically contaminated, it was then used to inoculate grape musts. Maintenance of pure cultures during multiple transfers was problematic due to the risk of contamination in addition to requiring a significant amount of time and resources.

Given the difficulty of preparation from stock cultures, enologists welcomed development of commercial wine yeasts starters in the 1950s. Early research indicated the potential for using compressed yeasts, a product similar to that used in the baking industry (Foy, 1994a). Because these products contained a high amount of moisture (70%), perishability at the winery before use was a problem. Unlike the baking industry, which requires yeast year-round, wineries required large amounts during a relatively short period of time at crush. This problem was solved in 1963 when compressed yeasts were successfully dried. Two years later, Red Star Yeast (Universal Foods Corporation) released the first commercial wine active dry yeast (WADY) in the United States. It has been estimated that active dry yeasts are used in all of the world's wine regions and by 85% or more of the producers in some areas (Foy, 1994a).

Even with the advantages of commercial cultures, the winemaking community remains divided with regard to the philosophy and practice of using starters. At one extreme are those winemakers that rely solely on yeasts and bacteria native to the winery in the hopes of creating unique products (Section 8.4). Others prefer to encourage the growth of some non-*Saccharomyces* yeasts early in alcoholic fermentation but eventually inoculate with *Saccharomyces*. Still others use *Saccharomyces* starters but at lower than recommended inoculum levels. The vast majority of winemakers employ active yeast starters prepared using manufacturers' recommendations.

8.3.2 Preparation of Starter Cultures

Viabilities of active dry yeasts vary but most formulations contain 1.1×10^{10} to 3.9×10^{10} CFU/g (Foy, 1994b). Given the low moisture content of active dry yeasts, these can be stored for longer periods of time but not indefinitely. For products still packaged under vacuum, Foy (1994b) estimated the monthly loss of fermentation activity at 0.5% to 1% when stored at 5°C/40°F and 1.5% to 2% when stored at 21°C/70°F. If stored at a high temperature (37°C/98°F), WADY can loose 75% to 80% of activity after only 16 months (Foy, 1994b).

The methodology of starter preparation and propagation is crucial to successful fermentation. Although tempting, the practice of simply spreading pellets into or over the surface of the must should be avoided. This generally results in the formation of clumps that are difficult to disperse. In addition, those cells trapped within the aggregate matrix are incompletely rehydrated, resulting in viable cell numbers that may be well below those expected. Rather, WADY should be rehydrated according to the supplier directions. In general, most recommend use of water or 1:1 diluted grape juice at 37°C/99°F to 40°C/104°F and rehydrate for 15 to 20min before inoculation. Monk (1986) noted that these rehydration conditions will result in near 100% viability compared with only 40% to 50% viability if the rehydration was at 15°C/60°F.

Actively growing yeasts, either recently rehydrated or from starter tanks, should not be transferred directly to chilled musts. As a general recommendation, yeasts should be acclimated to within 10°C/18°F of the must temperature before inoculation. Llauradó et al. (2005) found that rehydration of yeast cultures using normal techniques followed by subculturing at lower temperatures could avoid cold-shock and improve yeast performance during cold-temperature fermentations (13°C/55°F). In addition, caution should be exercised if agitation is used because this practice may result in decreased vitality and vigor of starter cultures.

Rather than rehydrate yeasts on a tank-by-tank basis, economics and time may dictate preparation of "starter" tanks in which larger volumes of yeast are propagated and used over a short period. In these cases, it is necessary to prepare relatively large volumes of sterile juice and, during the propagation phase, maintain yeast growth without contamination. Preparation of larger volumes of juice for expansion of starters may be accomplished by heating, cold-clarification, and/or filtration, with the latter two methods most commonly used by wineries. Clarified juice for starters may also be sterilized by the use of dimethyl dicarbonate (Section 5.2.2).

After the addition of yeast inoculum to the starter tank, 24 to 72 h may be required before the expanded starter reaches a sufficient cell number to add to the must at an initial inoculum of 1×10^6 to 3×10^6 CFU/mL. Growth should be followed microscopically, noting viability, percentage of budding cells, and any microbial contaminants. Ongoing starter tanks must also be closely monitored to ensure secondary contamination does not occur. In actively growing starters, budding cells should comprise 60% to 80% of the total cell number upon addition to must. By comparison, the budding cell number of recently rehydrated yeasts is generally only 2% to 5% of the total. Procedures for monitoring cell populations are presented in Chapter 14.

Some wineries have attempted to use other sources for a yeast inoculum including pomace, fermenting wine, or lees from another fermentation that is almost complete. Although seemingly economical, these practices dramatically increase the potential for contamination from spoilage microorganisms (*Acetobacter, Lactobacillus,* and others). In addition, the yeast lacks long-term vigor and vitality at this stage in its life cycle.

As is the case with reusing fermentation lees and pomace as starter inocula, the practice of reserving 5% to 10% of a starter for fresh sterile juice will lead to decreased yeast vigor and/or microbial contamination. Further, continual growth under semi-anaerobic conditions results in stressed cells that may neither initiate nor complete fermentation in a timely manner. Because nutrient exhaustion significantly reduces cell viability, starter tanks should be regularly monitored (°Brix and micro-scopically) and transferred before sugar is depleted.

One difficulty with using WADY is that these cultures are rarely, if ever, pure isolates. In fact, WADY may contain bacterial populations including lactic acid bacteria, contaminants that should not exceed 10^3 to 10^4 CFU/g (Foy, 1994b).

8.3.3 Strain Selection

Yeast strains isolated from wineries or research institutes that possess good qualities for winemaking have been commercialized. Hundreds of strains have now been isolated, and are marketed to winemakers by several inter-national companies. Unfortunately, the origins of some strains are no longer known. According to Foy (1994a), W.V. Cruess (University of California, Davis) originally brought the strain Montrachet (UCD 522) to the United States in the 1930s, presumably from the Montrachet region in France. However, information regarding the winemaker or scientist who initially isolated and cultured this strain has been lost.

Although winemakers often have strain preferences for particular applications, the issue continues to be one of debate. A generalized list of desirable traits of wine yeasts can be found in Table 8.2. Each strain possesses different characteristics including varying fermentation rate and production of H_2S (Ough and Groat, 1978; Jiranek et al., 1995c).

Table 8.2. Desirable characteristics and traits of wine yeasts.

- Initiate fermentation quickly
- Tolerant to low pH, SO_2
- Ferment at low temperatures
- Tolerate high temperatures
- Ferment to dryness
- Produce desirable bouquet
- Low nitrogen requirement
- Low H_2S production
- Low urea excretion
- Low foaming
- Good flocculation ability (aids clarification)
- Malolactic compatibility (if needed)

8.3.4 Temperature

During fermentation, *Saccharomyces* metabolize available sugar as a source of energy. However, some of the energy generated is lost as heat during fermentation. One consequence is that the temperature of the must increases during the active phase. According to Margalit (2004), fermentation of a 23°Brix must being conducted in a thermally isolated vessel could theoretically result in a temperature increase of as much as 26.5°C/47.7°F. Thus, temperature control during fermentation is vital to success.

White wines are generally fermented at lower temperatures (10°C/50°F to 18°C/65°F) for better aroma retention, whereas red wines are fermented at higher temperatures (18°C/65°F to 29°C/85°F) for increased color and tannin extraction (Ough and Amerine, 1966). Peynaud (1984) recommended slightly different fermentation temperatures, 18°C/65°F to 20°C/68°F for making white and rosé wines and 26°C/79°F to 30°C/86°F for red wines. Margalit (2004) suggested that white as well as rosé and blush wines be fermented at even cooler temperatures, 8°C/46°F to 14°C/57°F, whereas red fermentation should be conducted at 22°C/72°F to 30°C/86°F.

As expected, fermentation rates by *Saccharomyces* vary with temperature. Ough (1964) reported that fermentations are relatively slow at 10°C/50°F compared with those conducted at 15°C/60°F or 27°C/80°F. Temperature also affects the population balance between *Saccharomyces* and non-*Saccharomyces* yeasts. In red wine fermentations (20°C/68°F to 30°C/86°F), *Saccharomyces cerevisiae* represents the dominant species (Sharf and Margalith, 1983), partially due to the warmer temperature of fermentation. At lower fermentation temperatures such as those used in white wine production, non-*Saccharomyces* yeasts can proliferate to yield much higher populations (Section 6.2.2).

8.3.5 Immobilized Yeast

Because the presence of yeasts can represent a logistical "bottleneck" in postfermentation clarification, an alternative that has been studied is the use of immobilized microorganisms. Here, yeasts are "trapped" in calcium alginate beads or strands that are collectively packed into a synthetic mesh sleeve that is immersed into the juice/must. Relatively few yeasts ($<10^3/$ mL) escape the encapsulation matrix (Yokotsuka et al., 1993) but yet conduct an active alcoholic fermentation. Yajima and Yokotsuka (2001) reported that concentrations of some undesirable volatile compounds (methanol, ethyl acetate, and acetaldehyde) were lower in wines made using *Saccharomyces* immobilized in double-layer beads. Immobilized yeasts have also been proposed to be used in the production of sparkling wines (Fumi et al., 1988; Yokotsuka et al., 1997).

8.4 "NATURAL" FERMENTATIONS

There has been renewed interest among some winemakers with regard to the use of native microflora present on the fruit and in the winery to carry out alcoholic and malolactic fermentations. Stylistic distinction is the driving force that tempts winemakers to accept the risks involved in these "natural" fermentations. Perceived benefits include added complexity and intensity as well as a fuller, rounder palate structure. The latter may reflect the presence of small amounts of unfermented sugar, lower alcohol, and increased production of important sensory impact metabolites.

Non-*Saccharomyces* yeasts are known to produce various odor and flavor molecules that alter wine quality (Holloway and Subden, 1991; Gil et al., 1996; Lema et al., 1996). Besides delaying (or eliminating) inoculation of commercial cultures of *Saccharomyces*, winemakers can ferment at cooler temperatures (10°C/50°F to 15°C/59°F) to encourage non-*Saccharomyces* yeasts (Section 6.2.2).

Selected non-*Saccharomyces* yeasts, most commonly *Candida, Pichia, Kloeckera, Kluyveromyces,* and *Torulaspora,* have been investigated for their potential as commercial cultures that modify or enhance wine quality (Ciani and Maccarelli, 1998; Henschke et al., 2002; Toro and Vazquez, 2002; Jolly et al., 2003; Mamede et al., 2005; Sommer ct al., 2005). In one study, Jolly et al. (2003) inoculated *C. pulcherrima* with *Saccharomyces cerevisiae* in Chenin blanc musts. Although standard chemical analyses revealed little difference between wines, the authors noted that sensorially, wines produced from co-cultures were deemed superior in quality over those fermented by *Saccharomyces* alone. Similarly, Henschke et al. (2002) and Toro and Vazquez (2002) used two species of *Candida* (*C. stellata* and *C. cantarellii*) in culture with *Saccharomyces* to produce Chardonnay and Syrah wines, respectively. Henschke et al. (2002) reported that growth of a fructophillic strain of *C. stellata* resulted in wines with very low residual sugar, in agreement with the findings of Toro and Vazquez (2002). Rather than using a direct inoculation, Ciani and Ferraro (1998) studied immobilizing *C. stellata* into calcium alginate beads. However, Ciani et al. (2000) reported low growth and fermentation rates of *C. stellata* under fermentative conditions making commercialization of this yeast potentially difficult. Another non-*Saccharomyces* yeast, *K. apiculata,* is capable of producing considerable amounts of various volatile compounds including ethyl acetate, which can negatively impact wine quality (Ciani and Maccarelli, 1998; Plata et al.,

2003), although its presence may also enhance the quality of some wines (Mamede et al., 2005). Finally, Sommer et al. (2005) described the use of mixed active dry culture that included *Saccharomyces*, *Kluyveromyces*, and *Torulaspora*.

Some non-*Saccharomyces* yeasts can spoil wine through synthesis of various volatile odor and flavor compounds. Yet, another concern regarding these yeasts involves nutrient depletion. Like *Saccharomyces*, non-*Saccharomyces* yeasts require various nutrients such as nitrogen, vitamins, and minerals which may be depleted before *Saccharomyces* initiates fermentation. This could potentially adversely affect alcoholic fermentation.

8.5 FERMENTATION PROBLEMS

8.5.1 Sluggish/Stuck Fermentations

A vinification problem of tremendous economic importance encountered by winemakers is slow alcoholic fermentation rates, especially in the case of fermentation of high-sugar musts (Alexandre and Charpentier, 1998; Bisson, 1999; Bisson and Butzke, 2000). Premature cessation of yeast growth and fermentation results in a wine with unfermented sugars and an ethanol concentration lower than expected (Fleet and Heard, 1993). The problem may manifest itself as sluggish activity during middle and later phases of alcoholic fermentation, whereas in other cases, cessation of fermentative activity may be abrupt. From a commercial standpoint, sluggish or stuck wines are a problem due to their sweet taste, inferior sensory quality, and the potential for microbial spoilage.

Causes of sluggish (slow) and stuck (stopped) fermentations include nutritional deficiency, inhibitory substances, processing difficulties, as well as bacterial antagonism (Section 6.6.2). Although various factors can contribute to a sluggish or stuck fermentation, the exact cause(s) of a particular occurrence cannot always be identified.

8.5.1.1 Nutritional Deficiency

Nitrogen deficiency is well-known to limit yeast growth and slows fermentation rates (Agenbach, 1977; Ingledew and Kunkee, 1985; Monteiro and Bisson, 1992; Henschke and Jiranek, 1993; Jiranek et al., 1995b; Spayd et al., 1995). Henick-Kling (1994) reported yeast may utilize 1,000 mg/L of amino acids in agreement with Dittrich (1987). It has been estimated that a minimal assimilable nitrogen concentration of 140 to 150 mgN/L is necessary to complete fermentation (Agenbach 1977; Spayd et al., 1995). However, it is clear that these values do not represent absolute minimal required concentrations. For instance, Bisson and Butzke (2000) suggested that the 140 mg N/L value does not take into account the high sugar concentrations in many musts and made recommendations based on must composition (Table 8.3). Furthermore, some winemakers have not observed a relationship between nitrogen requirements and the sugar concentrations in musts. For example, Wang et al. (2003) reported that fermentations of a synthetic grape juice medium containing only 60 mg N/L yeast assimilable nitrogen and 24% sugar achieved dryness. In addition, it is known that nitrogen requirements vary with yeast strain (Ough et al., 1991; Manginot et al., 1998; Julien et al., 2000).

Unlike bacteria, *Saccharomyces* can accumulate large intracellular concentrations of amino acids. Depending upon the particular amino acid, stage of growth, and activity of necessary transport enzymes, these amino acids may be (a) directly incorporated into proteins, (b) degraded for either their nitrogen or carbon components, or (c) stored in vacuoles or cytoplasm for later use (Bisson, 1991).

Another "nutrient" important for yeast is the presence of oxygen. Without some initial oxygen, fermentation can slow because O_2 is needed for sterol synthesis (Section 1.4.2). Owing to the activity of grape oxidases (Section 7.5.1), the oxygen content of unsulfited musts can rapidly decrease. Aeration or pumping-over musts are methods employed by winemakers to supply additional oxygen to a deficient must, and addition of SO_2 at crush can help limit the activity of oxidases.

8.5.1.2 Inhibitory Substances

It is well established that high sugar concentrations in grape musts are inhibitory to yeast growth and therefore slow fermentation (Ough, 1966; Lafon-Lafourcade, 1983; Casey et al., 1984). More recently, Erasmus et al. (2004) noted lower maximum cell densities of *Saccharomyces* grown in 40% w/v sugar musts compared with populations grown in musts with only

Table 8.3. **Estimated amounts of nitrogen required by *Saccharomyces* in musts with different sugar concentrations as recommended by Bisson and Butzke (2000).**

Must ripeness (°Brix)	Yeast-assimilable nitrogen (mg N/L)
21	200
23	250
25	300
27	350

20% sugar. Nishino et al. (1985) attributed the slower fermentation rates of high-sugar musts to increases in osmotic pressure on yeasts.

Besides ethanol, the presence of fungicides on grapes may also slow fermentations. In support, Conner (1983) noted that of the 25 vineyard agrochemicals examined, five inhibited yeast. However, Pilone (1986) observed that triadimenol (Summit™), a fungicide with activity against powdery mildew and black rot, did not affect alcoholic fermentation by Montrachet UCD 522 or Pasteur Champagne UCD 595. Fungicides affect *Saccharomyces* in various ways including altered sterol content (Doignon and Rozès, 1992).

8.5.1.3 Other Factors

At temperatures of approximately 29°C/85°F, fermentation becomes sluggish, and complete cessation is possible at higher temperatures (37°C/100°F) (Ough and Amerine, 1966). The lethal effects of high temperature transients do not result from temperature alone; rather, the inhibition is coupled to lower ethanol tolerance. Temperature and ethanol tolerance of yeast varies with species/strain but also reflects intrinsic and extrinsic properties of the growth medium. Ough and Amerine (1966) also warn that infections by *Lactobacillus* can also be of concern at high temperatures. Finally, overclarification of musts prior to fermentation can cause problems, potentially due to decrease of nutrients and removal of oxygen physically trapped by the insoluble solids (Section 7.4.2).

8.5.1.4 Restarting Fermentations

If faced with a sluggish or stuck fermentation, winemakers have several potential solutions. Because populations of spoilage yeasts and bacteria may flourish in pomace and lees, wines should be racked-off prior to attempting the restart. For the same reason, free-run wine is favored over press wine to prepare the restart medium. Although it is tempting to use already fermenting juice/must for re-inoculation, Peynaud (1984) cautioned against this procedure, pointing out that the potential alcohol differences between fermenting wine and a fermentation that is stuck late in the cycle may be sufficient to "shock" the fermenting yeasts, thereby creating an even sweeter stuck fermentation. Cavazza et al. (2004) concluded that the major obstacles to regaining yeast activity were the concentrations of SO_2 and ethanol in the stuck wine. However, these authors reported successful restart of fermentations by direct inoculation of WADY at a concentration of 1 g/L.

Although other techniques exist, the following recommendation is generally useful in restarting stuck fermentations (M. Bannister, personal communication, 1995). The restart medium is prepared by removing 2.5% of the total volume of stuck wine and mixing this volume with an equal volume of water. Because the stuck wine may be microbiologically contaminated (Section 6.6.2), it is recommended that the wine be sterile-filtered (0.45 μm). Yeast nutritional supplements containing diammonium phosphate are then added at recommended levels. The winemaker may elect to use one of several commercial yeast formulations that contain vitamins and yeast hulls (ghosts), the latter known to be stimulatory to fermentation (Section 8.2). Sugar levels are then adjusted to approximately 5% w/v and the medium is warmed to 30°C/86°F prior to re-inoculation of yeast at a rate of 2 to 4 lb/1000 gal. Yeast must be rehydrated according to the manufacturer's instructions (Section 8.3.2). Once inoculated, the medium is slowly cooled to 20°C/68°F to 22°C/72°F over several hours. When the sugar (°Balling) of the restart medium decreases by approximately half, additional sterile-filtered stuck wine is added in increments of 20% v/v. Subsequent incremental additions of 20% are made each time the sugar concentration decreases by half or until all the stuck wine has been added. This process may take several weeks during which it is important to prevent MLF.

A similar technique to restart fermentations was outlined by Bisson and Butzke (2000). Here, the authors recommended racking and/or rough filtering the stuck wine, adding 30 mg/L total SO_2, and adjusting the temperature to 20°C/68°F to 22°C/72°F. The re-initiation medium for 1000 L of stuck wine is prepared by mixing 15 L of water at 20°C/68°F to 22°C/72°F with about 3000 g grape juice concentrate (65°Brix) and 30 g diammonium phosphate. Yeast (1 kg of *Saccharomyces bayanus*) is then rehydrated in 5 L of water at 38°C/100°F to 41°C/106°F for 15 to 20 min. The re-initiation medium is then gradually added to the yeast inoculum over a period of 30 min and allowed to ferment until about half of the sugar has been metabolized (only a few hours). At this time, the volume of the re-initiation medium is doubled by adding an equal volume of the stuck wine. The mixture should be aerated by either pumping-over or sparging with sterile compressed air at a rate of 10% of the tank volume per minute. Fermentation should be allowed to continue until one-half the sugar is metabolized, at which time, the volume is again doubled by addition of more stuck wine. Repeat the addition of stuck wine one more time to reach a total volume of about 160 L (this procedure may require between 12 h and 3 days to reach this stage). Add the fermenting medium (160 L) to the remaining stuck wine (860 L) along with 20 g diammonium phosphate. Bisson and Butzke (2000) warn not to allow the fermentation to reach dryness or a °Balling lower than the stuck wine. For stuck wines greater than 14.5% v/v ethanol, it may be necessary to reduce the ethanol content before attempting to restart fermentation.

During harvest and crush, it may not be feasible to deal with stuck fermentations immediately. In these cases, the wine should be stabilized against further biological deterioration by racking and sulfiting (30 to 40 mg/L) the free-run wine followed by storage at a low temperature until time is available to attempt the restart. Higher inoculum levels of yeast (8 to 10 lb/1000 gal) may be necessary using strains recommended by a supplier.

8.5.2 Hydrogen Sulfide

During and toward the end of alcoholic fermentation, H_2S may be released (Eschenbruch et al., 1978; Hallinan et al., 1999; Spiropoulos et al., 2000). Having an odor reminiscent of "rotten-eggs," H_2S has a very low sensory threshold of only a few parts per billion (Henschke and Jiranek, 1991). Hydrogen sulfide can also act as a precursor for other reduced sulfur compounds (mercaptans) that impart additional off-odors to wine (Lambrechts and Pretorius, 2000). Yeast strains differ in their ability to produce H_2S as exemplified by Montrachet, a strain known to produce higher amounts of H_2S (Guidici and Kunkee, 1994; Wang et al., 2003). As noted by Sea et al. (1998), consistently low-sulfide-producing strains of yeast are not commercially available, even with hundreds of strains examined.

Because nitrogen deficiency is a well-known contributor to hydrogen sulfide production, one strategy that can reduce the risk of formation is the addition of nutritional supplements to grape musts prior to and/or during fermentation (Vos and Gray, 1979; Guidici and Kunkee, 1994; Jiranek et al., 1995a; 1995b; Hallinan et al., 1999; Tamayo et al., 1999; Park et al., 2000; Spiropoulos et al., 2000). Jiranek et al. (1995a; 1996) concluded that excessive H_2S produced under nitrogen deficiency was due to ongoing reduction of sulfite by sulfite reductase even though the nitrogen-containing precursors that react with sulfide, O-acetylserine (OAS) or O-acetylhomoserine (OAH), were exhausted (Fig. 1.12).

However, nitrogen is not the only nutritional factor that influences H_2S evolution in grape musts as evidenced by Sea et al. (1998) who reported poor correlations between H_2S and must nitrogen concentrations. Metabolic depletion of OAS and OAH could be the result of a lack of pantothenic acid, a vitamin required for the synthesis of coenzyme A (CoA), which is necessary for formation of these precursors (Fig. 1.12). In agreement, pantothenic acid deficiency is known to increase H_2S production by *Saccharomyces* in synthetic media (Eschenbruch et al., 1978; Slaughter and McKernan, 1988). Wainwright (1970) and others have noted reduced H_2S formation in synthetic media containing amounts less than $150\,\mu g/L$.

Wang et al. (2003) reported that a complicated relationship exists between nitrogen and pantothenic acid that affects H_2S production (Fig. 8.2). Here, H_2S production decreased with an increase of nitrogen but only in the presence of $250\,\mu g/L$ pantothenic acid. If pantothenic acid was present at $50\,\mu g/L$ or less, the amount of H_2S evolved actually increased with an increase in available nitrogen. This observation had not been reported previously and casts doubt on the belief that addition of nitrogen to grape musts will always reduce H_2S problems (Tamayo et al., 1999).

Other factors are also known to impact H_2S in wine. For instance, Karagiannis and Lanaridis (1999) studied addition of sulfite, must turbidity, yeast strain, fermentation temperature, and lees contact on H_2S formation. Among other findings, the authors noted that more H_2S was present if the wine was left on lees for 2 months. Although elemental sulfur used in the vineyard can also be a source of H_2S (Acree et al., 1972; Eschenbruch, 1974), very high concentrations on the treated grapes may be required (Thomas et al., 1993).

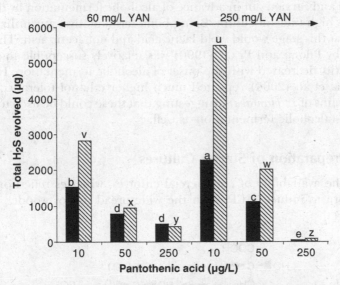

Figure 8.2.   Cumulative evolution of H_2S by EC1118 (■) or UCD 522 (▧) during fermentation of a synthetic grape juice containing variable concentrations of YAN and pantothenic acid. Within a given yeast, means with different letters are significantly different at $p < 0.05$ for EC1118 (letters a to e) and UCD 522 (letters u to z). Adapted from Wang et al. (2003) with the kind permission of Blackwell Publishing.

8.6 MALOLACTIC FERMENTATION

Numerous review articles have documented the enological importance of malolactic fermentation, or MLF (Kunkee, 1967b; Davis et al., 1985b; Wibowo et al., 1985; Edwards and Beelman, 1989; Henick-Kling, 1993; van Vuuren and Dicks, 1993; Lonvaud-Funel, 1999). During this process, L-malic acid is converted to L-lactic acid and CO_2. Although lactic acid bacteria can produce D-, L-, or DL-lactic acid from glucose (Section 2.4.1), only the L-isomer is produced during MLF (Fig. 8.3). One consequence of MLF is a reduction in wine acidity with an increase in pH of about 0.2 units (Bousbouras and Kunkee, 1971).

Malolactic fermentation can be carried out by a number of lactic acid bacteria but commercial strains of *O. oeni* such as ML-34, PSU-1, MCW, EQ-54, Viniflora, to name a few, have been used. In addition, there were some early attempts to use other heterofermentative strains of *Lactobacillus*, *L. brevis* strain Equilait and *L. hilgardii*, for induction of malolactic fermentation (Hayman and Monk, 1982; Caillet and Vayssier, 1984). A homofermentative species of *Lactobacillus*, *L. plantarum*, has also been used as a means to deacidify grape musts (Pilatte and Prahl, 1997; G-Alegría et al., 2004). Pilatte and Prahl (1997) noted that MLF can be initiated and carried out in advance of alcoholic fermentation by the addition of a high-titer inoculum. Being homofermentative, any utilization of glucose at this stage would yield lactic acid and not acetic acid. The strain studied by Pilone and Prahl (1990) was relatively susceptible to alcohol, and activity decreased with the onset of alcoholic fermentation. However, G-Alegría et al. (2004) reported much higher ethanol tolerance among other strains of *L. plantarum*, suggesting that these could be used to induce MLF postalcoholic fermentation as well.

8.6.1 Preparation of Starter Cultures

Before the availability of commercial cultures, wineries relied on native microflora to induce MLF. With the widespread use of wooden storage

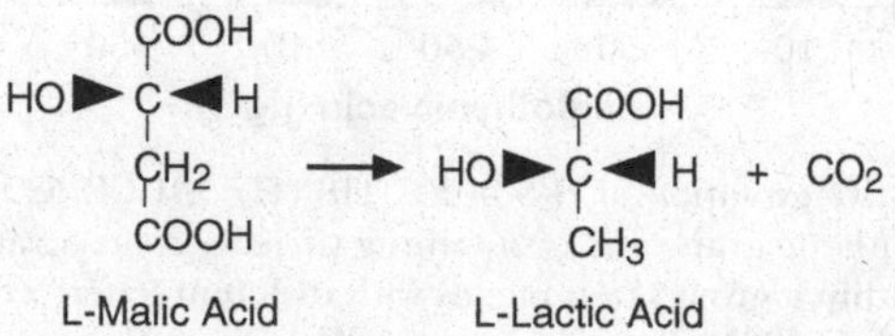

Figure 8.3.   Conversion of malic acid to lactic acid by malolactic fermentation.

tanks and barrels, a ready source of "in-house" inoculum was commonly available. Under these conditions, promotion of MLF is accomplished by maintaining a temperature of 21°C/70°F, avoid adding sulfites, and maintaining a pH greater than 3.2 (Olsen, 1994). Given that MLF can occur immediately after or several months after completion of alcoholic fermentation (Wibowo et al., 1985), there is a risk of spoilage because the wine is unprotected. Moreover, spontaneous MLF by unidentified lactic acid bacteria can produce unpredictable and/or undesirable flavor characteristics in wines (Zeeman et al., 1982; Bartowsky and Henschke, 1995). Because of this, Olsen (1994) advised regularly monitor the wine microscopically, check for off-aromas, and perform routine analysis such as quantifying malic and acetic acids.

Although some wineries continue the tradition of using native microflora, winemakers increasingly inoculate grape must or wine with lactic acid bacteria starter cultures to improve the success of MLF (Davis et al., 1985b; Fugelsang and Zoecklein, 1993; Henick-Kling, 1995). Many strains of *O. oeni* are available in lyophilized, cultures frozen concentrates, and liquid forms, although winemakers may elect to prepare their own starters by growing up the strains in diluted grape juice or wine (Beelman et al., 1980; Fugelsang and Zoecklein, 1993; Pompilio, 1993; Henick-Kling, 1995; Nielsen et al., 1996; Semon et al., 2001). Lyophilized starter cultures usually contain high populations of viable bacteria ($>10^8$ CFU/g) and are easy to ship and store. These cultures sometimes require rehydration and adaptation in diluted grape juice or wine prior to inoculation (Lafon-Lafourcade et al., 1983a; Krieger et al., 1990; Pompilio, 1993; Henick-Kling, 1995). Adaptation steps aid in restoring a loss of viability due to cell damage that can occur during freezing and lyophilization processes (Liu and Gallander, 1983; Henick-Kling, 1993). By comparison, direct inoculation formulations that do not require acclimation or rehydration are also available. Frozen concentrates and liquid cultures that have low cell densities may require additional expansion in the winery prior to inoculation.

Media for preparing malolactic starter cultures frequently contain grape or apple juice supplemented with other nutrients like yeast extract, peptone, and Tween 80 (Pilone and Kunkee, 1972; Kunkee, 1974; Costello, 1988; Champagne et al., 1989; Krieger et al., 1990, 1993; Henick-Kling, 1993). One preculture method is to inoculate pH adjusted (4.5) grape juice that has been diluted 1:1 with water and contains 0.5% w/v yeast extract. Once the culture reaches a cell density of 10^7 to 10^9 CFU/mL, it can be inoculated directly into wine at rates of 1% to 5% v/v. Other enologists prepare starter cultures using diluted wine because of reduced risk of contamination from spoilage microorganisms and increased ethanol tolerance of the malolactic bacteria (Hayman and Monk, 1982; Nault et al., 1995).

8.6.2 Strain Selection

Many strains of *O. oeni* are now commercially available to winemakers for inducing MLF in wine (Pompilio, 1993; Henick-Kling, 1995). Some are considered more desirable for inducing MLF than others due to their different and possibly unique malolactic activity and growth characteristics (Lafon-Lafourcade et al., 1983b; Fleet et al., 1984; Davis et al., 1985b; 1988; Izuagbe et al., 1985; Britz and Tracey, 1990; Rodriguez et al., 1990; Henick-Kling, 1993). Sought-after attributes among strains include growth at low pH, resistance to SO_2 and ethanol, production of low amounts of biogenic amines (Section 11.3.6), and compatibility with *Saccharomyces* (Liu and Gallander, 1983; Wibowo et al., 1988; Fugelsang and Zoecklein, 1993; Henick-Kling, 1993).

8.6.3 Timing of Inoculation

When MLF starter cultures are used in the winery, the winemaker will be faced with the decision as to the timing of bacterial inoculation. Although cultures can be inoculated simultaneously with yeast or early in the alcoholic fermentation, some winemakers inoculate after completion of the alcoholic fermentation (Webb and Ingraham, 1960; Kunkee, 1967b; 1974; Henick-Kling, 1993; Pompilio, 1993). In the survey of Fugelsang and Zoecklein (1993), 41% of red wine producers added starters during the course of alcoholic fermentation, 17% at the end, and 17% after pressing at the end of extended maceration. With the advent of direct-inoculation cultures, many winemakers now add cultures after completion of primary fermentation.

Some researchers suggest that early inoculation of malolactic bacteria is best for inducing MLF because rapid fermentation would allow winemakers to complete finishing operations sooner (Kunkee, 1967b; 1984; Davis et al., 1985b; Henick-Kling, 1993). This notion suggests that yeast may not yet have depleted nutrients also essential for malolactic bacteria. Furthermore, ethanol and SO_2, compounds known to be inhibitory to *O. oeni* (Britz and Tracey, 1990), would be present in lower concentrations. Early inoculation of malolactic bacteria has resulted in successful completion of both the alcoholic and malolactic fermentations (Beelman, 1982; Beelman et al., 1982; Beelman and Kunkee, 1985; Cannon and Pilone, 1993; Henick-Kling and Park, 1994; Huang et al., 1996; Nygaard et al.,

1998). In fact, a few researchers have reported successfully inducing simultaneous alcoholic and malolactic fermentations (Beelman, 1982; Beelman and Kunkee, 1985; Henick-Kling and Park, 1994).

One potential problem associated with early inoculation of malolactic bacteria involves antagonistic interactions between yeast and bacteria (Section 6.6.1). Antagonistic interactions can result in inhibition of malolactic bacteria by yeast resulting in delays or failed MLF (Beelman et al., 1982; Lafon-Lafourcade, 1983; King and Beelman, 1986; Cannon and Pilone, 1993). In support, a number of researchers have reported bacterial viability to decline from upwards of 10^7 CFU/mL to undetectable levels within a short period of time of bacterial inoculation (Beelman et al., 1982; Liu and Gallander, 1983; Wibowo et al., 1985; King and Beelman, 1986; Rodriguez et al., 1990; Cannon and Pilone, 1993; Henick-Kling and Park, 1994; Nygaard et al., 1998).

Additionally, early inoculation may result in production of acetic acid due to the presence of fermentable carbohydrates (Lafon-Lafourcade and Ribéreau-Gayon, 1984; Ribéreau-Gayon, 1985). However, Semon et al. (2001) did not observe excessive volatile acidities in wines inoculated early or during alcoholic fermentation with *O. oeni*, in agreement with other studies (Giannakopoulos et al., 1984; Beelman and Kunkee, 1985; Rodriguez et al., 1990; Edwards et al., 1991). Semon et al. (2001) noted that volatile acidities of Chardonnay wines inoculated with different strains of *O. oeni* prior to alcoholic fermentation were higher than those inoculated after completion of the fermentation, but not to undesirable concentrations (Table 8.4).

Table 8.4. Chemical analysis of Chardonnay wines inoculated with *O. oeni* strains EQ-54 or WS-8 before (day 0) or after (day 22) completion of alcoholic fermentations.

Strain of *O. oeni*	Day of bacterial inoculation	pH	Titratable acidity (g/100 mL)	Volatile acidity (g/100 mL)	Malic acid (g/L)	Lactic acid (g/L)
None	None	3.55[d]	0.72[a]	0.014[c]	4.49[a]	0.73[d]
EQ-54*	0	3.82[a]	0.49[b]	0.040[a]	1.70[bc]	4.31[bc]
	22	3.73[bc]	0.48[b]	0.026[b]	1.48[d]	4.64[a]
WS-8*	0	3.79[ab]	0.48[b]	0.040[a]	1.65[bcd]	4.17[c]
	22	3.70[c]	0.50[b]	0.030[b]	1.58[cd]	4.56[a]
WS-8[†]	0	3.72[c]	0.49[b]	0.034[ab]	1.81[b]	4.62[a]
	22	3.73[bc]	0.49[b]	0.028[b]	1.66[bc]	4.44[ab]

Means within a column with different superscript letters are significantly different (p < 0.05).
*Direct inoculation of a rehydrated lyophilized culture.
[†]Starter culture prepared using diluted grape juice medium.
Adapted from Semon et al. (2001) with the kind permission of the *Australian Journal of Grape and Wine Research*.

A third problem associated with early inoculation is the possibility of antagonism of yeast by malolactic bacteria. Huang et al. (1996) demonstrated that some lactic acid bacteria can inhibit yeast during vinification. Whereas one of their bacterial strains was determined to be a novel species of *Lactobacillus* (Edwards et al., 1998a), others were later shown to belong to *O. oeni* (Edwards et al., 1998b). Furthermore, some researchers have reported faster death rates of wine yeast when grown in culture with some *O. oeni* strains (Beelman et al., 1982; Lafon-Lafourcade et al., 1983a; Beelman and Kunkee, 1985).

To avoid the potential problems associated with early inoculation, some advocate inoculation of bacterial starter cultures after completion of the alcoholic fermentation (Ribéreau-Gayon, 1985; Krieger et al., 1990; 1993; Henick-Kling, 1993; Nygaard et al., 1998). Late inoculation has the advantage of minimizing undesirable interactions between yeast and bacteria, which ensures completion of alcoholic fermentation (Henick-Kling, 1993). Moreover, sugar concentrations are lower at this time, which reduces the risk of undesirable bacterial by-products being formed from carbohydrate metabolism (Lafon-Lafourcade et al., 1983a). Malolactic bacteria may also benefit from the nutrients released through yeast autolysis following alcoholic fermentation (Fornachon, 1968; Beelman et al., 1982; Henick-Kling, 1993). In fact, bacterial cultures inoculated into wine have maintained viabilities of ca. 10^6 CFU/mL and reached $>10^7$ to 10^9 CFU/mL rapidly to induce MLF (Krieger et al., 1990; 1993; Henick-Kling and Park, 1994; Liu et al., 1995a).

Unfortunately, late inoculation with bacterial starter cultures may not ensure successful bacterial growth and inducement of MLF. Some studies have reported rapid declines in bacterial populations after inoculation into wine with delayed or no MLF (Beelman, 1982; Beelman et al., 1982; Liu and Gallander, 1983; Krieger et al., 1993). The loss of bacterial viability after late inoculation has been attributed to a variety of conditions found in wine including high ethanol concentrations, low pH, presence of SO_2, presence of other antimicrobial compounds, or nutrient depletion by yeast (Kunkee, 1967b; Beelman et al., 1982; Krieger et al., 1993; Larsen et al., 2003).

Another factor that could influence bacterial growth is the presence of bacteriophages (Davis et al., 1985a; Henick-Kling et al., 1986a; 1986b; Arendt and Hammes, 1992). Bacterial starter cultures can be at risk for phage attack. Davis et al. (1985a) noted that phage survived in wines at pH 3.5 or greater but were inactivated at lower pH or by the addition of sulfur dioxide or bentonite. However, Henick-Kling et al. (1986a; 1986b) noted active phage in wines of pH 3.2 that were not affected by 50 mg/L SO_2. The extent of bacteriophage in wines is not known but potentially low given general success rates in inducing MLF.

8.6.4 Use of *Schizosaccharomyces*

Since *Schizosaccharomyces* utilizes L-malic acid, this yeast has been proposed as an alternative to using bacteria for deacidification of high-acid musts (Gallander, 1977; Snow and Gallander, 1979; Dharmadhikari and Wilker, 1998). Unfortunately, growth of some strains of *S. pombe* yields undesirable sensory characteristics (Gallander, 1977; Snow and Gallander, 1979; Yokotsuka et al., 1993). However, Yokotsuka et al. (1993) and Silva et al. (2003) successfully applied immobilized or encapsulated cells of *S. pombe*, which could be easily removed after consumption of malic acid but prior to synthesis of off-flavors. Furthermore, Thornton and Rodriguez (1996) proposed using a mutant of *Schizosaccharomyces malidevorans* rather than *S. pombe* as a means to deacidify wines.

8.7 POST-FERMENTATION PROCESSING

After alcoholic and malolactic (if applicable) fermentations are complete, wines undergo further processing that potentially influences the growth of microorganisms. One of the most important spoilage issues during storage of red wines, *Brettanomyces*, is discussed in Section 11.2.2.

8.7.1 Aging and Storage

Once alcoholic and malolactic fermentations are completed, wines may be aged in wooden (oak) barrels for different periods of time. Because wood is not an absolute barrier, ethanol, water, and oxygen will diffuse into or out of the barrel at different rates. Changes in the environmental conditions inside the barrel may support or hinder growth of many spoilage microorganisms such as *Acetobacter* (Section 11.2.1) and film yeasts (Section 11.2.3). If the relative humidity of a cellar is less than 60%, water is preferentially lost from barreled wines (ethanol content increases). Conversely, ethanol is lost at relative humidities greater than 60% (ethanol content decreases). The visual manifestation is development of headspace (ullage) in the barrel, which can allow entry of sufficient oxygen to support proliferation of oxidative microorganisms such as *Acetobacter* and film yeasts.

The evaporation process from a barrel may be slowed by the combination of humidity and temperature control. Several commercial mist systems are available for this purpose. Beyond this, however, effective seals coupled with a regular topping routine are essential to the barrel-aging program. In tightly bunged barrels, the diffusion of water and ethanol out of the barrel creates a partial vacuum in the developing headspace. To minimize oxygen incursion through imperfect seals between bung and barrel, wine-makers using traditional wooden bungs often store the barrels "bung over" or at the "two o'clock" position to keep the bung moist. Fortunately, silicon bungs, which provide an effective seal without needing to be kept moist, are now available. Using the latter allows the cellar worker to store barrels "bung up" thereby reducing the time and effort required for topping. Whereas adherence to topping schedules is crucial to cellaring, topping too frequently may cause more spoilage than this practice prevents by accelerating the incursion of oxygen.

Although the practice is not recommended, it may be necessary to store wine in partially filled containers. Rather than wooden barrels, stainless steel tanks that can be sealed effectively should be used. After transfer but prior to sealing, such containers should be topped with argon gas, although other less expensive gasses like nitrogen or carbon dioxide can also be applied. Because N_2 is far less soluble in wine than CO_2 (14 mg/L compared with 1500 mg/L), the former will not solubilize as readily and will remain on the surface of the wine longer. But because the atomic weight of N_2 (28 g/mole) is close to that of O_2 (32 g/mole), effective displacement of the oxygen is questionable. Given its molecular weight, CO_2 (44 g/mole) should provide a more effective environmental barrier. However, its solubility in juice/wine reduces this barrier to only a short period of time. Due to these limitations, argon (Ar) has become the preferred blanketing gas among winemakers. With an atomic weight of 40 g/mole, Ar is heavier than either nitrogen or oxygen and is not as soluble as carbon dioxide.

To minimize microbial problems, Ough and Amerine (1966) recommended cellar temperatures of 13°C/55°F to 18°C/65°F for storage of white and red dry wines.

8.7.2 Adjustment of Volatile Acidity

Adjustment of volatile acidity in salvageable wines has historically been achieved through blending with other wines that contain a much lower level of acetic acid. Direct attempts to chemically neutralize the acetic acid are not feasible because of preferential reduction in the stronger fixed (tartaric and malic) acids. Similar problems have been encountered using conventional anion exchange of the wine. However, recent application of reverse osmosis, coupled with ion-exchange technology, has proved to be successful (Smith, 2002). In this case, only the permeate, which contains acetic acid, ethanol, and water, passes through an anion exchange column that removes acetic acid (Fig. 8.4). Once the acetic acid is removed, perme-

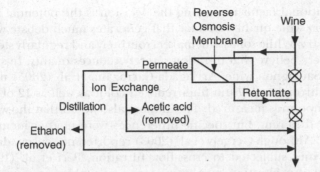

Figure 8.4.   Removal of acetic acid and/or ethanol using reverse osmosis and other processing technologies.

ate and retentate are recombined. This system can also be used to remove ethanol from wines by distilling the permeate.

8.8 BOTTLING

Once final blends are prepared, wines are ready to be bottled. In cases where the wine has been determined by chemical and microbiological examination to be prone to instability, the winemaker is faced with the decision of how to stabilize the product such that post-bottling microbiological activity is prevented. In this regard, microbial activity is either inhibited or prevented through the use of specific preservatives or sterilants (Section 5.2) or by physical removal using filtration (Section 5.3). Although heating wines to a specific point to destroy all microorganisms is possible, caution should be exerted when applying the historical practice of "hot bottling" because quality can rapidly deteriorate. Interestingly, Malletroit et al. (1991) did not detect any influence of pasteurization on the sensory quality of Chenin blanc wine bottled in 375 mL bottles.

The decision whether to sterile filter/package should be partially based on the inherent chemical and microbiological parameters of the wine. Here, some wines are less likely to exhibit post-bottling microbiological instability than others. As an example, a red wine containing 14.5% v/v ethanol, <2 g/L reducing sugars, <30 mg/L malic acid, and >50 mg/L total SO_2 is less likely to be a candidate for sterile filtration than a white wine that contains 11.5% v/v ethanol, >5 g/L reducing sugars, >0.5 g/L malic acid, and little SO_2. Another example would be wines that have some "Brett-character" before bottling. If unfiltered, there is a high risk that these wines could experience a significant microbiological bloom after bottling.

An additional factor in making the decision is the potential impact of filtration on wine quality, a factor that generates much debate within the wine industry. While some winemakers routinely and regularly sterile-filter wines, others believe that filtration severely reduces quality. In studying a Cabernet Sauvignon wine, Arriagada-Carrazana et al. (2005) noted that filtration through a 0.65 μm filter reduced color as well as 12 of the 100+ aroma compounds measured. The authors also noted that the wines were sensorially different, but specific differences were neither identified nor quantified. Although Gergely et al. (2003) reported no sensory differences in some wines subjected to cross-flow filtration, Peri et al. (1988) concluded that ultrafiltration membranes were unsuitable for wine processing due to losses of polymeric anthocyanins.

Finally, wineries may utilize cost considerations to make this decision. In these cases, the value of the final product does not justify the added cost of sterile packaging. Rather than sterile filtering, the winemaker may opt for the inclusion of chemical preservatives in order to limit risk of microbiological spoilage after bottling.

Regardless of whether the wine is to be sterile-filtered or stabilized with chemical preservatives, post-bottling microbiological examination should be carried out to ensure effectiveness of the treatment and operation. Bottles should be collected off the line at (minimally) 1 h intervals over the course of the bottling run. As populations of microorganisms in the bottled product should be low or undetectable, wines should be examined using membrane filtration (Section 14.5.3). Critical times for sampling are shortly after start-up of the bottling line and after return from employee breaks. At these times, hydraulic surges, resulting from reactivation of pumps, may produce sufficient force to push previously retained microorganisms through filter membranes. If viable wine spoilage microorganisms are recovered, the integrity of filter membranes should be examined. As pointed out by Loureiro and Malteito-Ferreira (2003), the outlet side of the sterile filter, filler (especially bell gaskets and rubber spacers), the corker (cork jaws and hopper), bottle sterilizer, bottle mouth, and environmental air are critical points for contamination. These surfaces should be periodically sampled for microbial growth (Section 9.8). Spraying 70% v/v ethanol on corker jaws and when resupplying the cork hopper will help minimize the potential for microbial contamination as well.

CHAPTER 9

WINERY CLEANING AND SANITIZING

9.1 INTRODUCTION

Microorganisms arising on the grape and during transportation not only survive but can also proliferate at any stage during the winemaking process (Chapters 7 and 8). At the winery, the presence of fruit residue on processing equipment surfaces for prolonged periods may promote rapid proliferation, leading to population blooms. Because many of these microorganisms can cause spoilage, winemakers should prevent uncontrolled growth early in the process thus reducing "pass-through" threats during later stages of processing. Corrective action requires that all fruit-contact surfaces, beginning with mechanical harvesters, picking lugs, and gondolas and through to fermenters, hoses, and tanks, receive regular and thorough cleaning and sanitizing treatments.

The terms "cleaning" and "sanitizing" represent two different but very important unit operations within a winery. Here, cleaning refers to the removal of mineral and organic material or debris from equipment surfaces, whereas sanitizing implies reduction or elimination of microorganisms through such means as addition of chemicals or heat (e.g., steam).

Properly conducted, cleaning and sanitizing programs limit the build-up of mineral and organic debris (grape "bloom," tartrates, biofilms, etc.), which serve as reservoirs for microbial proliferation and re-infection. It must be remembered that (a) improperly cleaned equipment cannot subsequently be sanitized and (b) cleaning equipment does not imply that the equipment has been sanitized. Winery personnel must, therefore, thoroughly clean equipment surfaces prior to sanitizing.

The terminology applied to describe chemicals used in sanitation programs can be confusing. Frequently, antimicrobial compounds are interchangeably and incorrectly referred to as "sterilants," "sanitizers," and/or "disinfectants." By definition, a sterilant kills all (100%) viable microorganisms, associated spores, and viruses on contact, whereas a disinfectant destroys or eliminates living (vegetative) cells but not necessarily bacterial spores. By comparison, a sanitizer is an agent that reduces viable cell populations to acceptably lower numbers.

9.2 SAFETY ISSUES

The hazard potential surrounding winery cleaning and sanitation is high owing to the use of (a) strong oxidants, caustics, and/or acidic chemicals, (b) high-pressure hot water and/or steam, and (c) slippery floors. Because of these, personal protective equipment (PPE) including gloves, goggles, appropriate footwear, and waterproof aprons are essential (Section 19.2.7). Given the slippery properties of detergents when discharged on floors, boots with nonskid soles should be used. Ideally, pants should not be tucked into boots as this could allow hot water or chemical directly into the boot. It is critical that employees read the product labels and understand the procedures for their use. The location of Material Safety Data Sheets (MSDS), sheets that contain the necessary health and safety information for employees working with the agent, should also be readily available and clearly identifiable (Chapter 19). Finally, dry chemicals should always be added to cold water rather than to hot.

9.3 WATER QUALITY

As part of implementing a sanitation program, careful consideration should initially be given to the quality of in-house water, particularly its chemical and sensory properties. In general, water should be potable, free from suspended particulates, and low in compounds that impart odor and flavor (e.g., "earthy" or "musty" odors). Furthermore, water should be analyzed two to four times a year depending on the source (well or city). Routine testing includes pH, alkalinity, calcium hardness, iron, silica, total dissolved solids, and a standard plate count for microorganisms.

Additional attention should be paid to the water hardness. Due to the presence of calcium, magnesium, and other alkali metals, hard water interferes with the effectiveness of detergents, particularly bicarbonates, and contributes to the formation of precipitates or "scale" on equipment. Besides diminishing the appearance of equipment, these precipitates serve as sites for accumulation of organic debris and microorganisms, making sanitation difficult if not impossible. Perhaps the least expensive method to alleviate this problem is through installation of a water softener.

Absent a water-softener system, many detergent formulations include special adjuncts that can help mitigate water hardness. Known as chelating agents, these chemicals physically remove metals from solution, thus softening the water. One group of chemicals, the polyphosphates, are widely used due to their ability to chelate calcium and magnesium and prevent precipitation. Specific examples of polyphosphates are sodium hexametaphosphate (e.g., Calgon®) and sodium tetraphosphate (Quadrofos). With the exception of trisodium phosphate, these chemicals are noncorrosive. Another compound used as a chelator is ethylenediamine tetraacetic acid (EDTA). Although more expensive than polyphosphates, EDTA has the advantage of being relatively heat stable. Because chelating agents bind water hardness minerals (metals), these will improve the subsequent effectiveness of cleaners.

9.4 PRELIMINARY CLEANING

In any sanitation process, the first step is to remove as much of the first-level or visible debris as possible. This is can be accomplished either manually or by mechanical cleaning systems such as spray balls or tank and barrel washers. High pressure water (i.e., 4000 to 8000 kPa or 600 to 1200 lb/in^2) without added chemicals can often readily remove organics and buildup at a fraction of the cost associated with the use of an equivalent amount of detergent or sanitizer needed to accomplish the same goal. Cleaning with high-pressure is most effective when the spray is directed at an angle to the surface being cleaned. In addition, using warm water (38°C/100°F to 43°C/109°F) in a high-pressure delivery system further improves cleaning operation while decreasing both the amount of water used and time required. It is recommended that workers not use hot water for preliminary cleaning because application of heat to organic materials adhering a surface debris may "cook-on" the debris, thereby requiring greater effort and costs to effect removal.

To prevent compromising the integrity of the protective oxide coating on stainless steel, only soft-bristle brushes should be used in cases where scrubbing is required. Once stainless steel surfaces are scratched, these rapidly become susceptible to oxidation and corrosion. As such, fiber or metal "scratch" pads or brushes should never be used for removal of tenacious deposits.

9.5 DETERGENTS, CLEANERS, AND SURFACTANTS

Once the majority of debris and film has been removed, the surface should be cleaned. Commonly, detergents or cleaners are used to solubilize any deposits at this stage. Each detergent has unique properties of action for the most effective application. Generally, increasing the concentration beyond recommended levels provides little additional benefit and is not cost effective. Another variable is the contact time necessary between the detergent and application surface, which varies with the agent(s) uses and the mode of application. On the one hand, high-pressure spray requires less contact time than foams or gels. However, gels allow additional contact time with the surface and can be used in low-pressure systems (Wirtanen and Salo, 2003).

Several different components may be present in a particular detergent formulation. These include alkalies, acids, surfactants, and chelating agents, in varying proportions. Such formulations are often referred to as "built detergents" or "built cleaners" and, depending on composition, may achieve multiple cleaning goals in a single application. Alternatively, formulations can be prepared in the winery using individual chemicals and ingredients. It is important to consult specific suppliers for information related to proper use of these chemicals because incorrect or improper mixing can create gases highly toxic to workers.

9.5.1 Alkali

The most commonly used detergents include strong alkalies or caustics such as NaOH (caustic soda or lye) or KOH (caustic potash). Although both NaOH and KOH have excellent detergent properties and remove fats and proteins, KOH has better rinse ability than NaOH. One method is to use 1% to 2% w/v sodium hydroxide in heated water (75°C/167°F to 80°C/176°F) for a cleaning time of 15 to 20 min (Wirtanen and

Salo, 2003). These chemicals are corrosive even to stainless steel if recommended application levels are exceeded.

Other alkali compounds can be used in a winery. Sodium *ortho-* and *meta-*silicates (Na_2SiO_3) are less caustic than NaOH, possess better detergent properties, and are less corrosive toward equipment. Where the relative amount of organic material is not heavy, mild alkalies such as sodium carbonate (soda ash) or trisodium phosphate find application. Sodium carbonate (Na_2CO_3) is an inexpensive and frequently used detergent, but regular use contributes to precipitate formation when prepared in hard water. Phosphates will help soften water by chelating calcium and magnesium, thus facilitating better cleaning while reducing mineral deposits (Section 9.3).

9.5.2 Acids

Phosphoric acid (H_3PO_4) solubilizes minerals and thus, is a commonly used adjunct in many formulations. Other acids are also used as cleaning compounds, but these tend to be more corrosive to metal equipment. For example, nitric acid is effective in removing stubborn mineral deposits but tends to prematurely degrade gasket material.

9.5.3 Surfactants

Also known as wetting agents, surfactants are organic molecules that structurally have both hydrophilic ("water-loving") and hydrophobic ("water-fearing") portions. The term "surfactant" is a contraction of "surface active agent" because these molecules reduce the surface tension of water. Surfactants help to suspend debris and microorganisms adhering to surfaces by facilitating contact between the detergent and the surface being cleaned (Wirtanen and Salo, 2003). Of the three types of surfactants available (Fig. 9.1), nonionic forms have the broadest range of properties because these can be either extremely good wetting agents or emulsifiers depending on chemical structure. Nonionic surfactants also vary widely in their ability to foam, a characteristic that can be important for cleaning some pieces of equipment.

9.6 RINSES

Once the cleaning cycle using detergents is completed, equipment surfaces should be thoroughly rinsed to remove residual chemicals and debris. Although hot or cold water is commonly used for this initial rinse, a mild

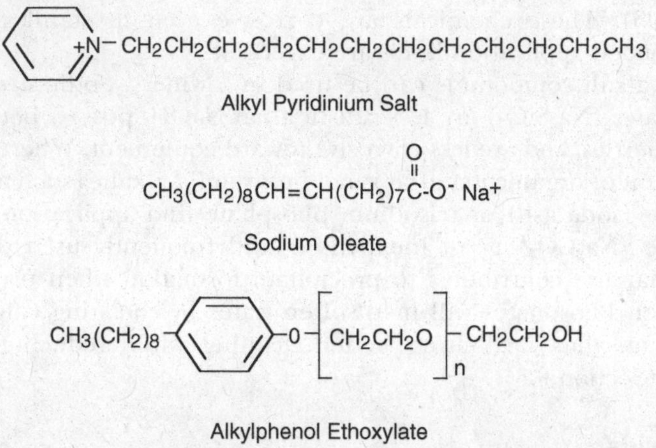

$$+N-CH_2CH_2CH_2CH_2CH_2CH_2CH_2CH_2CH_2CH_2CH_2CH_3$$

Alkyl Pyridinium Salt

$$CH_3(CH_2)_8CH=CH(CH_2)_7C\text{-}O^- Na^+$$

Sodium Oleate

$$CH_3(CH_2)_8\text{—}\phenyl\text{—}O-[CH_2CH_2O]_n-CH_2CH_2OH$$

Alkylphenol Ethoxylate

Figure 9.1.   Examples of cationic (alkyl pyridinium), anionic (sodium oleate), and nonionic (alkylphenol ethoxylate) surfactants.

acid (e.g., citric) rinse will neutralize alkaline detergent residues. In addition, acid rinses will help reduce mineral deposits. However, acid solutions are corrosive toward stainless steel and other metals at the pH of maximum effectiveness (2.5). As already noted, phosphoric acid is generally preferred because of relatively low corrosiveness.

9.7 SANITIZERS

Once deposits and debris are removed and surfaces are visibly clean, the equipment can then be sanitized. Sanitizing agents may include the halogens (e.g., iodine), hot water, ozone, peroxides, quaternary ammonium compounds (QUATS), or acidulated sulfur dioxide. Table 9.1 compares commonly used chemical sanitizers with respect to their relative advantages and disadvantages.

All chemicals used in a sanitation program must be approved including their intended-use concentration. Any deviation from prescribed and approved formulas is potentially unlawful and may also be a safety concern. It is therefore very important to follow recommendations provided by suppliers.

Once the sanitation process has been completed, surfaces are rinsed to remove residual sanitizer and drained. For instance, tank and hose sanitation is typically followed with a citric acid rinse to neutralize any residual alkali. Ideally, heat-sterilized water should be used for the final rinse.

Table 9.1. General characteristics of sanitizers.

Sanitizer	Advantages	Disadvantages
Iodophores	Noncorrosive, easy to use, broad activity spectrum	Expensive, possible flavor/odor concerns
Quaternary ammonium compounds (QUATS)	Effective, nontoxic, prevents regrowth, supports microbial detachment, noncorrosive, nonirritating	Inactivated in low pH and by salts (Ca^{2+} and Mg^{2+}), ineffective against Gram-negative bacteria
Peroxyacetic acid	Effective in low concentration, broad microbial spectrum, penetrates biofilms, nontoxic (forms acetic acid + H_2O)	Corrosive, unstable
SO_2 (acidulated)	Cheap, readily available, most microbes sensitive depending on pH and strain	Effectivness depends on pH and microorganisms, can be corrosive at high concentrations
Ozone	Similar effect as chlorine, decomposes to O_2, no residues, decomposes biofilms	Corrosive, inactivated easily, reacts with organics, safety monitoring needed

Adapted from Wirtanen and Salo (2003) with the kind permission of Springer Science and Business Media.

9.7.1 Iodine

Formulations including iodine and acid (commonly phosphoric acid) are called iodophors. Iodophors are very effective in that a concentration of $25\,mg/L$ results in similar effects to $200\,mg/L$ chlorine (Jennings, 1965). Iodophors are considered to be broad spectrum with demonstrated effectiveness against a variety of bacteria, yeasts, and molds. After sanitation, residual iodine can be detected using inexpensive test strips.

Iodophors are not as readily degraded by organics and are not irritating when applied at recommended levels. Iodophor product labels normally have a "titratable iodine" statement, which represents the minimum amount of iodine available in the product. These formulations are most effective at pH $\leq$ 4 where the concentration of I_2 is maximum. However, I_2 volatilizes at $>49°C/120°F$ so these formulations cannot be used with hot water. Iodophors may foam excessively, stain polyvinylchloride, and can be expensive.

9.7.2 Quaternary Ammonium Compounds

Quaternary ammonium compounds (QUATS) have the basic structure of a nitrogen covalently bound to four alkyl or aromatic groups such as that illustrated in Fig. 9.2. QUATS function by disrupting microbial cell

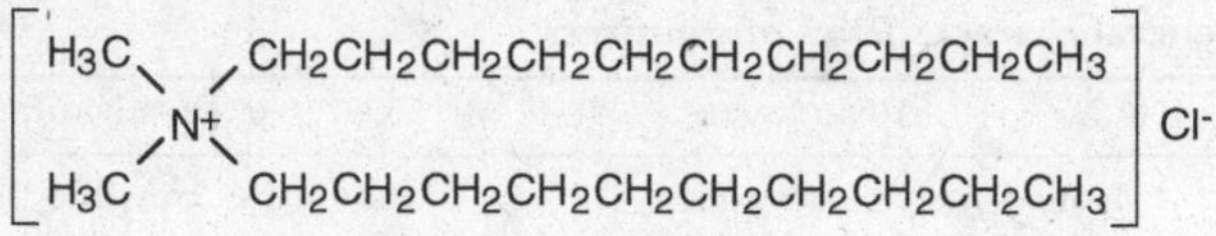

$$\left[\begin{array}{c} H_3C \diagdown \quad \diagup CH_2CH_2CH_2CH_2CH_2CH_2CH_2CH_2CH_2CH_3 \\ N^+ \\ H_3C \diagup \quad \diagdown CH_2CH_2CH_2CH_2CH_2CH_2CH_2CH_2CH_2CH_3 \end{array}\right] Cl^-$$

Figure 9.2.   Chemical structure of didecyldimethylammonium chloride, a type of QUAT.

membranes but have differential activity toward bacteria, yeasts, and molds. Gram-positive bacteria (*Oenococcus, Pediococcus,* and *Lactobacillus*) are most sensitive, whereas Gram-negative microorganisms (*Acetobacter* and *Gluconobacter*) are less so. QUATS also have limited activity against bacterial endospores and fungal spores and none against bacteriophage.

QUATS have several advantages over other sanitizers. Formulations have extended activity over a broad pH range, possess residual activity if not rinsed away, are heat stable, and noncorrosive. In addition, activity is not compromised by hard water or poorly prepared surfaces. At typical application levels of 200 to 400 mg/L, a thorough post-application rinsing is required.

In wineries, QUATS commonly find application in controlling mold growth on tanks, walls, floors, and in drains. Here, the formulation is sprayed onto the surface and left without rinsing. Depending on environmental conditions and the extent of mold growth, a single application may last for several weeks.

9.7.3 Acidulated Sulfur Dioxide

During vinification, SO_2 has historically represented one important tool used by winemakers to control microbiological growth. Due to health concerns (Yang and Purchase, 1985), some wineries have reduced usage dramatically, with a possible increased risk of microbial spoilage.

Acidulated SO_2 may be used as an effective sanitizing agent especially for hoses. Because the antimicrobial activity of SO_2 is pH dependent (Section 5.2.1), the sanitizing agent (100 mg/L SO_2 or 200 mg/L potassium metabisulfite) is usually made up in acidulated solution by inclusion of 3 g/L citric acid. Due to volatility and corrosive properties, SO_2 solutions should only be used in a well-ventilated area away from metal surfaces. For the same reasons, employees should be cautioned to avoid direct contact or inhalation of SO_2. Although wineries commonly prepare this sanitizer in acidulated, hot water (60°C/140°F), this practice serves to increase the volatility of SO_2 as well as to increase safety risks. When not in use, SO_2 solutions should be sealed to minimize volatilization.

9.7.4 Peroxides

Peroxides, or "proxy" compounds, are characterized by having at least one pair of highly reactive covalently bonded oxygen atoms (–O–O–). The group includes hydrogen peroxide and peroxyacetic acid, which are strong oxidizing agents.

Hydrogen peroxide (H_2O_2), available in concentrations ranging from 3% to 30% v/v, breaks down to generate toxic singlet or superoxide (O_2^-) oxygen. Solutions of 3% can be purchased at pharmacies as a topical antiseptic for treatment of abrasions, whereas the highly concentrated form (30%) must be obtained from chemical supply companies. H_2O_2 has limited winery application but can be legally used to remove SO_2 during the manufacture of sparkling wine cuvees, although this practice can result in potential problems of oxidation.

At concentrations >5% v/v, hydrogen peroxide becomes a strong irritant that can cause burns and blisters on exposed skin. Staff working with the agent should be trained and cautioned in its use. Unless stored in a sealed container, H_2O_2 rapidly breaks down with time. Even when stored properly, chemical decomposition occurs. Thus, it is best to replace laboratory peroxide (3% v/v) on a regular basis.

Sodium percarbonate is a stabilized powder containing hydrogen peroxide. The product is widely used as the active component in laundry detergent and all fabric bleach as well as denture cleaners, pulp and paper bleaching, and wine barrel treatment. The highly reactive product has an available oxygen equivalent to 27.5% H_2O_2 and, like peroxide, breaks down to the reactive form, oxygen as well as water and sodium carbonate upon full reaction.

In the wine industry, sodium percarbonate is sold under the trade name Proxycarb^{TM} and is widely used to treat barrels believed to be contaminated with spoilage microorganisms and/or to neutralize offensive odors that may be present. Like other barrel treatments, 100% kill is unlikely given the porous nature of wood. It is probable that viable populations can be sequestered in areas where the compound cannot reach during a cleaning cycle. As with the use other peroxide-based cleaners/sanitizers, employees should trained in its application and safe use.

Peroxyacetic acid (PAA), sometimes referred to as "peracetic acid," is also a highly reactive oxidant with antimicrobial properties similar to hydrogen peroxide. As a sanitizer and sterilant, PAA has several desirable characteristics over H_2O_2 including better stability at application concentrations (100 to 200 mg/L), improved compatibility with hard water, and reduced foaming. Compared with chlorine, PAA is biodegradable and exhibits reduced corrosive properties.

Like hydrogen peroxide, concentrated PAA (40%w/v) is a highly toxic oxidant. As such, employees must receive special training prior to use. In diluted form, its best applications include barrel and bottling line sanitation and sterilization.

9.7.5 Chlorine and Chlorinated Formulations

Chlorine or chlorinated cleaners have a long history of use in the wine and food industries. The active form, hypochlorous acid (HOCl), is a powerful oxidant and antimicrobial agent. Although chlorine sanitizers are relatively inexpensive and easy to use, activity will prematurely degrade if organic debris reflecting inadequate preliminary cleaning is present. Moreover, the use of chlorine-based sanitizers may damage stainless steel and aluminum surfaces causing localized pitting (corrosion) on the surfaces and welds of processing equipment. Subsequent pitting and corrosion makes these surfaces difficult to clean and sanitize (Frank and Chmielewski, 2001). Although such pitting can also result from exposure to high local concentrations of sulfur dioxide, the problem can commonly be traced back to residual chloride after use of chlorine-based sanitizing agents.

Another major concern surrounding the use of chlorine in wineries is the potential formation of 2,4,6-trichloroanisole (TCA). This compound can be produced by various microorganisms from chlorinated precursors (Section 4.6.3) and imparts a "musty" or "corkiness" off-odor to wines (Lee and Simpson, 1993). Use of chlorine and/or chlorine-based sanitizers in a winery can be involved in the formation of environmental TCA (Peña-Neira et al., 2000). Due to these concerns, it is advisable to reduce or even eliminate chlorine use in the winery. If chlorine-based products are to be used, these should never be mixed with acidic products (toxic gas and/or explosives can be produced).

9.7.6 Hot Water and Steam

Hot water (>82°C/180°F) and steam are ideal sterilants. Both have excellent penetrative properties, are generally noncorrosive, and effective against all juice and wine microorganisms. For instance, Wilker and Dharmadhikari (1997) compared various treatments for sanitizing barrels infected with acetic acid bacteria and noted that the hot water treatment used (85°C/185°F to 88°C/190°F for 20min) was most effective (Section 11.2.1). However, both hot water/steam may more rapidly degrade gaskets compared with other techniques.

Common applications for hot water/steam within a winery are at the bottling line and for stainless steel tanks. Hot water can be employed to sanitize lines based on a guideline of >82°C/180°F for more than 20 min as measured at the most distant point in the line. When steam is used to sanitize tanks, the recommendation is to continue until condensate from valves reaches >82°C/180°F for 20 min. In both cases, the temperature should be monitored at the point most distant from the steam source (i.e., the end of the line, fill spouts, etc.). The practice of dismantling valves, racking arms, and other parts for immersion in containers of hot water may not yield an adequate time and temperature relationship necessary for sanitization.

9.7.7 Ozone

Ozone (O_3) is one of the most potent sanitizers available and is finding increased use in the food industry. Though a very strong oxidant, ozone is also unstable with a half-life of only 20 to 30 min depending on conditions (Khadre et al., 2001).

Ozone is most commonly dissolved in water rather than applied as a gas. Because O_3 rapidly degrades to O_2, it cannot be stored and must be generated on demand. This is accomplished by specific equipment that exposes a stream of dry air to either ultraviolet light (185 nm) or electrical discharge. O_3 is best used in clean-in-place (CIP) operations as the bottling line or for treating in-house water for off-odors or discoloration.

Although special equipment is needed to use ozone, it has several advantages that may justify the additional cost. For instance, the sanitizer is effective against bacteria, fungi, viruses, protozoa, as well as bacterial and fungal spores (Khadre et al., 2001). In addition, Hampson (2000) observed that ozone was less corrosive against stainless steel (316L) than chlorine. Furthermore, Greene et al. (1994) noted only slight differences between several gaskets (Buna N, white Buna N, EPDM [ethylene propylene diene monomer], polyethylene, silicone rubber, Teflon, and Viton) treated with ozone.

From a health and safety point of view, ozone is a strong irritant, and uncontrolled exposure may result in inflammation of eyes, nose, throat, and lungs. Limits for ozone exposure have been set by the U.S. Occupational Safety and Health Administration (OSHA). Here, the legal maximum concentration for an 8 h continuous exposure is 0.1 mg/L, whereas the limit for short-term exposure is 0.2 mg/L for 10 min (Khadre et al., 2001). Where ozone is used, the staff should be well trained and use proper ozone safety monitors.

9.8 SANITATION MONITORING

Although the cleaning and sanitation processes will significantly lower microbial populations, there may be survivors depending on the degree of debris buildup and the effort expended. For instance, microorganisms deposited on equipment can adhere to the surface, grow, and multiply to form a colony of cells known as a "biofilm." Eventually, these colonies become large enough to trap soil, debris, nutrients, and other microorganisms (Kumar and Anand, 1998). In this regard, Nel et al. (2002) observed biofilm formation by *O. oeni* on stainless steel.

It is well-known that biofilms exhibit increased resistance to antimicrobials and are difficult to remove (Kumar and Anand, 1998; Wirtanen and Salo, 2003). As a commonly used procedure for clean-in-place systems, washes of caustic (75°C/167°F for 30 min) and acid (75°C/167°F for 30 min) have proved successful in removing biofilms (Parkar et al., 2004). Recently, detection of biofilms in hard-to-examine areas using remote sensors has been successful (Tamachkiarow and Flemming, 2003). In general, it is better to design an effective cleaning and sanitation program to remove microorganisms and debris as a means to limit biofilm formation (Kumar and Anand, 1998).

Because of the probability of residual microorganisms on equipment, it is important to evaluate the effectiveness of any sanitation program. Sampling sites on winery equipment should be selected to include all points that are liable to shelter microorganisms. Of importance would be to sample direct-contact surfaces such as crushers, stemmers, presses, the interior of pipelines, conveyors, and tanks. Other areas to sample include where indirect contamination could occur such as condensation from ceilings or equipment, aerosols, and lubricants.

Although microbiological methods can be used to evaluate surface cleanliness (Section 14.2), each protocol has common problems that make quantification difficult. For instance, sampling equipment surfaces can be difficult if the surface is pitted or irregular. Furthermore, the results may not be available for hours if not days (i.e., plate counts), and it is not possible to delay processing while waiting for results. A less scientific method to detect poorly cleaned or sanitized equipment would be the presence of "slippery" surfaces or "off-odors." Although sometimes adequate for fermenters and storage tanks, areas such as bottling lines require a more complete microbiological evaluation.

9.9 SCHEDULES AND DOCUMENTATION

Each winery should develop a series of standard operating procedures (SOPs) for each operation of the cleaning and sanitizing program. The SOPs must include specific protocols, schedules, and methods to document implementation of the approved protocols and schedules.

As starting points, it is recommended that crushers/stemmers, and presses be cleaned out and thoroughly rinsed between lots and a complete sanitation cycle performed every 8 h. These practices should also be applied to belt conveyors and hoppers where sugar can accumulate. Tanks should be cleaned and sanitized after use and then rinsed with high-pressure water before the next filling. Bottom and side clean-outs doors and values should be left open to provide adequate water drainage and air circulation. At this point, a "green flag" can be attached to the tank, indicating that the tank is ready to be filled. When not in use, hoses should be sanitized with SO_2 and citric acid (Section 9.7.3) and, after the operation is complete, hung to drain and dry. Hoses should be stored on sloped drain racks or hung upright to facilitate drainage rather than left on processing floors.

Wineries will apply different procedures and schedules for cleaning and sanitizing, and it is very important to document the approved procedures.

Cleaning and Sanitation Schedule For _____ (date) _____

Equipment and Description	Procedure Number	Frequency	Times			Notes	Employee Initials
Crusher (line 2)							
Rinse	9.01	After every lot					
Clean	9.02	Every 8 hours					
Sanitize	9.03	Every 8 hours					
Destemmer (line 2)							
Rinse	9.01	After every lot					
Clean	9.02	Every 8 hours					
Sanitize	9.03	Every 8 hours					
Press (#2)							
Rinse	9.01	After every load					
Clean	9.05	Every 8 hours					
Sanitize	9.06	Every 8 hours					

Assigned Employee: _____ Approval of Supervisor: _____

Figure 9.3.   An example of a cleaning and sanitizing schedule.

This documentation allows the winery to accurately describe what procedures, frequency, and chemicals were used to ensure that operations were conducted in a consistent manner. At the conclusion, representative surfaces should be sampled for microbiological populations and/or excess sanitizer. All data should be collected for future examination and potential changes. These records not only become part of the overall quality points program (Chapter 10) but also will serve as documentation if a legal issue arises.

One of the easiest means to document cleaning and sanitizing would be to prepare and use a standard form (Fig. 9.3). These forms describe cleaning and sanitizing procedures (time, frequency, temperatures, types and concentrations of chemicals, etc.) approved for each processing step. The employee and supervisor then initials or sign the form, indicating that each step has been completed according to the approved procedures.

CHAPTER 10

QUALITY POINTS PROGRAM

10.1 INTRODUCTION

Hazard analysis and critical control points (HACCP) is a systematic approach to prevent, reduce, or minimize risks associated with foods (Anonymous, 2002). Basically, the program identifies potential problems and determines which critical junctures in production must be controlled to eliminate or reduce the risk of a food safety hazard to an acceptable level. Originally created to meet the needs for food safety in support of the space program in the 1960s, HACCP programs are now applied to agricultural production, food preparation and handling, food processing, food service, and distribution. In fact, the programs have been successfully implemented in breweries (Señires and Alegado, 2005).

Although HACCP-based approaches are frequently used to address food microbiological hazards, wines do not normally present a risk of human illness or death from the ingestion of pathogenic microorganisms. As such, HACCP programs are not necessarily applicable to wine production. However, a HACCP-based approach can be used in the winery to improve process control, which, in turn, can lead to higher and more consistent quality. Therefore, wineries should consider implementing a quality points (QP) program based on general HACCP principles.

10.2 DEVELOPING QP PROGRAMS

Prior to development of a successful QP program, a number of other programs must already be in place at the winery. These programs include good manufacturing practices (GMPs) including production records (Chapters 7 and 8); pest control strategies (birds, insects, and rodents; pesticide applications; integrated pest-management strategies); maintenance schedules and records; sanitation protocols, schedules, and records (Chapter 9); water quality evaluation and monitoring records (Chapter 9); strategies for handling consumer complaints; employee training programs (Chapter 19); preferred supplier and supplier guarantee programs (ingredient specifications; grower contracts); inventory control and product storage; supply chain management, traceability, and recall or market withdrawal policies and procedures.

Because facilities, production operations, and staffs are different, the QP program for each facility and product will be unique. However, a simple and concise QP program provides the basis for just about any situation faced during wine production. As discussed in the following sections, the general steps for the winemaking staff to develop a QP program are as follows:

(a) Develop an operational sequence or flow diagram, beginning with the raw material (grapes) and continuing through to the bottled wine and case goods storage (Section 10.3).

(b) Conduct an analysis of each step in processing and identify potential quality issues (critical quality points) at each juncture (Section 10.4).

(c) Establish critical limits for each critical quality point (Section 10.5).

(d) Develop monitoring procedures and records (Section 10.6).

(e) Establish corrective actions when the process deviates from a critical limit (Section 10.7).

(f) Establish procedures for record-keeping and documentation (Section 10.8).

(g) Develop a verification plan to show that the QP program has been effectively implemented and monitors important quality factors (Section 10.9).

10.3 PROCESSING AND FLOWCHART

A "standard" flowchart or diagram for wine processing can be used as a starting point in preparing a QP program (Fig. 10.1). As each facility varies in its design and objectives, a more detailed diagram may be needed to describe all the steps involved as the grapes, must, or wine move through various stages. If a winery handles its own distribution or sales including tasting rooms, these steps should also be included on the flow diagram.

Each QP program must include descriptions (criteria) at each step for acceptable fruit quality (sugar concentrations, titratable acidity, microbial contamination, etc.), SO$_2$ additions (when and how much), press times and pressures, must clarification procedures, yeast inoculation rates, concen-

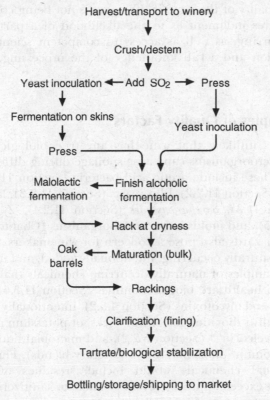

Figure 10.1. Abridged wine processing scheme that can be used to establish a quality points program. Much more detail is normally needed for an effective flowchart.

trations of nutrients to add, fermentation temperatures, malolactic inoculations (or not), schedule and number of rackings and/or topping off, clarification agents and rates, temperatures and times for tartrate stabilization, and methods for biological stabilization.

10.4 QUALITY FACTOR ANALYSIS

Under HACCP, a food safety hazard is one that is reasonably likely to occur for which a prudent processor would establish controls, based on experience, best industry practices, and the best available technical or scientific information. Hazards are divided into three groups: (a) biological, (b) chemical, and (c) physical. Similarly, a quality factor can be defined as any biological, chemical, or physical agent that is reasonably likely to cause a decrease in quality of the wine if the agent is not being controlled. This analysis requires judgment as to the likelihood of a particular quality problem occurring, as well as access to competent scientific expertise and information and a full knowledge of the processing, storage, and distribution.

10.4.1 Examples of Quality Factors

Although it is unlikely that wine has any true biological hazards, a number of microorganisms can cause spoilage during different stages of vinification. These include acetic acid bacteria (Section 11.2.1), *Dekkera/ Brettanomyces* (Section 11.2.2), film yeasts (Section 11.2.3), lactic acid bacteria (Section 11.3), *Saccharomycodes* (Section 11.2.4), *Zygosaccharomyces* (Section 11.2.5), and molds/other microorganisms (Chapter 4).

Chemical hazards also pose a concern for winemakers. These can be divided into naturally occurring, intentionally added, and unintentional/ incidental. Examples of naturally occurring chemicals that could pose a risk to human health are biogenic amines (Section 11.3.6), ethyl carbamate (11.3.2), and mycotoxins (Section 4.5.2). Intentionally added chemicals include sulfur dioxide (gas, sulfur wicks, or potassium metabisulfite), tartaric acid, Velcorin™ (Section 5.2.2), and microbial nutrient formulations (diammonium phosphate or proprietary blends). Finally, unintentional/incidental chemicals would include residues of agricultural chemicals that exceed approved levels, cleaners or sanitizers (Chapter 9), and the inadvertent transfer of lubricants from equipment.

Any potentially harmful extraneous matter not normally found in wines would also be considered to be a physical hazard. Perhaps the most important would be glass, pieces of which could be introduced during bottling. Other potential contaminants include wood or metal fragments, diatomaceous earth, packaging material, or cork dust that escaped removal during bottling. Under a conventional HACCP program, filth such as the presence of fruit flies or plant debris is not considered a hazard. However, controlling these factors should be considered within a QP because these can adversely affect the value and marketability of wines.

10.4.2 Preventive Measures

Preventive measures are developed and implemented as a way of reducing product hazards and maintaining market quality. For instance, a winemaker may decide to add additional nitrogen to a specific grape must prior to or during alcoholic fermentation (Section 8.2). The decision regarding the concentration to add becomes more complicated due to the availability of different nutritional supplements and conflicting scientific information as to when to add the nutrients (before or after yeast inoculation and in what amounts). In this case, it is important for the winemaking staff to agree before crush as to what preventive measures should be taken if the amount of nitrogen in grape musts be below a specified concentration. Another example of a preventative measure would be the use of SO_2 or lysozyme at the crusher as a means to reduce the risk of *Lactobacillus* spoilage.

10.4.3 Critical Quality Points

Critical quality points (CQPs) are actions that can be taken at specific processing steps to control a factor thus ensuring product quality and safety. CQPs will vary between wineries and products within a winery. Except for those involving product safety which must be controlled, specific CQPs should be developed to reflect the overall goals and philosophies of the owners and staff.

Some examples of different CQPs and their specific microbiological problems along with possible corrective measures are found in Table 10.1. As an example, detection of one cell of *Saccharomyces* in a bottle of sweet, unfortified wine may be enough to cause refermentation in the bottle. In this case, a preventative measure that could be taken would be assurance that the bottling line was correctly sterilized and that sterile filtration equipment functioned properly. Additional measures would be inclusion of sorbates to limit potential yeast growth. Known as "hurdle technology," this concept relies on implementation of several preventative measures as

Table 10.1.   Examples of microbiological issues and potential preventive measures associated with selected Critical Quality Points.

Critical quality point	Microbiological issue	Potential corrective actions
Harvest/transport to winery	• Excessive *Botrytis*, rots, or acetic acid bacteria	• Alter vineyard practices, use SO_2, set rejection limits of moldy fruit
Alcoholic fermentation	• Sluggish/stuck fermentation	• Adjust nutrients, use proper yeast rehydration schemes and amounts, dilute very high °Brix musts
	• *Lactobacillus* infection	• Lower pH, use SO_2 or lysozyme
Maturation in barrels or bulk storage	• Undesirable or failed malolactic fermentation	• Lower ethanol, raise pH, examine starter culture, lower SO_2
	• Film or slime formation	• Avoid ullage, use SO_2, sterile filtering
	• Off-odors (volatile acidity, acetaldehyde, ethyl acetate, "mousy," and/or "geranium"	• Avoid ullage, use SO_2, sterile filtering, avoid sorbate in sweet wines
	• Undesirable levels of *Brettanomyces*	• Use SO_2 or Velcorin™, lower pH, sterile filtration, maintain cool cellar and hygiene
Bottling	• Microbial contamination	• Use SO_2, sterile filtration

a means to reduce the risk of spoilage as much as possible. This approach has the further advantage that if one preventative measure is not correctly implemented, other actions will hopefully maintain a low risk of spoilage.

To identify CQPs in a winery, the staff should use the flowchart as the starting point and then develop a plan for dealing with raw materials and ingredients used, activities at each step of process, equipment, potential design flaws, as well as storage and distribution conditions. CQPs are specific for the process and product being made and are dependent on the winery layout, formulations, equipment, ingredients used, and sanitation programs.

10.5 CRITICAL LIMITS

A critical limit (CL) is an objective measurement taken at a particular point in the process. A critical limit can be a (a) process specification, (b) measurement on a process or finished product sample, or (c) a simple yes/no decision. An example of a process specification would be fermentation temperature because excessive temperatures (high or low) may result in a sluggish or stuck alcoholic fermentations. Supplementing a grape must with nitrogen is another example of a product measurement because this decision is based on an analysis performed at the winery (e.g., determination of yeast assimilable nitrogen). Finally, an example of a "yes/no" decision would be to sterile filter wine prior to bottling or not.

Establishment of critical limits usually involves a test measurement on the product as it is being processed or a determination of an operating parameter for a piece of equipment. Specific limits or acceptable minimum/ maximum levels must be determined and verified. For example, some wineries capable of measuring yeast assimilable nitrogen have established a minimum concentration necessary to complete fermentation. If the actual value for a given must lot is lower than the required concentration, then the winemaker should add a specific amount of nitrogen. Other critical limits could be the amount of rot in grapes, sugar concentrations, temperatures, times, pH, acidity (titratable and/or volatile), filter flow rates or the presence of foreign objects. Sources of information needed to determine a CL include scientific publications, regulatory guidelines, competent experts, and experimental studies.

10.6 MONITORING PROCEDURES

Monitoring implies that the winery will conduct a planned sequence of observations or analyses to assess whether a CQP is under control. Monitoring procedures identify what should be monitored (temperature, pH, volatile acidity, SO_2, sensory observations, etc.) including the appropriate CL along with the analytical methods to be used (thermometer, pH meter, Cash still, Ripper or aeration/oxidation SO_2 methods, smell and/or taste, etc.). The procedures should also specify the frequency of analysis, the individual conducting the test, development of monitoring records, and how monitoring is to be verified. Many analytical methods are available that can measure a selected CL to ensure the process is in control and spoilage issues are prevented, reduced, or minimized. Various microbiological methods can be found in Chapters 12 through 16, and chemical procedures are available through Ough and Amerine (1988) and Zoecklein et al. (1995).

Besides being fast and accurate, monitoring must be cost effective. As such, it is appropriate to include financial constraints in decisions regarding what method to use, what analyses are to be performed, and the frequency of monitoring. At times, wineries may elect to send wines to commercial analytical laboratories for testing that cannot be performed in-house. One example is 4-ethyl phenol, an indicator of *Brettanomyces* (Section 11.2.2). In that this analysis is expensive, it should only be performed on suspect wines rather than on a routine basis.

In general, measurement of microbiological populations are avoided as CLs because results are commonly not available for hours if not days. Enumeration of microorganisms (Chapter 14) should be incorporated into sanitation programs, GMPs, or as part of verification. If microbiological criteria will be applied, it is advisable to have uniform analytical procedures including methods of sampling, diluents, culture media, and incubation conditions. As pointed out by Loureiro and Malfeito-Ferreira (2003), methods applied by wineries vary widely, which makes comparing results difficult.

10.7 CORRECTIVE ACTIONS

Once a critical limit has been violated, it is important to determine what corrective action(s) should be taken (Table 10.1). Preferably, there should be a number of feasible corrective actions developed in anticipation of problems. For example, if bottles are missing the alcohol consumption warning label, the corrective action is to apply the correct label.

Commonly, the actual cause for a loss of process control cannot be ascertained with complete certainty, but corrective actions may be possible. For instance, a sudden increase in volatile acidity may indicate the growth of certain species of *Lactobacillus* (Sections 6.6.2 and 11.3). Here, corrective actions could include (a) lowering pH, (b) addition of SO_2, and/or (c) use of lysozyme to limit the infection. Given that many grape musts can have a pH of greater than 3.8, the addition of tartaric acid can lower must pH to make SO_2 more effective due to the formation of molecular SO_2. In this case, decreasing the pH from 3.8 to 3.7 will result in far less free SO_2 needed (40 mg/L vs. 50 mg/L) to yield 0.5 mg/L molecular SO_2 (Section 5.2.1).

10.8 RECORD-KEEPING AND DOCUMENTATION

Record-keeping confirms that production is within specifications and provides technical data for each production lot. As traceability becomes more important in the marketplace, having good records of source control and distribution are more critical than ever before. Verification steps within a

QP require record review, including calibration records for key measuring instruments and processing equipment.

It is important to maintain a set of monitoring records that are recorded (preferably in ink) by the operator taking the measurement and at the time the measurement is taken. These records should be reviewed by a supervisor within a couple of days and before any product has left the facility. A frequent review of these records can identify possible production errors while corrections are still possible.

One very important safety-related QP for a winery is labeling. Mandatory warning labels for alcohol use are required by federal law in the United States and in other jurisdictions. Because some consumers are sensitive to sulfites, an approved sulfite warning must appear on the label if sulfites are present in excess of 10 mg/L. Furthermore, all wine labels must be approved by the regulatory agencies (e.g., U.S. Alcohol and Tobacco Tax and Trade Bureau).

10.9 VERIFICATION

Verification encompasses those activities, other than monitoring, that establish the validity of the safety or quality plan. For instance, this step indicates that a program has been implemented and that it is effective for a given operation. Verification includes a variety of elements such as calibration of equipment and instruments used in monitoring, chemical testing, and microbiological samples, validation of experimental test methods, and record review protocol. A quality program should be verified when formulations or production methods change, key ingredients change, or storage, distribution, or product end-use changes because any of these factors could affect wine safety or quality. Verification is conducted when a plan is first adopted and then is reviewed and modified when quality or safety factors change, and again prior to the next vintage.

The term "validate" is used when a quality program is first adopted to determine if it has been properly implemented and will effectively control the identified safety and quality factors. In essence, validation is the "bridge" between the development and implementation of the plan. As part of the initial implementation plan, it is important to integrate the quality program with existing sanitation, production, maintenance, procurement, distribution, marketing (including existing recall and market withdrawal programs), and employee training.

CHAPTER 11

WINE SPOILAGE

11.1 INTRODUCTION

Because microbological activity can develop quickly and without warning, early identification of potential spoilage problems is critical to implementing corrective remedies. However, identifying the causative microorganisms and therefore the appropriate correction is not always simple because a given microorganism can bring about multiple spoilage problems, and a specific wine fault can be caused by different microorganisms. This chapter summarizes spoilage microorganisms and their control as well as sources of important wine faults.

11.2 SPOILAGE MICROORGANISMS

11.2.1 *Acetobacter*

Control of *Acetobacter* in wines should be initially accomplished by limiting populations entering the winery on fruit and by implementing and maintaining an adequate sanitation program (Chapter 9). Because mold-damaged fruit will have higher populations of acetic acid bacteria than sound fruit (Section 6.5.1), use of the former can lead to increased *Acetobacter* infections in wines during post-fermentation storage. Therefore, the risk of pass-through infections can be reduced by maintaining agreed upon quality standards of the harvested fruit.

Although adherence to fruit quality standards clearly represents the winemaker's first line of defense against microbial spoilage, on-site decisions can play a pivotal role in containment or proliferation of undesirable species. In that *Acetobacter* are obligate aerobes (Chapter 3), frequent topping of barrels to minimize headspace and limit introduction of oxygen is an important control measure (Section 8.7.1). SO_2 is also used, but the concentration required depends on the species/strain and wine pH (Du Toit and Pretorius, 2002; Du Toit et al., 2005). These authors recommended a concentration of 0.7 to 1 mg/L molecular SO_2 as necessary to inhibit all species of *Acetobacter* and *Gluconacetobacter*. Because *A. pasteurianus* can survive under anaerobic conditions in wine even with SO_2 (Du Toit et al., 2005), limiting oxygen and use of SO_2 should be used in conjunction with other winemaking practices including juice clarification to physically remove bacteria, lowering pH, use of a high inoculum of *Saccharomyces*, pre-fermentation additions of SO_2, appropriate temperatures, and good cellar hygiene (Du Toit and Pretorius, 2002; Du Toit et al., 2005). Joyeux et al. (1984b) pointed out the importance of maintaining cellar temperatures of 10°C/50°F to 15°C/59°F during barrel aging as another means to limit populations blooms.

Despite the winemakers best efforts, isolated instances of contamination and spoilage can occur. This is more frequently seen in facilities with a large barrel inventory where barrels are missed during the topping rotation. Although these barrels should be removed from production as soon as possible, some cooperage may be sanitized and reused depending on the degree of contamination. Wilker and Dharmadhikari (1997) compared four treatments of sanitizing wood infected with acetic acid bacteria: (a) chlorine (pH 7.0; 250 mg/L for 24 h), (b) acidified sulfur dioxide (pH 3; 250 mg/L for 24 h), alkaline/SO_2 (soak in 1.25 g/L potassium carbonate for 24 h followed by a 300 mg/L SO_2 rinse), and (d) hot water (initially at 85°C/185°F to 88°C/190°F for 20 min). After storing a susceptible wine in the barrels for 92 days, the hot water treated staves were the only ones that did not redevelop an *Acetobacter* infection (<10 CFU/mL). An alternative to hot water would be the use of ozone (Section 9.7.7).

Normally, the low concentration of oxygen in bottled wines prevents secondary growth and spoilage due to *Acetobacter*. However, Bartowsky et al. (2003) described an unusual situation where *Acetobacter* were able to spoil bottled wines. In this case, the unfiltered wines were stored vertically, which was believed to have resulted in incursion of enough oxygen into the bottle to support bacterial growth.

11.2.2 Dekkera/Brettanomyces

Various sensory descriptors have been used to characterize *Dekkera/ Brettanomyces*–tainted wines. These range from "cider," "clove," "spicy," "smoky," "leather," and "cedar" to "medicinal," "Band-Aid®," "mousy," "horsy," "wet wool," "rodent-cage litter," "barnyard," or even "sewage." These odors are due to synthesis of a number of volatile compounds including 4-ethyl guaiacol and 4-ethyl phenol (Fugelsang et al., 1993; Licker et al., 1999). 4-Ethyl guaiacol has been sensorially described as "clove" or "spice" and 4-ethyl phenol is "smoky" or "medicinal." Given the wide range in sensory descriptors of infected wines, more "sensory-neutral" strains of *Brettanomyces* may exist, and use of these strains may improve the quality of some wines without imparting off-odors or flavors.

The sensorial impact of *Brettanomyces* depends on the wine as well as preferences of the winemaker and consumer. For instance, Loureiro and Malfeito-Ferreira (2003) noted that some consumers would find wines objectionable if the concentration of 4-ethyl phenol exceeded 620μg/L whereas others would not. If present at concentrations less than 400μg/L, the authors suggested that this compound contributes complexity by imparting sensory descriptors of "spice," "leather," "smoke," or "game." In contrast, Licker et al. (1999) described a "high Brett" wine that contained 3000μg/L 4-ethyl phenol, a "medium Brett" wine as having 1700μg/L, and a "no Brett" wine with 690μg/L. Volatile phenol production varies with the strain of *Brettanomyces* as found by Fugelsang and Zoecklein (2003) who reported 120μg/L 4-ethyl guaiacol and 440μg/L 4-ethyl phenol produced by one strain but <10μg/L by another.

Biochemically, 4-ethyl guaiacol and 4-ethyl phenol originate from ferulic acid and *p*-coumaric acid, respectively. The reaction is a two-step process with an initial decarboxylation of the hydroxycinnamic acids catalyzed by cinnamate decarboxylase and the reduction of the vinyl phenol intermediates by vinyl phenol reductase (Fig. 11.1). Although the specific coenzyme involved remains unknown, one possible metabolic benefit of the second reaction to *Brettanomyces* could be reoxidation of NADH. Under low oxygen conditions such as those found in wines, the availability of NAD$^+$ can be limited so that carbohydrate metabolism is inhibited (Section 1.5.1). Reduction of the vinyl phenols to the ethyl phenols would allow the cell to increase the availability of NAD$^+$ and thus maintain metabolic functions.

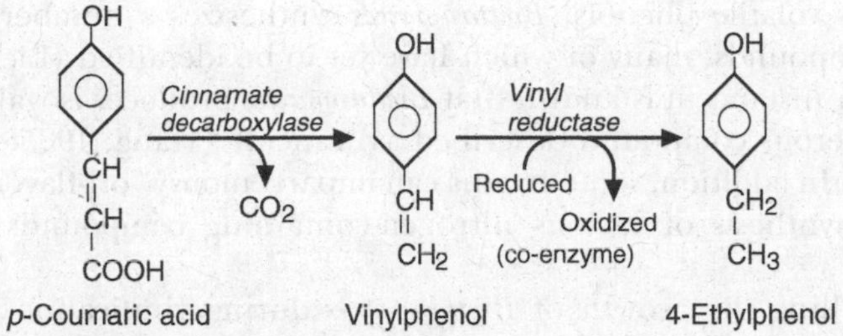

OH   OH   OH

Cinnamate decarboxylase   *Vinyl reductase*

CH
‖
CH CO_2 CH   Reduced CH_2
| ‖   Oxidized |
COOH CH_2 (co-enzyme) CH_3

p-Coumaric acid   Vinylphenol   4-Ethylphenol

Figure 11.1. Formation of volatile phenolics from hydroxycinnamic acids. Copyright © Society of Chemical Industry. Adapted from Chatonnet et al. (1992) and reprinted with the kind permission of John Wiley & Sons Ltd. on behalf of the SCI.

Although many wine microorganisms like *Acetobacter, O. oeni, L. hilgardii, L. plantarum, L. brevis, P. pentosaceus, P. damnosus,* and *Saccharomyces* can synthesize 4-vinyl guaiacol or 4-vinyl phenol from ferulic and *p*-coumaric acids, respectively, most are not able to reduce the vinyl intermediates to 4-ethyl guaiacol or 4-ethyl phenol (Chatonnet et al., 1992; 1995; Shinohara et al., 2000). Given this, *Brettanomyces* could theoretically reduce vinyl phenols synthesized by other wine microorganisms and thereby benefit from the growth of other microorganisms (e.g., lactic acid bacteria). In support, Dias et al. (2003b) found that *Brettanomyces* could use 4-vinylphenol as a precursor to 4-ethylphenol in the absence of *p*-coumaric acid.

As only *Brettanomyces* produces significant amounts of ethyl phenols, wineries have attempted to use measurement of 4-ethyl phenol as an indicator for infections. However, synthesis of ethyl phenols varies with strain (Fugelsang and Zoecklein, 2003). Some lactic acid bacteria, most notably *L. plantarum* (Chatonnet et al., 1992; 1995; Cavin et al., 1993) as well as *Pichia guilliermondii* (Dias et al., 2003a) are reported to produce small amounts of ethyl phenols. Because of these reasons, direct comparisons between concentrations and viable cell concentration can be difficult.

Processing methods are currently being sought to remove 4-ethyl guaiacol and 4-ethyl phenol from wines after an infection of *Brettanomyces*. For example, Ugarte et al. (2005) noted some success using reverse osmosis coupled to an adsorption system to remove these volatile phenols from wine. Perhaps a simpler method was that described by Chassagne et al. (2005) where yeast lees were used to remove volatile phenols. Compared with active dry yeasts, the authors reported that yeast lees from fermentation were especially strong at removing 4-ethyl phenol. Sorption of these volatile phenols depended on wine conditions, in particular pH, temperature, and ethanol concentration.

Besides volatile phenols, *Brettanomyces* synthesizes a number of odor-active compounds, many of which have yet to be identified (Licker et al., 1999). For instance it is known that *Brettanomyces* produces isovaleric acid, an odoriferous compound described as "rancid" (Wang, 1985; Licker et al., 1999). In addition, some strains can impart "mousy" off-flavors to wine through synthesis of various nitrogen-containing compounds (Section 11.3.3).

Controlling the growth of *Brettanomyces* during vinification is not an easy task. The yeast is relatively tolerant to SO_2 as indicated by Sponholz (1993) who suggested the use of 100 mg/L total or 30 to 50 mg/L free SO_2 for control in wine.

Maintenance of 0.4 to 0.6 mg/L molecular SO_2 is effective in limiting growth. Du Toit et al. (2005) suggested that *Brettanomyces* may enter a "viable-but-not-culturable" state in the presence of SO_2 (Section 6.1). As such, the yeast may escape detection by conventional microbiological methods (i.e., plating) prior to bottling, potentially resulting in sudden "blooms" during bottle aging if conditions permit.

Other control measures have included fining and filtration (Section 5.3), and recent research has investigated the use of dimethyl dicarbonate (Section 5.2.2). Although the research looks promising (*Brettanomyces* was inhibited at a concentration of 25 mg/L), Daudt and Ough (1980) only used one culture (species not reported). Furthermore, the combined use of DMDC and SO_2 against *Brettanomyces* has not yet been explored although a synergy between DMDC and SO_2 against *Saccharomyces* was noted by Ough et al. (1988a). Another option is temperature control. Lowering the cellar temperature to less than 13°C/55°F can also be used to slow the growth of *Brettanomyces*. Heat resistance of the yeast has also been reported (Couto et al., 2005).

Once established in wood cooperage, elimination of these yeasts are difficult partially due to the physical properties of wood. Unlike polished stainless steel or glass, inside surfaces of barrels are difficult to clean due to the many irregular cracks and crevices where particulates, including spoilage yeasts, can settle.

11.2.3 Film Yeasts

The visual manifestation of oxidative yeast activity is the formation of a film, sometimes referred to as "mycoderma." The film results from repeated budding of mother and daughter cells that, rather than separating, remain attached, forming chains that branch and rebranch to eventually cover the surface of the wine (Section 1.2.2.4). Initially, the yeasts can appear as floating "flowers." If allowed to continue, growth may rapidly develop into a thick pellicle, which appears "mold-like." Baldwin (1993) described the film as a chalky or filamentous white substance that was dry enough to appear "dusty." *Candida vini* (formerly *Candida mycoderma*) is a relatively common film yeast capable of producing a thick pellicle. Besides formation of a film, these yeasts can synthesize sensorially active compounds such as ethyl acetate and acetoin among others (Clemente-Jimenez et al., 2004).

Because film formation by certain non-*Saccharomyces* yeasts reflects oxidative growth, the best preventative measure is to maintain topped tanks and barrels, thereby depriving the yeasts of air (oxygen) needed for growth. Baldwin (1993) suggested that addition of dry ice to barrels of wine and subsequent release of CO_2 may also help limit the influx of O_2. Because some non-*Saccharomyces* yeasts (e.g., *Pichia membranefaciens* and *Candida krusei*) are resistant to molecular levels of more than $3\,mg/L$, reliance on SO_2 is generally ineffective once a film has formed in the barrel (Thomas and Davenport, 1985). Furthermore, one of the major metabolites of film yeasts is acetaldehyde, which can effectively bind SO_2 and decrease its antimicrobial properties (Section 5.2.1). Some winemakers have had success placing a few grams of potassium metabisulfite onto a small plastic Petri dish that is allowed to float on wine in a barrel (Baldwin, 1993).

Use of lower cellar temperatures ($<15°C/60°F$) can slow the growth of film yeasts because the alcohol content and temperature interactively inhibit growth. As support, Dittrich (1977) reported no growth of film-forming yeasts in wines of 10% to 12% alcohol when stored at $8°C/47°F$ to $12°C/54°F$, whereas growth was observed in other wines up to 14% alcohol at warmer temperatures.

11.2.4 *Saccharomycodes*

Saccharomycodes ludwidii is referred to as the "winemaker's nightmare" because the yeast is highly resistant to SO_2, approximately five times that of *Saccharomyces* (Stratford et al., 1987). Ciani and Maccarelli (1998) hypothesized that the extraordinary resistance of this yeast could be due to its ability to produce high concentrations of acetaldehyde, which binds SO_2. As such, the best known control method is sterile filtration (Section 5.3). Fortunately, the yeast has rarely been reported in either cellar aged or bottled wines possibly due to slow growth or poor competition with other yeasts (Fleet, 2003).

Saccharomycodes spoils wines by formation of cloudiness and/or sediment as well as off-odors (Du Toit and Pretorius, 2000; Loureiro and Malfeito-Ferreira, 2003). Ciani and Maccarelli (1998) studied 27 different strains of *Saccharomycodes* grown in grape musts and found high concentrations of acetaldehyde (46.7 to 124 mg/L), acetoin (211 to 478 mg/L), and ethyl acetate (141 to 540 mg/L).

11.2.5 *Zygosaccharomyces*

Another potential post–alcoholic fermentation problem is *Zygosaccharomyces*. This yeast causes spoilage by forming gas, sediment, and/or cloudiness in bottled wines (Loureiro and Malfeito-Ferreira, 2003). Synthesis of other compounds, namely succinic, acetic, and lactic acids, as well as acetaldehyde and glycerol has also been reported (Rankine, 1967; Oura, 1977; Zeeman et al., 1982; Nykanen, 1986; Herraiz et al., 1990; Moreno et al., 1991; Mateo et al., 1992; Lema et al., 1996). Thomas (1993) estimated that for yeasts such as *Z. bailii,* as few as one viable cell in a bottle of wine is sufficient for spoilage.

Zygosaccharomyces is osmophilic, resistant to ethanol, SO_2, sorbate, and other commonly used preservatives. Some species can even grow at temperatures as low as 2.5°C/36.5°F (Warth, 1985; Thomas, 1993). Infections can often be traced to the addition of contaminated grape juice concentrates used to increase the sugar concentration of must prior to alcoholic fermentation or adjustment of wine sweetness before bottling. At low storage temperatures, the microorganism is capable of slow and often unnoticed growth. Warming of the product during shipment and/or changes in chemistry upon blend formulation may stimulate previously repressed populations.

The yeast can become a problem in wineries using poor sanitation at bottling. For instance, sublethal doses of chemical sterilizing agents or hot water steam that do not meet time and temperature requirements for cell destruction are often causes for product contamination during bottling. Once introduced into the winery, difficult to sanitize areas serve as the principle source for the yeast. In one extraordinary case, Rankine and Pilone (1973) reported an established population present in an in-line pressure gauge on the filtrate side of a sterile filter and concluded that this represented a reservoir for continued contamination of sweetened bottled wines. In this case, the configuration of the pressure gauge allowed substantial resident populations of *Zygosaccharomyces* to escape steam sterilization.

11.3 WINE FAULTS

11.3.1 Volatile Acidity

Though normally thought of as referring to the amount of acetic acid in a wine, the sensory perception of "volatile acidity" is not exclusively the result of this acid. As an example, volatile acidity produced by lactic acid bacteria is often sensorially different from that resulting from acetic acid bacteria. In the latter case, volatile acidity is often perceived as a mixture of acetic acid and ethyl acetate, whereas with lactic acid bacteria, the ethyl acetate component is either missing or present at very low levels (Henick-Kling, 1993).

Because the "volatile character" or "acetic nose" is commonly due to ethyl acetate, some winemakers believe this metabolite rather than acetic acid should be a legal indicator of wine spoilage. Ethyl acetate is present in wines at concentrations ranging from less than 10 mg/L to greater than 1200 mg/L (Drysdale and Fleet, 1989b; Vianna and Ebeler, 2001; Rojas et al., 2003; Peinado et al., 2004; Mamede et al., 2005) and is described sensorially as the odor of "airplane glue" or "fingernail polish remover."

Although there are no legal limits on the concentration of ethyl acetate, the concentrations of acetic acid are regulated (Table 11.1). The maximum legal limits for acetic acid in wine in the United States are 0.9 to 1.7 g/L depending on the wine. Processing methods are available to reduce acetic acid from wines above the legal limit (Section 8.7.2).

Aside from potential sensory implications, acetic acid is well-known to be inhibitory to *Saccharomyces* (Doores, 1993), influencing both growth and fermentative abilities (Pampulha and Loureiro, 1989; Ramos and Madeira-Lopes, 1990; Kalathenos et al., 1995; Edwards et al., 1999a). In agreement, Rasmussen et al. (1995) added 4 g/L acetic acid midway through fermentation and noted that sugar utilization drastically slowed. In fact, certain lactobacilli have been implicated in slowing alcoholic fermentations, potentially due to acetic acid production (Section 6.6.2). As an example, the heterofermentative species *L. kunkeei* can produce between 3 and 5 g/L acetic acid in a wine (Huang et al., 1996; Edwards et al., 1999a).

Table 11.1. Legal limits for volatile acidity in wines worldwide expressed as acetic acid.

Wine type	USA (g/L)	California (g/L)	EU[a] (g/L)
Table (red)	1.40	1.20	1.20
Table (white)	1.20	1.10	1.08
Dessert (red)[b]	1.70		
Dessert (white)[b]	1.50		
Export	0.90		

[a]European Union. Maximum concentrations can vary given specific country regulations.
[b]Wines produced from unameliorated juice of >28°Brix.

11.3.2 Ethyl Carbamate

Ethyl carbamate (urethane) has been found in a number of fermented foods such as alcoholic ciders, beers, breads, olives, sake, soy sauces, wines, and yogurts as well as some unfermented foods like orange and grape juices (Ough, 1976; Canas et al., 1989). Of concern is the fact that this compound is a suspected carcinogen (Zimmerli and Schlatter, 1991). Even though ethyl carbamate is produced in small quantities, its concentration in wine is subjected to international regulation and therefore must be carefully managed. At present, the United States wine industry has established a voluntary target level of <15 µg/L for table wines and <60 µg/L for dessert wines.

Urea and citrulline, arising from the metabolism of arginine, serve as precursors to the formation of ethyl carbamate (Fig. 11.2). With regard to urea utilization, yeast strains exhibit variability in terms of both uptake and excretion during fermentation (Ough et al., 1991; An and Ough, 1993). For this reason, the addition of urea as a nitrogen source for *Saccharomyces* is no longer legal (Kodama et al., 1994). Although N-carbamyl amino acids can also act as precursors (Ough et al., 1988b), their concentration in wines appears to be very low and so probably do not

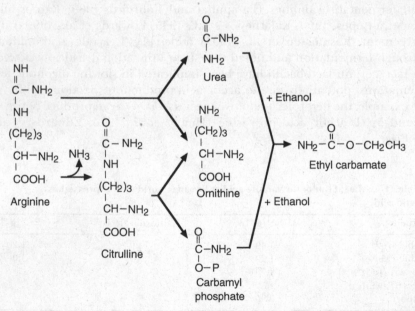

Figure 11.2.   Formation of ethyl carbamate from citrulline or urea. Adapted from Ough (1976) and Ribéreau-Gayon et al. (2000).

contribute to the formation of ethyl carbamate (Huang and Ough, 1993). Finally, diethyl dicarbonate, a potent antimicrobial compound once used by wine industry, has been banned in the United States because the additive can, in theory, react with ammonia to form ethyl carbamate (Fig. 11.3).

In yeasts, arginine is largely converted to ornithine and urea, rather than to carbamyl phosphate (Ough et al., 1988c). Conversely, citrulline is a major by-product of the degradation of arginine by lactic acid bacteria (Section 2.4.2). Strains of *O. oeni* and *L. buchneri* have been shown to excrete citrulline and carbamyl phosphate (Liu et al., 1994; Mira de Orduña et al., 2000; 2001). Even though these bacteria produce the necessary precursors, Tegmo-Larsson et al. (1989) reported that malolactic fermentation did not affect the concentrations of ethyl carbamate in wine. However, more recent information (Uthurry et al., 2006) suggests that some lactic acid bacteria, specifically *O. oeni* and *L. hilgardii*, can contribute to ethyl carbamate formation.

Although many factors influence the formation of ethyl carbamate in wines, there are general guidelines that can be used to limit its formation. For instance, excessive nitrogen applications in the vineyard can lead to increases in the wines (Ough et al., 1989b). Furthermore, some have argued for the use of yeast or malolactic bacterial strains that will not metabolize arginine or release urea (Ough et al., 1988c; 1991; Mira de Orduña et al., 2001). In an innovative approach, Ough and Trioli (1988) successfully applied an acid urease produced from *Lactobacillus fermentatum* as a post-fermentation treatment of wines to remove urea. In that formation of ethyl carbamate increases exponentially as a function of storage temperature (Stevens and Ough, 1993), wines should not be exposed to excessively high temperatures during storage and/or transport.

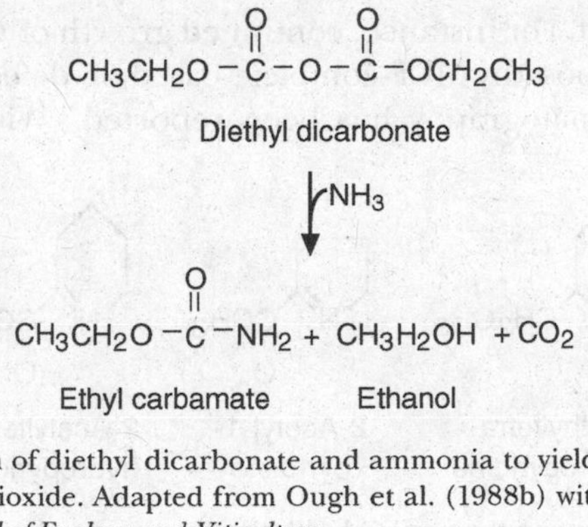

$$CH_3CH_2O-\overset{O}{\underset{\|}{C}}-O-\overset{O}{\underset{\|}{C}}-OCH_2CH_3$$

Diethyl dicarbonate

$\downarrow NH_3$

$$CH_3CH_2O-\overset{O}{\underset{\|}{C}}-NH_2 + CH_3H_2OH + CO_2$$

Ethyl carbamate   Ethanol

Figure 11.3. Reaction of diethyl dicarbonate and ammonia to yield ethyl carbamate, ethanol, and carbon dioxide. Adapted from Ough et al. (1988b) with the kind permission of the *American Journal of Enology and Viticulture*.

11.3.3 Mousiness

Growth of some strains of heterofermentative lactobacilli (*L. brevis, L. hilgardii*, and *L. cellobiosus*) as well as *Brettanomyces* may lead to the "mousy" defect in wines (Heresztyn, 1986; Grbin and Henschke, 2000; Costello et al., 2001). This type of spoilage is characterized by the formation of an offensive odor reminiscent of rodent-cage litter as well as a noticeable lingering aftertaste (Costello and Henschke, 2002). According to Sponholz (1993), the unpleasant odor becomes distinct after rubbing a sample of the mousy wine between the fingers.

An initial report by Craig and Heresztyn (1984) indicated that 2-ethyl-3,4,5,6-tetrahydropyridine may be involved in the problem, although the authors were not able to detect the compound in wines determined to be "mousy." More recently, Costello and Henschke (2002) noted that 2-ethyltetrahydropyridine, 2-acetyl-1-pyrroline, and 2-acetyltetrahydropyridine all contributed to this defect (Fig. 11.4).

This fault is associated with wines rather than musts because synthesis of these compounds requires the presence of ethanol (Heresztyn, 1986). As the aroma threshold in wine is very low, $1.6\,\mu g/L$ (Riesen, 1992), very little growth of these bacteria is required to potentially spoil a wine. Sponholz (1993) suggested that mousiness is not a common problem, but low-acid wines with insufficient SO_2 can be more prone to spoilage. Lay (2003) reported that different cultures of *Brettanomyces* formed a distinctive mousy taint in the presence of lysine or ammonium phosphate under both aerobic and anaerobic conditions.

11.3.4 Post-MLF Bacterial Growth

When utilizable levels of L-malate and viable bacterial populations remain, the wine is at risk for continued microbial activity, resulting in spoilage of the bottled product. For instance, continued growth of *Oenococcus* or *Pediococcus* in wines supposedly MLF-complete based on detection of malic acid using paper chromatography has been reported (Wibowo et al., 1988;

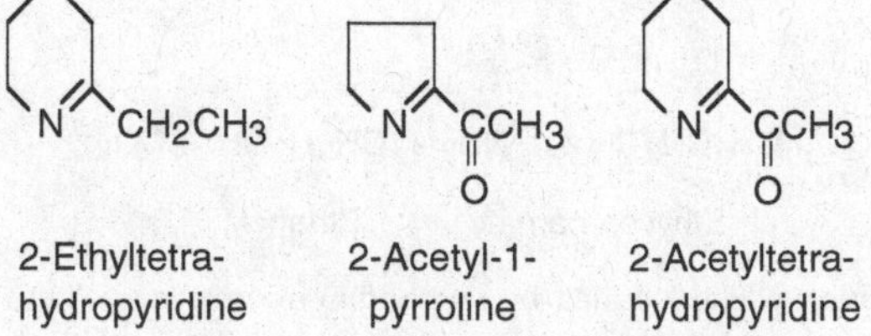

2-Ethyltetra-hydropyridine	2-Acetyl-1-pyrroline	2-Acetyltetra-hydropyridine

Figure 11.4.   Compounds associated with the occurrence of mousiness in wines.

Edwards et al., 1994). Paper chromatography (Section 15.4.9) has a high detection threshold for malic acid (100 to 200 mg/L), a concentration higher that what winemakers consider to be stable for a wine. In a survey conducted by Fugelsang and Zoecklein (1993), most of the winemakers responding (66%) considered L-malate levels of <15 mg/L as indicative of completion of MLF. The remaining 34% considered <30 mg/L as complete and "stable" with respect to continued bacterial activity. The decision to accept one concentration compared with another depends on inherent wine chemistry (pH, free and total SO_2, and ethanol) as well as the capability of the winery to sterile bottle.

An unusual example of secondary spoilage is that produced by *Lactobacillus fructivorans* (formerly *L. trichodes*). This bacterium has an extraordinary high tolerance to alcohol as evidenced by the fact that the species has been isolated from high-alcohol (>20% v/v) dessert wines (Fornachon et al., 1949; Gini and Vaughn, 1962). Visually, the bacterium appears as mycelial growth in the bottled wine, hence its nicknames of "cottony bacillus" or "Fresno mold." With modern winemaking techniques, particularly use of sulfur dioxide and attention to sanitation, reported incidences of infection have dramatically decreased. According to Kunkee (1996), the microorganism has re-emerged in the port cellars of Portugal where its growth has been correlated with reduced usage of sulfur dioxide.

To minimize the potential for secondary growth, winemakers should consider racking, SO_2 addition, and acidulation where the pH has increased above 3.5. In this case, acid adjustment should be made with tartaric acid rather than malic or citric acids, compounds that provide a source of carbon for bacteria. Despite its relative low cost, citric acid should not be used for wines except immediately prior to sterile bottling due to the possibility of excessive diacetyl (Section 2.4.5).

11.3.5 Geranium Odor/Tone

Sorbic acid is a short-chain unsaturated fatty acid used as a chemical preservative against *Saccharomyces* in sweetened wines at bottling (Section 5.2.4). Lactic acid bacteria are not inhibited by the preservative. In fact, some species can reduce the acid to sorbic alcohol, a compound that undergoes rearrangement at wine pH to yield 3,5-hexadien-2-ol (Fig. 11.5). Upon subsequent reaction with ethanol, 2-ethoxyhexa-3,5-diene is formed (Crowell and Guymon, 1975), a compound that has an odor reminiscent of "crushed geranium leaves" and has a sensory threshold of approximately 100 ng/L (Riesen, 1992). The original German term for this defect was "geranienton," or "geranium tone," and should not be

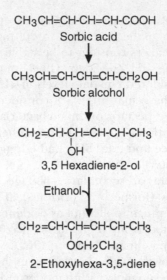

CH₃CH=CH-CH=CH-COOH
Sorbic acid

↓

CH₃CH=CH-CH=CH-CH₂OH
Sorbic alcohol

↓

CH₂=CH-CH=CH-CH-CH₃
|
OH
3,5 Hexadiene-2-ol

Ethanol↘

↓

CH₂=CH-CH=CH-CH-CH₃
|
OCH₂CH₃
2-Ethoxyhexa-3,5-diene

Figure 11.5.   Microbiological formation of "geranium" odor or tone.

confused with geraniol, a monoterpene important for the odor of some grape cultivars.

Edinger and Splittstoesser (1986) reported that all strains of *O. oeni* studied reduced sorbic acid to sorbic alcohol but noted that strains of *Pediococcus* and *Lactobacillus* could not. Based on these findings, winemakers wishing to stabilize juice with sorbate for longer term storage should consider the potential presence and activity of *O. oeni*. In the absence of measures to limit bacterial growth (SO_2 or sterile filtration), reduction of the sorbic acid may occur and upon blending into wine, chemical rearrangement and esterification with ethanol can then produce the "geranium" odor. To minimize the potential for this problem, sweet reserves should be well clarified, and preferably filtered, sulfited, and stored at low temperature until used.

11.3.6 Biogenic Amines

Biogenic amines are formed from the decarboxylation of amino acids by certain lactic acid bacteria. As such, these compounds are found in a variety of fermented foods such as cheese, dry sausage, sauerkraut, miso, and soy sauce (Stratton et al., 1991; Lonvaud-Funel, 2001). In wine, histamine, tyramine, putrescine, cadaverine, phenylethylamine, and others have been identified (Zee et al., 1983; Baucom et al., 1986; Ough et al., 1987; Vidal-Carou et al., 1991; Bauza et al., 1995; Soufleros et al., 1998;

Arena and Manca de Nadra, 2001; Moreno-Arribas et al., 2003). The decarboxylations of histidine and ornithine to form histamine and putrescine, respectively, are illustrated in Fig. 11.6. Putrescine can arise from the decarboxylation of ornithine, which, in turn, is formed from arginine (Fig. 2.6). However, there is an alternative pathway in which arginine is first decarboxylated to yield agmatine, a compound that breaks down to urea and putrescine. Evidence for this alternative pathway in lactic acid bacteria is lacking (Moreno-Arribas et al., 2003).

From a health perspective, consumption of excessive amounts of foods containing biogenic amines may result in headaches and other symptoms (Rivas-Gonzalo et al., 1983; Jarisch and Wantkle, 1996; Silla Santos, 1996). However, Ough (1971) reported concentrations of histamine in fruit, dessert, and table wines from California averaging 2.39 mg/L, a concentration below the 8 mg/L thought to induce headaches. Glòria et al. (1998) and Soufleros et al. (1998) noted higher concentrations of amines in wines from Oregon and Bordeaux, respectively. Currently, the biogenic amine content in wines is not regulated in the United States. Given emerging regulations from other wine-producing regions (European Union), it is possible that policies may be in place one day that provide for rejection of non-compliant wines.

Soufleros et al. (1998) noted that biogenic amines are formed from their precursor amino acid during and after spontaneous malolactic fermentation. Spoilage bacteria such as *Pediococcus* and *Lactobacillus* had been implicated (Delfini, 1989; Moreno-Arribas et al., 2000; 2003; Arena and Manca de Nadra, 2001), however, yeast strain used for alcoholic

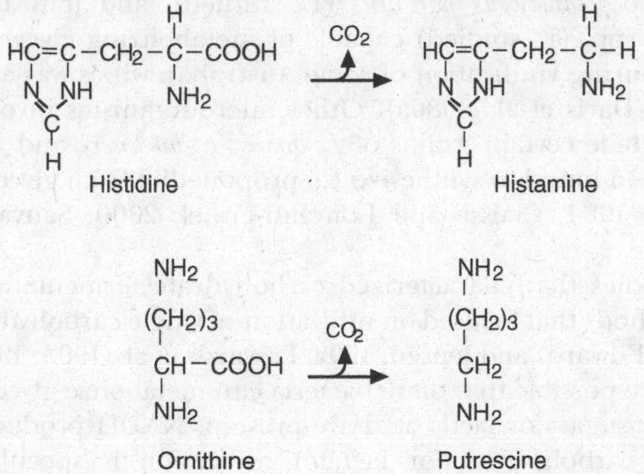

Figure 11.6. Formation of histamine and putrescine from the amino acids histidine and ornithine.

fermentation may also impact formation (Goñi and Azpilicueta, 2001). Others have found only *O. oeni* to be able to produce these amines (Coton et al., 1998). It is also clear that not all strains within a species can form biogenic amines (Lonvaud-Funel, 2001). For instance, Lonvaud-Funel and Joyeaux (1994) found a strain of *O. oeni* that produced these amines, whereas Moreno-Arribas et al. (2003) did not observe this activity in five other strains. To limit the synthesis of biogenic amines, Lonvaud-Funel (2001) advocated the use of malolactic starter cultures that do not decarboxylate amino acids.

11.3.7 Acrolein

Some lactic acid bacteria can metabolize glycerol to form 3-hydroxypropionaldehyde (3-HPA), a compound that is either (a) reduced to 1,3-propanediol, (b) oxidized to 3-hydroxypropionic acid, or, to a lesser extent, (c) undergoes chemical dehydration to yield acrolein (Sobolov and Smiley, 1960; Kandler, 1983; Slininger et al., 1983; Claisse and Lonvaud-Funel, 2000). Although acrolein is not synthesized by bacteria directly, a small portion of the 3-hydroxypropionaldehyde will form the compound through a chemical dehydration reaction (Fig. 11.7). Having little sensory impact itself, acrolein reacts with anthocyanins or phenolics to yield intensely bitter compounds (Sponholz, 1993; Louw, 2001). Little is known regarding which wine phenolics are involved in these reactions.

It has been generally thought that the ability to metabolize glycerol is limited to a few microorganisms. In a survey of a number of lactic acid bacteria, Davis et al. (1988) reported one strain of *Oenococcus oeni* (71 studied), two *Pediococcus parvulus* (17 studied), and four unidentified *Lactobacillus* spp. (13 studied) capable of metabolizing glycerol. Glycerol utilization during vinification of some Australian wines was attributed to *P. parvulus* (Davis et al., 1986a). Other microorganisms involved in this problem include certain strains of *L. brevis*, *L. buchneri*, and *L. collinoides* that have been found to synthesize 1,3-propanediol from glycerol (Schütz and Radler, 1984; Claisse and Lonvaud-Funel, 2000; Sauvageot et al., 2000).

Many studies that characterized carbohydrate fermentation patterns applied methods that focused on utilization of single carbohydrates (Davis et al., 1988; Edwards and Jensen, 1992; Edwards et al., 1993; 1998a; 2000). It is therefore possible that these bacteria can metabolize glycerol to acrolein if other sugars or lactic acid are present. NADH produced as a by-product of carbohydrate (or lactate) utilization is speculated to be reoxidized to NAD^+ from the reduction of 3-HPA to 1,3-propandiol (Schütz

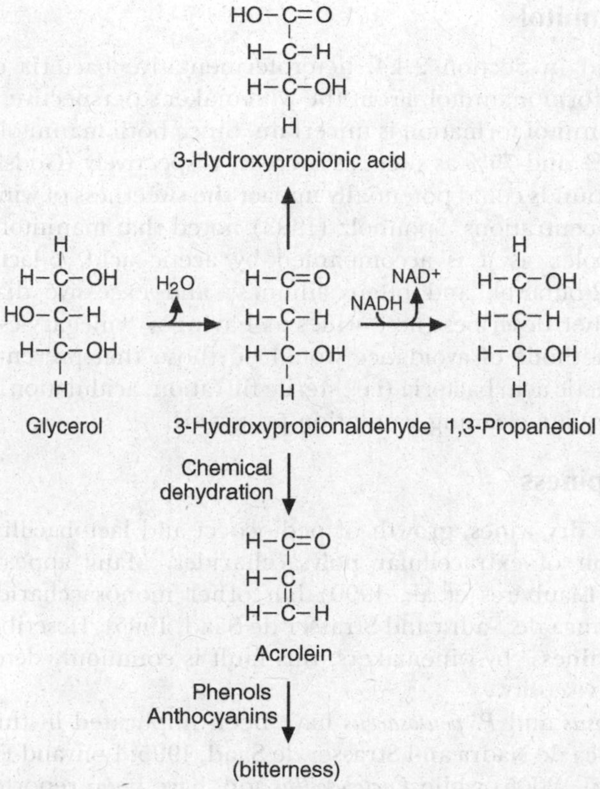

$$
\begin{array}{c}
\text{HO}-\overset{|}{\text{C}}=\text{O} \\
\text{H}-\overset{|}{\text{C}}-\text{H} \\
\text{H}-\overset{|}{\text{C}}-\text{OH} \\
\overset{|}{\text{H}}
\end{array}
$$

3-Hydroxypropionic acid

↑

$$
\begin{array}{c}
\overset{\text{H}}{|} \\
\text{H}-\overset{|}{\text{C}}-\text{OH} \\
\text{HO}-\overset{|}{\text{C}}-\text{H} \\
\text{H}-\overset{|}{\text{C}}-\text{OH} \\
\overset{|}{\text{H}}
\end{array}
\quad \xrightarrow{\text{H}_2\text{O}} \quad
\begin{array}{c}
\text{H}-\overset{|}{\text{C}}=\text{O} \\
\text{H}-\overset{|}{\text{C}}-\text{H} \\
\text{H}-\overset{|}{\text{C}}-\text{OH} \\
\overset{|}{\text{H}}
\end{array}
\quad \underset{\text{NADH}}{\overset{\text{NAD}^+}{\longrightarrow}} \quad
\begin{array}{c}
\overset{\text{H}}{|} \\
\text{H}-\overset{|}{\text{C}}-\text{OH} \\
\text{H}-\overset{|}{\text{C}}-\text{H} \\
\text{H}-\overset{|}{\text{C}}-\text{OH} \\
\overset{|}{\text{H}}
\end{array}
$$

Glycerol   3-Hydroxypropionaldehyde   1,3-Propanediol

Chemical
dehydration ↓

$$
\begin{array}{c}
\text{H}-\text{C}=\text{O} \\
\text{H}-\overset{|}{\text{C}} \\
\text{H}-\overset{||}{\text{C}}-\text{H}
\end{array}
$$

Acrolein

Phenols
Anthocyanins ↓

(bitterness)

Figure 11.7. Metabolic formation of 1,3-propanediol and 3-hydroxypropionic acid from glycerol through the intermediate 3-hydroxypropionaldehyde (3-HPA) by some lactic acid bacteria. Acrolein is formed through a chemical dehydration of 3-HPA. Adapted from Sobolov and Smiley (1960), Slininger et al. (1983), and Schütz and Radler (1984).

and Radler, 1984; Claisse and Lonvaud-Funel, 2000). As NADH becomes limiting, excess 3-HPA could be chemically dehydrated to yield acrolein. Thus, it is plausible that the growth of other species of *Pediococcus* or *Lactobacillus* during post–malolactic fermentation could produce enough acrolein to impart bitterness to wine even though the strain(s) could not utilize glycerol as a sole carbon source. Besides different strains, acrolein formation is favored at (a) higher alcoholic fermentation temperatures and (b) in high °Brix musts, the latter resulting from increased yeast synthesis of glycerol and high pH, which favors bacterial growth (Louw, 2001).

11.3.8 Mannitol

As described in Section 2.4.4, heterofermentative bacteria can reduce fructose to form mannitol. From the winemaker's perspective, the importance of mannitol formation is uncertain. Since both mannitol and erythritol are 40% and 75% as sweet as sucrose, respectively (Godshall, 1997), these compounds could potentially impact the sweetness of wine if in high enough concentrations. Sponholz (1993) noted that mannitol spoilage is highly complex as it is accompanied by acetic acid, D-lactic acid, *n*-propanol, 2-butanol, and often sliminess and excessive diacetyl. The author further describes these wines as having a "vinegary-estery" taste. The best methods of avoidance would be those that prevent unwanted growth of lactic acid bacteria (i.e., sterile filtration, acidulation, use of SO_2, and an adequate cleaning/sanitation program).

11.3.9 Ropiness

In low-acid, dry wines, growth of pediococci and lactobacilli may result in formation of extracellular polysaccharides. Many appear to be β-D-glucans (Llaubéres et al., 1990) but other monosaccharides may be present (Manca de Nadra and Strasser de Saad, 1995). Described as "ropiness" or "oiliness" by winemakers, this fault is commonly detected as an increase in viscosity.

P. damnosus and *P. pentosaceus* have been implicated in this defect in wines (Manca de Nadra and Strasser de Saad, 1995; Lonvaud-Funel, 1999; Walling et al., 2005) while *Lactobacillus* spp. have been reported to cause ropiness in ciders (Duenas et al., 1995). Factors that affect bacterial growth also impact ropiness. For instance, exopolysaccharide production by *P. damnosus* is enhanced above pH 3.5 (Walling et al., 2005). Although ropy strains of pediococci are thought to be more tolerant of alcohol, acid, and SO_2 than other strains of *Pediococcus* (Lonvaud-Funel and Joyeux, 1988; Walling et al., 2005), synthesis of these polysaccharides is not a response to ethanol stress (Walling et al., 2005).

This problem occurs in wines either during alcoholic fermentation or after bottling (Du Toit and Pretorius, 2000). Ropiness can initiate in the bottom of cooperage and eventually spread throughout the vessel. Surprisingly, low levels of glucose (50 to 100 mg/L) may be enough to allow formation of the polysaccharide. This may partially explain why ropiness may be observed several months post-bottling. Walling et al. (2005) concluded that nonagitated wines in which the pH is high and glucose and a nitrogen source are present are most at risk for this problem. If detected early enough, addition of SO_2 and lowering pH can limit bacterial growth (Du Toit and Pretorius, 2000).

11.3.10 Tartaric Acid Utilization

Except for precipitation as the insoluble salt potassium bitartrate (KHT), winemakers generally regard tartaric acid as being microbiologically stable. However, there are species of *Lactobacillus* that can degrade the acid (Wibowo et al., 1985). In fact, the French term *tourne* is used to describe the loss of tartaric acid due to microbial infection (Ribéreau-Gayon et al., 2000). Wibowo et al. (1985) reviewed reports where decreases of 3% to 30% were observed during malolactic fermentation of wines. Microbial decomposition of tartrates during recovery operations in California have also been observed (K.C. Fugelsang, personal observation).

Early research by Krumperman and Vaughn (1966) reported five species of *Lactobacillus*, including *L. brevis* and *L. plantarum*, that can utilize tartaric acid. Apparently, pathways to metabolize tartaric acid differ between heterofermentative and homofermentative bacteria. As described by Radler and Yannissis (1972), *L. brevis* (heterofermentative) converts tartaric to oxaloacetic acid and through a series of intermediates, succinic acid, acetic acid, and CO_2. By comparison, *L. plantarum* (homofermentative) decarboxylates oxaloacetic acid to pyruvic acid, which is subsequently reduced to lactic acid or decarboxylated to acetic acid and CO_2. More recently, a novel species of *Candida* has been isolated from wine lees that can degrade tartaric acid (Fonseca et al., 2000), but it remains unknown if this yeast causes spoilage in wines.

SECTION III

LABORATORY PROCEDURES AND PROTOCOLS

CHAPTER 12

BASIC MICROSCOPY

12.1 INTRODUCTION

When a winery considers investment in laboratory equipment, a compound microscope should be a priority. Microscopic capabilities allow winemakers to quickly monitor the progress of alcoholic and malolactic fermentations and to tentatively determine the source of microbiological problems. This chapter outlines basic microscopy as well as techniques to view wine microorganisms.

12.2 MICROSCOPES

Costs for a good quality brightfield binocular scope equipped with 10×, 40×, and 100× objectives range widely, with many priced at $1000 to $3000. A microscope with dual action, commonly both brightfield and phase-contrast, can cost $3000 or more. Additional features such as binocular design and camera attachments will increase costs as well. Some brightfield microscopes can be upgraded with phase-contrast, but the cost may be prohibitive. Buyers of used microscopes must be aware that manufacturers may no longer provide parts or service for that particular model. It is therefore wise to contact the manufacturer prior to buying a used microscope to be certain parts and service are available.

The quality of a microscope depends on three elements; magnification, resolution, and contrast. In addition, the microscope may have additional features such as fluorescence capabilities.

12.2.1 Magnification

The first and most obvious goal in microscopy is to magnify the image of the microorganism so that details can be easily seen by the human eye. Total magnification is the product of the contributions of the objective lens and the ocular (eyepiece). For example, if an image is viewed with an ocular of 10× and an objective of 100× (oil immersion), a total magnification of $(10) \times (100) = 1000\times$ is achieved.

Although magnification is theoretically limitless, it is practically limited by the point at which further useful information is lost. Most wineries are satisfied with using microscopes that have three objectives of 10×, 40×, and 100×, the latter being oil immersion. Of the microscopes available in the marketplace, many have "achromatic" objectives, which yield a sharp image by correcting chromatic (color) aberrations. Although more expensive than achromatic objectives, "planachromatic" types correct for these aberrations but also yield a flatter image, which is recommended for higher quality applications such as photography. Some enologists have also had success using 15× oculars rather than the standard 10× because small images (e.g., wine bacteria) will appear larger.

12.2.2 Resolution

The ability to visually separate two objects, a property called resolution, is as important as magnification. One factor that influences resolution is the refractive index of the medium that lies between the objective lens and the cover glass on the microscope slide. Under normal conditions, the medium is air but others can be used. Because increasing the refractive index of the medium will yield better resolution, immersion oil is placed between the objective and the cover slip. Because immersion oil cannot be used with 10× and 40× objectives, all wineries should have a microscope with an oil immersion 100× objective. Using this objective, the resolving power of a compound microscope is approximately 0.2 μm with an overall magnification of 1000×, enough to view even the smallest of microorganisms.

12.2.3 Contrast

If the microorganism cannot be differentiated from the matrix in which it is suspended, the ability to magnify and resolve is of little value. Contrast is created by the use of dyes or stains or by manipulation of the optical system.

Staining achieves contrast by reaction between the stain and the microorganism (or other material) because the amount of light absorbed by the microorganism would be greater relative to the unstained matrix. Staining is useful in developing information relative to specific intracellular structures or cell wall characteristics (i.e., Gram stain). Unfortunately, the process of staining often distorts cell structure, which may or may not be a problem. Although dyes such as methylene blue do not cause aberrations in cellular morphology, these chemicals are eventually toxic to the microorganism. Alternatively, negative (or background) stains, such as nigrosin, create contrast because these are unable to penetrate the cell. Here, the background appears dark whereas the microorganism is clear. Such dyes are often effectively used to obtain information regarding cell shape and the presence of a capsule. As phase-contrast optical systems are not designed to view color, brightfield microscopes are used to view stained or dyed preparations.

The other method to gain contrast is to use a technique known as phase-contrast. This technique relies on the fact that light passing through a medium that is denser (the cell) relative to the surrounding medium (water or wine) is retarded and diffracted. The degree to which light is diffracted is defined by the refractive index. Microscopes equipped with a phase-contrast optical system are designed to enhance the differences in refractive index between the microorganism and surrounding medium, thereby allowing the microorganism to be visualized. The significant advantage of phase microscopy over brightfield is the ability to observe living cells without staining. Because an image viewed under phase-contrast is not colored, this technique should not be used for the examination of stained or dyed microorganisms.

12.2.4 Fluorescence Microscopy

Fluorescence microscopy is used to estimate microbial population density and viability based on the fact that some molecules absorb light of a specific wavelength and emit light of different wavelengths (fluorescence). Compounds that fluoresce can be either present in nature (e.g., chlorophyll) or chemically synthesized (fluorophores). Fluorophores absorb blue light resulting in formation of excited but unstable electrons that rapidly shed the excess energy upon return to ground state (Fig. 12.1). As the energy level of electron decreases en route to ground state, green light that is of lower energy and longer wavelength than blue light is emitted. However, some fluorophores absorb green light and emit red light, which is even lower in energy and of longer wavelengths.

The specific dyes used in fluorescence microscopy are directly taken up by the cells and may react indiscriminately with organic material or be incorporated and concentrated in specific subcellular organelles. Still other techniques utilize immunochemical methodology (Section 16.3.2) whereby a fluorescent dye react with specific moieties of the viable cell membrane. In these cases, the dye can be used to distinguish between viable and dead cells. These methods have been applied toward detection of viable-but-non-culturable cells because these cells cannot be cultured using standard microbiological media (Section 6.1).

Splittstoesser (1992) described a method using fluorescent dye (acridine orange) known as direct epifluorescence filter technique (DEFT), a method also applied by Divol and Lonvaud-Funel (2005). Divol and Lonvaud-Funel (2005) used a different substrate, fluoresceine diacetate, which is hydrolyzed by viable cells to form a fluorescent product, fluoresceine. However, Atlas and Bartha (1981) observed that cell population values can differ substantially between (epifluorescence and direct plating) methods (10^9 vs. 10^7 CFU), possibly due to the presence of viable-but-non-culturable cells. In addition, Meidell (1987) reported interference

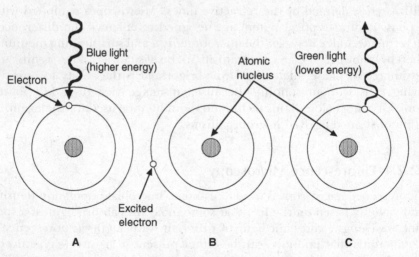

Figure 12.1.   Excitement of an electron by blue light (A) and its movement to a higher energy state (B) prior to returning to the original state and emitting green light (C).

by preservatives such as sorbic acid that, upon adhering to cells, fluoresce intermediate shades.

Fluorescent microscopes differ from either brightfield or phase-contrast types in that these use a mercury lamp as a source of white light, and an in-line filter selects specific wavelengths of light based on the fluorphore used. In order to visualize the emitted light, a second filter is used to separate the much brighter excitation light. As such, fluorescing materials appear bright against a dark background.

More recent technology has been termed laser scanning confocal microscopy (LSCM). These instruments utilize one or more lasers as the light source in order to narrow the excitation light bandwidth from 20–30 nm of classic fluorescence microscopes to 2–3 nm. The laser beam rapidly scans the area, and the image is developed on a monitor. LSCM has the advantage of clearer images facilitating observation of fine detail, including the possibility for a three-dimensional perspective. However, such systems can be expensive.

12.3 USING MICROSCOPES

12.3.1 Components

The various components of a microscope illustrated in Fig. 12.2 are defined below.

Arm: Physical support for the microscope and the part that is held when moving the microscope to different areas (A).

Body tube: Holds the eyepiece lenses and nosepiece in the correct position and orientation (B).

Condenser: Condenses light rays into a "pencil-shaped" cone of light thereby allowing more light to enter the microscope. The position of the condenser can be raised/lowered to change the amount of light entering the microscope (C).

Coarse-adjustment knob: Moves the stage up and down in large increments to permit quick focusing on the sample of interest (D).

Diaphragm: Increases/decreases the amount of light passing through the slide containing the sample to be examined (E).

Eyepiece (ocular) lenses: Used to view samples and normally possess a magnification of 10× (F).

Fine-adjustment knob: Moves the stage up and down in small increments to permit fine focusing on the sample of interest (G).

Magnification: The ratio of the apparent size of the sample as seen through the microscope to the actual size of the sample seen by the

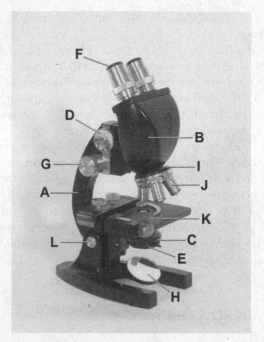

Figure 12.2.   Components of a microscope.

unassisted eye at a distance of 25 cm. Total magnification = (ocular magnification) × (objective magnification).

Mirror: Used to direct light into the microscope from a separate, unattached light source. Most modern microscopes will have a direct light source rather than a mirror (H).

Nosepiece: Objective lenses are screwed into the nosepiece, which allows changing objectives quickly by rotating its base (I).

Objective lenses: Most microscopes contain three objectives, which act to further magnify the image. These objectives can be low (10×), high (40×), and oil immersion (100×). Objectives can be switched by simply rotating the nosepiece. Commonly the highest magnified objective, the oil immersion objective requires a drop of immersion oil to be placed between the objective and the sample in order to provide clear images of the sample (J).

Parfocal: Microscopes are designed so once an image is in focus using the low (10×) objective, other objectives can be used with a minimum of refocusing.

Stage: This platform holds the sample to be examined. Most microscopes have controls to move the stage so that more of the sample can be completely examined (K).

Tilt stage adjustment knob: Allows the microscope to be tilted to/from the microbiologist to assist viewing. Be careful of liquid samples running over the stage if the stage is tilted too far (L).

12.3.2 General Use

1. Place the prepared slide onto the microscope stage and turn on the light source. Adjust the mirror (if so equipped) for the light beam to shine directly into the mirror.
2. Move the condenser all the way up to the stage and open the diaphragm completely. Newer microscopes do not have a diaphragm adjustment.
3. Using the low magnification objective (10×), attempt to focus on the image using the coarse-adjustment knob. To help with adjustment, mark an "X" on a clean slide using an ink pen and place the slide onto the stage. As the image is viewed, the condenser and diaphragm can be adjusted to obtain optimal light and image.
4. Once the object is in focus, higher magnifications can be achieved by rotating the nosepiece to the proper objective. Oil immersion objectives require immersion oil to be placed between the objective and the slide (usually only one drop). Higher magnifications will also require more light to pass through the condenser.
5. Be sure to clean all objectives with the appropriate microscopy cleaning fluid and lens paper. Never use Kimwipes® or similar lab tissues because these will scratch the lenses and objectives. Given the overall investment in a microscope, avoid using substitute materials when cleaning.
6. For phase-contrast microscopes, it is crucial to maintain adjustments of the phase and annular rings. As every microscope is different in this process, consult the manufacturer for specific instructions regarding this adjustment.

12.3.3 Calibration

At times, wine microbiologists may need to determine the size of the microorganisms being viewed to help to determine the identity of the microorganism in question. To do this, a separate ocular micrometer is placed into the eyepiece of a microscope but must be calibrated for each objective. Both ocular and stage micrometers are required.

1. Place the ocular micrometer into the eyepiece and the stage micrometer onto the center of the stage. Viewed together through the eyepiece of the microscope, the micrometers appear as shown in Fig. 12.3.

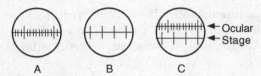

Figure 12.3.   Views of ocular (A), stage (B), and both ocular and stage (C) micrometers used in microscopy. The lines on the stage micrometer are exactly 10 μm apart.

2. Use the low objective (10×) to focus onto the scale of the stage micrometer.
3. Once focused, rotate the objective to be calibrated into place. Be sure to use oil with the oil immersion objective.
4. Using the control knobs, move the stage until the two micrometer scales are close to overlap as viewed in Fig. 12.3C. It can be easier to super-impose the scales, unlike the drawing.
5. Count the number of ocular micrometer lines that exactly occupy the space between two (2) stage micrometer lines. With four ocular lines per space on the stage micrometer lines, one space of ocular micro-meter would be: 10 μm ÷ 4 = 2.5 μm.
6. With this objective, the distance between two lines in the ocular micro-meter are 2.5 μm. Given the initial calibration of one space = 2.5 μm, the approximate length of a rod-shaped microorganism that occupies 3 spaces would be 7.5 μm (3 × 2.5 = 7.5).

12.4 PREPARING SMEARS

In order to see a yeast or bacterium under a microscope, sometimes it is necessary apply a stain such as methylene blue (yeast) or to perform a Gram stain (bacteria). Before staining, a slide smear of the microorganism must be prepared. Whereas dye can be added to wet mounts (Section 12.5), smears rely on preparations that are dried.

12.4.1 From Liquid Media

1. Using a transfer loop, place one or two loops of the liquid containing the microorganism on the microscope slide.
2. Spread the liquid over a large area.
3. Allow the smear to dry at room temperature. Avoid the temptation to accelerate the drying process by exposing the preparation directly to heat from the flame.

4. Once the smear is dry, heat fix the microorganisms to the slide by attaching a clothespin to a short side of the slide and passing the slide through an open flame.
5. The smear is now ready for staining and visualization.

12.4.2 From Solid Media

1. Place one or two loops of water on the slide.
2. Transfer a small amount of the microorganism with an inoculating needle. Mix with the water on the slide and spread over a large area.
3. Allow the smear to dry at room temperature. Do not expose the smear to heat from a flame.
4. Once the smear is dry, heat fix the microorganisms to the slide by attaching a clothespin to a short side of the slide and passing the slide through an open flame.
5. The smear is now ready for staining and visualization.

12.5 PREPARING WET MOUNTS

Frequently, a wine microbiologist does not have time to isolate microorganisms from a sample. Therefore, it is easier to examine a sample under a wet mount using a microscope equipped with phase-contrast. Normally, a minimal population of 10^4 cells per milliliter is necessary to be seen microscopically. As such, several microscopic fields should be examined and/or samples should be centrifuged to concentrate microorganisms prior to microscopic evaluation.

1. Using either a laboratory loop or small pipette, place a drop of juice/wine on a clean microscope slide. It is frequently necessary to centrifuge samples to concentrate the microorganisms, especially if the number of microorganisms in the juice/wine is low. Centrifuge 10 mL of sample for 15 min at $3000 \times g$ in a plastic, conical centrifuge tube using a laboratory centrifuge. Remove the liquid by decanting and mix the sediment using a small glass rod or a Pasteur pipette prior to placing a drop onto a slide.
2. Place a cover slip over the suspension and examine the slide using low and oil immersion powers. For short-term storage, seal the edge of the cover slip with petroleum jelly to prevent drying of the mount.

12.6 PREPARING MOLD SLIDE CULTURES

Molds are identified based largely on morphological properties. The latter are dramatically influenced by the medium on which the mold is grown. Further, microscopic observation is complicated by the fact that molds easily fragment when disturbed. Thus, preparation of wet mounts, as used in bacterial and yeast identification, is of minimal value. Rather, slide cultures are typically employed for the observation of molds. The procedure described below permits visualization of the structure directly or after staining with minimal mechanical damage to the microorganism. A similar technique (Johnson's slide culture method) was described by Harrigan (1998).

1. Prepare lactophenol–cotton blue dye by mixing equal volumes of component A and component B.
   a. Component A: Slowly mix 10 g phenol in 10 mL of distilled water. Once dissolved, add 10 mL lactic acid and 10 mL glycerol.
   b. Component B: Prepare a saturated solution of soluble aniline blue (cotton blue) and transfer 10 mL to a solution of 10 mL glycerol and 80 mL distilled water.
2. Pour approximately 10 mL of sterile molten potato dextrose agar into a Petri plate and allow to solidify.
3. Prepare a sterile (autoclaved) growth chamber (glass Petri plate, sterilized tubing, sterile microscope slide and cover slip) as shown in Fig. 12.4.
4. Using a sterile spatula, cut the agar into cubes of suitable size that fit within the dimensions of the microscope slide cover slip.

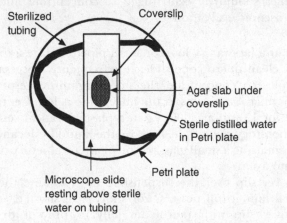

Figure 12.4.   Apparatus to microscopically observe molds.

5. Using a sterile needle, collect a portion of mycelial mass and aseptically inoculate the corners of the agar cube.
6. Using sterilized forceps, place the sterile cover slip over the top of the mold-inoculated agar cube.
7. Transfer the slide into the Petri plate and incubate at room temperature until growth is observed (12 to 36 h). The slide culture may be removed for microscopic examination during this period. To reduce the likelihood of contamination, the number of examinations should be limited and restricted to high-dry (i.e., 40×) or lower magnifications.
8. When suitable growth has developed, prepare a wet mount by carefully lifting the cover slip (with adhering mold) from agar substrate and transfer to a fresh clean microscope slide that contains a drop or two of the dye described above. Wet mounts can be preserved by sealing the edges of the cover slip to the slide with clear fingernail polish.

CHAPTER 13

MEDIA PREPARATION AND CULTURE TECHNIQUES

13.1 INTRODUCTION

Successful cultivation of microorganisms for growth and identification requires use of various types of media, either liquid (referred to as broths) or solid by inclusion of agar. For winemakers, a growth medium may be as simple as diluted sterile grape juice for the activation and expansion of yeast starter cultures or as complicated as that necessary to grow lactic acid bacteria.

Historically, microbiologists prepared growth media by mixing and cooking undefined components and then adding gelatin as the solidification agent. Prepared as a 15% w/v solution, gelatin (unfortunately) liquefies near 26°C/79°F and therefore is not suitable for most applications where cultures need to be incubated at higher temperatures. To overcome these problems, most modern media formulations use agar. Agar is a complex polysaccharide prepared from seaweed and is incorporated at concentrations ranging from 1% to 2% w/v. Neither yeast nor bacteria metabolize agar, and the polysaccharide does not suffer from temperature-related liquefaction problems noted with gelatin.

The availability of a variety of preformulated dehydrated media has dramatically shortened the time required for preparation. To prepare these media, frequently only distilled water is added and then the medium is sterilized, usually by autoclaving. Most suppliers handle a variety of specialized media and, depending on demand, can prepare others as needed. Even when specialized media cannot be obtained, individual ingredients are normally available. Major suppliers of microbiological media include Difco, BBL (Baltimore Biological Laboratories), and Oxoid. For price conscious individuals, "house-brand" formulations of routine media are also available from most major suppliers at competitive prices.

Dehydrated media are hygroscopic and, thus, have a limited shelf life once opened (Flowers et al., 1992). In general, unopened containers should be used within 1 year of receipt and, once opened, the contents should be used within 6 months. Media should be stored under cool (<30°C/86°F), dry conditions, preferably out of direct sunlight, and should be discarded if clumping/caking or off-colors and odors develop. Since media in opened bottles will deteriorate relatively rapidly, it is better to purchase smaller quantities depending on expected use.

13.2 PHYSICAL/CHEMICAL REQUIREMENTS FOR CULTIVATION

Several criteria must be met in order to successfully cultivate microorganisms. For instance, the medium must contain a utilizable source of organic carbon (commonly a sugar) for energy, nitrogen for the synthesis of amino acids, proteins, enzymes, and nucleic acids, as well as various other compounds including vitamins and minerals. Furthermore, media pH, incubation temperature, and atmospheric conditions (presence or absence of oxygen) must be optimized. Microorganisms isolated from grape juices or wines have widely varying requirements for nutrients (Chapters 1, 2, 3, and 4).

13.2.1 Carbon and Nitrogen

Microorganisms require sources of utilizable carbon and nitrogen for growth and metabolism. By changing the medium composition through selective exclusion or inclusion of ingredients, specific microorganisms or groups of microorganisms can be isolated from an otherwise diverse population. This is true even when the microorganism(s) of interest are present in relatively low populations. An example would be the use of lysine agar in which the amino acid lysine represents the sole utilizable source of nitrogen. Although *Saccharomyces* cannot grow on this nitrogen source due to the formation of an inhibitory intermediate, most non-*Saccharomyces* yeasts can use this amino acid as a sole nitrogen source.

13.2.2 Oxygen

Oxygen requirements and tolerances vary depending on the microorganism in question. Some have absolute requirements for oxygen (obligate aerobes), whereas others are rapidly killed by even minute amounts (obligate anaerobes). Although winemakers are generally not concerned with obligate anaerobic species, aerobic organisms (e.g., molds) play important roles in grape growing and winemaking. Of great importance to winemakers are those microorganisms that grow under conditions where oxygen is present or absent (facultative anaerobes).

In winemaking, there are two groups of facultative anaerobes. Fermentative yeasts, such as *Saccharomyces*, are metabolically able to grow either fermentatively in juice where oxygen is limiting or oxidatively at the surface of wine as a film yeast where wines are exposed to air. However, *Saccharomyces* cannot grow for extended periods under anaerobic conditions due to the eventual lack of certain cellular components only synthesized under aerobic conditions (Section 1.4.2). Besides fermentative yeasts, lactic acid bacteria possess a fermentative metabolism and do not require oxygen for growth. These bacteria can grow under aerobic conditions but grow better under atmospheres of reduced oxygen ("microaerophilic"). Fermentative species found in wine can be normally cultivated in low-oxygen environments by use of stab cultures, agar overlay techniques (pouring sterile agar over solidified agar already containing the microorganism of choice), or by use of low-oxygen incubators such as candle jars or GasPak® systems (Section 18.5).

13.2.3 Hydrogen Ion Concentration (pH)

Hydrogen ion concentration plays an important role in cultivation and identification of microorganisms and may be used to selectively promote growth of some over others. Depending on composition of the medium and the microorganism involved, dramatic pH shifts may occur when yeast or bacteria grow. For example, growth in a medium rich in amino acids and peptides may liberate ammonia (increase pH), whereas growth on sugars produces acids (decrease pH), depending on the buffering capacity of the medium. Both situations are eventually inhibitory and potentially toxic to the microorganism.

The most frequent technique for reducing the impact of moderate acid or alkali production during growth on media is through incorporation of a buffer. Phosphate buffers are often the agents of choice when preparing microbiological media because these can be formulated to function around pH 7.0. Furthermore, phosphate buffers are relatively nontoxic to microorganisms at the low concentrations used (<5 g/L, as potassium phosphate). Although the specific compounds and relative concentrations vary depending on desired final pH, a typical buffer is prepared as a mixture of a weakly basic salt (e.g., potassium monohydrogen phosphate, or K_2HPO_4) and the weakly acid salt (e.g., potassium dihydrogen phosphate, or KH_2PO_4). When combined in equal molar concentrations, the resultant buffered solution is approximately pH 6.8.

Under some situations, the microorganism may utilize sugar or ethanol to produce acetic acid in such large enough amounts that the buffering capacity of the medium is overwhelmed. Such is the case with the growth of *Brettanomyces/Dekkera* or *Acetobacter* whereby the acid shift results in premature cell death of the species during cultivation. The problem may be limited by inclusion of calcium carbonate ($CaCO_3$) in these media. Because the salt is insoluble, it does not initially affect pH of the medium. However, acid produced by these microorganisms results in the salt dissolving and eventually yielding carbon dioxide:

$$CO_3^{2-} (+ H^+) \rightarrow HCO_3^- (+ H^+) \rightarrow H_2CO_3 \rightarrow CO_2 + H_2O$$

Microbiologists also use this chemical reaction as a rapid screening tool for acid-producing microorganisms. Here, colonies that produce organic acids can be readily visualized by a clearing or "halo" around a colony as contrasted against an opaque background of the medium containing $CaCO_3$.

13.2.4 Moisture and Water Activity

Yeast, bacteria, and molds all require minimal levels of moisture for growth. In general, most wine-borne yeast and bacteria need more moisture than molds. During solidification of cultivation media, agar binds water and lowers free moisture (a_w) or water activity. As such, it is important to adhere closely to the supplier's recommendations regarding the amounts of agar to include in formulations because a_w could be reduced too far to support growth.

Moisture may be further lost during the process of pouring plates. For instance, immediately pouring agar upon removal from the autoclave promotes condensation on the inside cover of Petri plates. This loss of water yields very hard agar due to a lower a_w. To avoid this problem, it is recommended that media be transferred to a 45°C/113°F to 50°C/122°F water bath and the internal temperature allowed to equilibrate ("temper") prior to pouring. When large numbers of plates are to be poured at one time, the flask containing agar may require frequent return to the water bath to prevent premature solidification of the agar.

Storage of pre-poured media can also result in water loss over time. Although pre-poured media should be stored at refrigeration temperatures, this increases the problem of dehydration. Because of these issues, pre-poured media should be used as soon after preparation as possible. If agar plates must be stored, returning the plates to the original plastic shipping sleeves slows desiccation. Where Petri plates are incubated for more than 1 week, an open container of water placed in the incubator will increase the relative humidity within the chamber. Finally, sealing the lid and bottom of a Petri plate with a sealing wrap material (e.g., Parafilm®) will also slow the process of dehydration.

13.2.5 Incubation Temperature and Conditions

Temperature optime for most microorganisms are commonly between 15°C/59°F and 40°C/104°F. Whereas many microbiological laboratories employ incubators for control of growth temperatures, most wine microorganisms grow well at room temperatures (20°C/68°F to 25°C/77°F). Although some lactic acid bacteria have higher optimal temperatures, 30°C/86°F, these normally will grow, albeit slower, at room temperatures. Normally, Petri plates are incubated upside down (inverted), which minimizes agar contamination from airborne microorganisms.

13.2.6 Selective Agents

There are many ingredients that can be added to a medium that select for growth of one microorganism or group over others present in a mixed culture. For instance, cycloheximide (Actidione®) has been used to select against *Saccharomyces*, which is generally inhibited at concentrations of 10 to 20 mg/L. Whereas *Brettanomyces/Dekkera* spp. survive at 50 mg/L, *Kloeckera* can tolerate even higher amounts, approaching 100 mg/L (Pfaff et al., 1978). Table 13.1 illustrates the various selective agents that can be used to isolate grape or wine microorganisms.

Cycloheximide may be added to a medium either before or after autoclaving. However, steam sterilization will decrease the activity of the ingredient by approximately half. In that *Saccharomyces* is generally inhibited at <20 mg/L, sufficient concentrations of the active component should be present in a medium before autoclaving. Alternatively, solutions of

Table 13.1.   Selective agents used in media for the cultivation of microorganisms from grape juices and wines.

Agent	Inhibited	Not inhibited
Cycloheximide (20 to 100 mg/L)	*Saccharomyces*	Non-*Saccharomyces* yeasts Lactic acid bacteria Acetic acid bacteria
Pimaricin (50 mg/L)	*Saccharomyces*	Non-*Saccharomyces* yeasts Lactic acid bacteria Acetic acid bacteria
Oxytetracycline (100 mg/L)	Lactic acid bacteria Acetic acid bacteria	*Saccharomyces* Non-*Saccharomyces* yeasts
Propionate (0.1% to 0.2% w/v)	Molds	*Saccharomyces* Non-*Saccharomyces* yeasts Lactic acid bacteria Acetic acid bacteria
Streptomycin (25 mg/L)	Lactic acid bacteria	Acetic acid bacteria

cycloheximide can be sterile filtered (0.45 µm) and then aseptically added to a sterilized medium.

Cycloheximide is a health hazard, and extreme caution should be used when adding this ingredient to a medium. Skin contact and/or inhalation of dust should be completely avoided. Refer to the Material Safety Data Sheets (MSDS) for specific instructions (Section 19.2.2) regarding proper disposal of residual powder.

13.3 STERILIZATION OF LABORATORY MEDIA AND SUPPLIES

The goal of sterilization is to kill or physically remove all of the living microorganisms and reproductive spores (mold conidiospores, yeast asco-spores, or bacterial endospores). Sterilization of laboratory media and equipment may be accomplished through exposure to (a) boiling water, (b) combined high temperature and pressure (autoclaving), (c) dry heat, (d) physical removal (filtration), or (e) chemicals.

13.3.1 Boiling Water

Boiling water for 5 to 10 min is enough to kill viable microorganisms but not all bacterial spores (Harrigan, 1998). In the absence of an autoclave, laboratory hardware like membrane-filter housings and utensils may be sterilized by immersion in boiling, distilled water (100°C/212°F) for

20 min. However, this sterilization treatment has drawbacks in that it is difficult to work with wet instruments.

13.3.2 Steam Sterilization

Most laboratory media can be steam sterilized using an autoclave, the exceptions being solutions of vitamins and some sugars that degrade at high temperatures. Although a range of features are available, autoclaves operate much like pressure cookers used in home canning and, while not recommended, the latter have been used for this purpose. In an autoclave, the environment becomes saturated with superheated, pressurized, steam ($121°C$ at 103.4 kPa or $250°F$ at 15 lb/in^2) and all air has been displaced. Saturated steam and pressure prevents the water phase of media from boiling while permitting internal temperature to rise well above boiling. For media volumes of 1 L or less, sterilization is defined as $121°C/250°F$ for 15 min while longer autoclave times are needed for larger volumes.

Given the conditions described above, flasks and other media-containing vessels should not be filled to capacity prior to autoclaving. As a general rule, flasks and containers should be filled using the general guidelines presented in Table 13.2. When screw-capped containers are used, it is recommended that the cap be tightened and then loosened one-half turn prior to autoclaving. Once the exhaust cycle is complete, the cap should immediately be tightened. This recommendation applies for glass as well as plastic containers. The latter undergoing distortion if pressure is not allowed to equilibrate during the autoclaving cycle. Finally, many laboratories stopper flasks with cheesecloth-wrapped cotton stoppers rather than using metal caps or purchasing the more expensive screw-cap flasks. To minimize the problem of condensate soaking the cotton during the sterilization cycle, a suitably sized piece of aluminum foil

Table 13.2.   Recommended presterilization fill volumes for selected containers.

Container Type	Volume/size	Maximum autoclave volume
Erlenmeyer flasks	250 mL	100 mL
	500 mL	200 mL
	1000 mL	400 mL
Bottles	100 mL	40 mL
Test tubes	18 × 150 mm	15 mL

Adapted from Pawsey (1974) with the kind permission of R. Pawsey.

should be placed over the inserted plug and crimped around the neck of the flask.

Upon completion of the sterilization cycle, steam is exhausted and the pressure drops to atmospheric levels. The rate of pressure decrease varies with the autoclave but most are equipped with fast and slow exhaust cycles. Fast exhaust is used for glassware, tubing, and utensils, while slow exhaust cycles are used with liquid media. Once the cycle is complete, media should be removed from the autoclave as soon as possible because prolonged heating will cause heat-sensitive components to degrade.

Although tempting, one should never attempt to open the autoclave door before the exhaust cycle is complete. The discharge of steam into the immediate area may result in serious burns. Furthermore, the rapid reduction in pressure will cause flasks to boil-over causing burns while creating a difficult mess to clean up. When removing liquids and media from an autoclave, care should be taken not to agitate the contents as this action will cause the superheated liquid to boil-over. To minimize risks of burns resulting from boil-overs, using insulated laboratory gloves are recommended.

It is good practice to verify that sterilization conditions are being met during the autoclave cycle. Depending on the type of microbiological work being performed, the autoclave should be tested on a yearly basis by qualified personnel. Various types of heat- or pressure-sensitive test tapes (indicator strips) and liquids are available to ensure that sterilizing conditions are being met. Tapes may be attached directly to media containers or placed in packaged supplies before autoclaving. Upon achieving necessary sterile conditions, the tape will change color.

13.3.3 Dry Heat

In comparison to high-pressure steam sterilization, dry heat is relatively inefficient and is usually reserved for sterilizing non-liquid items such as empty glassware or utensils. In general, glass pipettes stored in metal canisters should be held at 170°C/338°F for 2 h while other glassware (flasks, etc.) can be sterilized at this temperature after 1 h. Gaskets or other heat-sensitive inserts must be removed prior to heating.

Drying ovens should never be used to re-melt solidified agar media because too much water will evaporate from the medium. Re-autoclaving media for a complete sterilization cycle will also degrade heat-sensitive nutrients in the medium. Rather, solidified media can be melted by placement in a boiling water bath or a microwave, the later performed with hand mixing every few minutes.

13.3.4 Sterile Filtration

Media that contain thermally labile compounds are sterilized by sterile filtration. Unlike filtration systems using depth filters, these membranes trap microorganisms on surfaces because the pore sizes are too small for the microorganisms to pass through. Such filters are described as being "absolute" as opposed to nominal. For instance, an absolute filter with a specification of $0.45 \pm 0.02\,\mu m$ will not have pores greater than $0.47\,\mu m$ in diameter (Section 5.3.2).

Most membranes are made of porous cellulose acetate and are available in a range of porosities, from $0.22\,\mu m$ to several micrometers. Sterilization and the complete removal of the smallest microorganisms (bacteria) requires the use of sterile membranes of porosities of $0.45\,\mu m$ or less. Filter housings and clamps should be packaged in aluminum foil and then heavy-grade wrapping paper (butcher paper) prior to autoclaving and not opened until use. Alternatively, suppliers make disposable filter units that facilitate sterilization of small volumes of media.

One drawback in using small-porosity filters is premature plugging or clogging. Some suspensions used in media are hazy or cloudy (e.g., the liver extract used in media for the cultivation of lactic acid bacteria) and these can quickly clog a $0.45\,\mu m$ membrane. In these cases, it may be necessary to prefilter through Whatman no.1 filter paper or centrifuge the medium or ingredients to remove haziness or cloudiness prior to sterile filtration.

13.3.5 Chemical Sterilization

Because microorganisms are airborne, laboratory surfaces should be sterilized prior to beginning work. To do this, there are many aerosol disinfectants available that are simply sprayed onto the area and wiped away. Some laboratories rely on solutions of chlorine (bleach) to assist in maintaining a sterile environment. Although ultraviolet light can be used as a surface sterilant, most laboratories rely on wiping down the area with a liquid disinfectant (Table 13.3). Several disinfectants that find application in the winery (Chapter 9) can also be used in a laboratory such as iodine or QUATS. Other laboratory disinfectants are the phenol-based formulations like *o*-phenylphenol (Lysol®) and hexachlorophene (PhisoHex®).

Another common disinfectant used in a wine laboratory is 70% v/v ethanol. Ethanol is typically used to sterilize surfaces as well as utensils (forceps, scalpels, hockey sticks, etc.) that are repeatedly used during the

Table 13.3. Comparison of various laboratory disinfectants.

Characteristic	Chlorine (Chlorox®)	Iodophor (Wescodyne®)	Ethanol	QUATS (Roccal®-D)	Phenol (Lysol®)	Oxidizing agents (Virkon-S®)	Chlorhexidine (Nolvasan®)
Concentration (v/v)	0.01% to 5%	0.5% to 5%	70%	0.1% to 2%	0.2% to 3%	0.2% to 3%	0.2% to 3%
Sensitive to hard water	Yes	No	Yes	No	Yes	Yes	Yes
Sensitivity to organics	Yes	Yes	Yes	Yes	No	Yes	Yes
Sensitivity to detergents	No	Yes	No	Yes	No	No	No
Residual effects	No	No	Some	Some	No	No	Yes
Active against bacteria	Yes	Yes	Yes	Yes	Yes	Yes	Yes
Activity against yeasts/molds	Yes	Some	Yes	Some	Some	Some	Some
Activity against viruses	Yes	Yes	Some	No	No	Yes	No

course of work. Utensils are commonly dipped into a beaker containing 70% v/v ethanol and then ignited. Once the alcohol has burned off, the tool is cooled prior to use. Given its flammability, alcohol and the open flame should not be in the immediate proximity of each other. Additionally, once the ethanol is ignited on an instrument, it must be held downward so that any excess alcohol flows off the end. If tilted upwards, the alcohol may run onto hands and clothing causing burns. It should also be noted that higher concentrations of ethanol (i.e., 95% v/v) are not nearly as effective in killing microorganisms as 70% v/v.

In packaging of sterile Petri plates and other plastic disposable supplies that cannot be heat sterilized, gases like ethylene oxide or ozone are used. Given its toxic and explosive properties, ethylene oxide is not widely used in routine laboratory work.

In winery laboratories, sterilization of grape juice to prepare yeast or bacteria starter cultures can be difficult. Heating the juice is a concern due to the development of "cooked" characters that may be transferred to the wine. Although grape juices can be sterile filtered, premature plugging of membranes can be a significant problem. An alternative method is the use of dimethyl dicarbonate, a sterilant sold under the trade name of Velcorin™ (Section 5.2.2). At concentrations of 250 to 400 mg/L, this "cold" sterilant accomplishes the goal of reducing microbial populations in grape juice within several hours. Residual DMDC will rapidly break down into methanol and carbon dioxide (Porter and Ough, 1982). However, DMDC is not easy to work with, and care must be taken in handling due to its toxicity.

13.4 STORAGE OF PREPARED MEDIA

Whether dealing with solid or liquid media, it is recommended to prepare an amount that can be used within one week. In the interest of efficiency, some laboratories prepare larger volumes that are stored for future use. However, when media are stored for a period of time, it is optimally packaged in screw-capped containers and held under refrigeration. As broths and dilution blanks are prepared using loose-capped test tubes, it is recommended that the original volume of a representative tube be marked and note the volume upon use. Media that experience losses of more than 10% in volume should be discarded without use. Table 13.4 summarizes maximum storage time for broth and solid media stored in several types of containers at 4°C/39°F.

Table 13.4. Suggested storage times for prepared media.

Container type	Storage time at 4°C/39°F
Prepoured agar plates in plastic bags	2 weeks (longer if ordered directly from supplier companies)
Agar in bottles or screw-capped flasks	3 months
Agar or broth in loose-capped tubes	1 week
Agar or broth in tight-capped tubes	3 months

Adapted from Clesceri et al. (1989) with the kind permission of the American Public Health Association.

13.5 MEDIA FOR YEASTS AND MOLDS

Numerous media are used for the isolation, detection, and/or enumeration of yeasts from grape juices and wines. Wort medium is non-selective for yeast and is commonly used to enumerate *Saccharomyces* (King and Beelman, 1986). An incubation period of 2 to 4 days at 22°C/72°F to 25°C/77°F is needed for colonies to appear. Similarly, grape juice agar is non-selective and will support the growth of many yeasts, bacteria, and molds.

WL nutrient broths and agars (Anonymous, 1984) represent general cultivation media for yeasts and molds including non-*Saccharomyces* yeasts, *Saccharomyces*, *Zygosaccharomyces*, and *Brettanomyces*. In fact, these media can be used to monitor yeast population diversity during fermentation due to unique colony morphologies of various wine yeasts (Pallman et al., 2001). Various colony morphologies are described in Section 15.2.

"WL" is the acronym for Wallerstein Laboratories, which originally marketed the medium. WL, sometimes called WL–Nutritional (WLN), does not contain cycloheximide and so is used for determination of total viable yeast populations. This WL medium contain a pH indicator, bromcresol green, which permits rapid screening of acid-producing colonies. In addition, this medium contains casitone, a specially prepared pancreatic digest of casein available from Difco.

A related medium, WL–Differential (WLD) or WL–Cycloheximide (WLC), contains cycloheximide that is selective for *Brettanomyces* against *Saccharomyces*. It is critical to know the specific concentration of cycloheximide present in a given WLD formulation because some suppliers only add 4 mg/L, a concentration below the 10 mg/L needed for inhibition of *Saccharomyces* (Vilas, 1993). Many enologists add 50 mg/L cycloheximide to make WL selective for *Brettanomyces*. Although no evidence exists, it is possible that some strains of *Brettanomyces* are more sensitive to the antibiotic than previously thought and so may not grow well on media with the agent added.

Besides WLD, *Brettanomyces* can be cultivated on other selective media. *Brettanomyces* medium A (Vilas, 1993) and *Brettanomyces* medium B contain cycloheximide as the selective agent while *Brettanomyces* medium C has both cycloheximide and ampicillin (B. Watson, personal communication, 1998). *Brettanomyces* medium D (*Dekkera/Brettanomyces* differential medium, or DBDM) was recently developed by Rodrigues et al. (2001) as a selective medium for *Brettanomyces* against *Saccharomyces* and other yeasts, although some strains may not be able to utilize ethanol as a sole carbon source. Because of slow growth, a long incubation time (7 to 14 days) is generally required to visualize colonies of *Brettanomyces*. However, some researchers have reported seeing colonies after 3 days using *Brettanomyces* medium C. Given that *Brettanomyces* produces significant quantities of acetic acid, calcium carbonate (2% w/v) may be added similar to media used to cultivate acetic acid bacteria (Section 13.7).

Lysine agar is used for cultivation of non-*Saccharomyces* yeasts where lysine serves as the sole source of nitrogen (Morris and Eddy, 1957). Because of the large number of ingredients needed, it may be easier to purchase this medium from a commercial supplier rather than preparing in-house.

Presumptive identification of a yeast suspected to be *Zygosaccharomyces* can be made by growing the isolate on the *Zygosaccharomyces* medium supplemented with 1% v/v acetic acid. Whereas *Saccharomyces* will not grow, most species of *Zygosaccharomyces* will (*Z. bailii* and *Z. bisporus*). Some species, however, will not (*Z. rouxii*). Other selective media for *Zygosaccharomyces* are available (Makdesi and Beuchat, 1996).

Several media are used for cultivation of molds. Dichloran rose bengal chloramphenicol agar (DRBC) contains rose bengal, which restricts the spread of mycelial growth across the surface of media, as well as chloramphenicol, which inhibits bacteria (Mislivec et al., 1992). This medium is recommended as general-purpose for enumeration of yeasts and molds and is available from Oxoid. It is important to enumerate molds using a spread plate method rather than with pour plates. Spread plates allow maximum exposure to oxygen. This method also avoids heat stress caused by addition of molten agar to cultures (Mislivec et al., 1992).

13.5.1 Wort Medium

1. Mix 150 g diastatic diamalt with 850 mL distilled water.
2. Steam for 10 min at 100°C/212°F and filter through cheesecloth. Add 20 g agar to solidify the medium.
3. Autoclave at 121°C/250°F for 15 min.

13.5.2 Grape Juice Medium

1. Add 250 mL of fresh or reconstituted grape juice to a 1 L volumetric flask and dilute to volume with distilled water. It is important that the grape juice not have any preservatives such as SO_2 or sorbate.
2. Sterilize by autoclaving at 121°C/250°F for 15 min.
3. If solidified agar is needed, add 20 g of agar to approximately 800 mL of the liquid medium. Boil to dissolve all ingredients, and then add back the remaining 200 mL of medium prior to autoclaving.

13.5.3 WL Medium

1. Mix and dissolve the following ingredients.

Yeast extract	4 g
Casitone	5 g
Glucose	50 g
Potassium dihydrogen phosphate (KH_2PO_4)	0.55 g
Potassium chloride (KCl)	0.425 g
Calcium chloride ($CaCl_2$)	0.125 g
Magnesium sulfate ($MgSO_4$)	0.125 g
Ferric chloride ($FeCl_3$)	0.0025 g
Manganese sulfate ($MnSO_4$)	0.0025 g
Bromcresol green	0.022 g
Distilled water	800 mL
Agar (if desired)	15 g
Final pH = 5.5	

2. Dilute medium to volume (1000 mL) and autoclave at 121°C/250°F for 15 min.
3. If desired, WLD/WLC medium can be made by adding cycloheximide as a sterile-filtered solution to the tempered medium but prior to solidification. The final concentration of cycloheximide should be between 20 and 100 mg/L to inhibit *Saccharomyces*; commonly 50 mg/L is used.

13.5.4 *Brettanomyces* Medium A

1. Mix and dissolve the following ingredients.

Peptone	50 g
Yeast extract	30 g

Glucose	100 g
Malt extract	30 g
Distilled water	800 mL

2. Dilute medium to volume (1000 mL) and autoclave at 121°C for 15 min.
3. If desired, cycloheximide can be added as a sterile filtered solution to the cooled media before solidification. Final concentration in the medium should be between 20 and 100 mg/L to inhibit *Saccharomyces*.

13.5.5 *Brettanomyces* Medium B

1. Mix and dissolve the following ingredients.

Wort agar	48 g
Bactoglycerol (or equivalent)	2.4 mL
Cycloheximide	0.050 g
Distilled water	800 mL
Agar	20 g

2. Once dissolved, bring to volume with distilled water (1000 mL) and sterilize by autoclaving at 121°C/250°F for 15 min.
3. Once sterilized, place the flask containing the medium into a 50°C/122°F water bath and allow the temperature to equilibrate prior to pouring plates.

13.5.6 *Brettanomyces* Medium C

1. Mix and dissolve the following ingredients.

Peptone	20 g
Yeast extract	10 g
Glucose	20 g
Cycloheximide	0.003 g
Agar	17 g
Distilled water	800 mL

2. Dilute medium to volume (1000 mL) and autoclave at 121°C/250°F for 15 min.
3. Dissolve 1 mg thiamin and 100 mg ampicillin in 20 mL distilled water.
   a. Thiamin is best added as 1 mL of a 1 mg/mL sterile filtered stock solution stored at 4°C/39°F.
   b. Ampicillin should be added as 1 mL of a 100 mg/mL sterile filtered stock solution stored at −20°C/−4°F.
4. Sterile filter the thiamin/ampicillin solution and add to the autoclaved medium that has been cooled to 50°C/122°F.
5. Pour the medium into Petri dishes as pour plates.

13.5.7 *Brettanomyces* Medium D

1. Mix and dissolve the following ingredients.

Yeast nitrogen base	6.7 g
Ethanol	60 mL
Cycloheximide	0.010 g
p-Coumaric acid	0.10 g
Bromcresol green	0.022 g
Distilled water	500 mL
Final pH = 5.4	

2. Sterile filter solution through an 0.45 μm absolute membrane into a previously sterilized container.
3. Suspend 20 g agar in 500 mL distilled water and autoclave at 121°C/250°F for 15 min.
4. Mix the sterile filtered nutrients with the sterilized agar just prior to pouring the plates.

13.5.8 Lysine Medium

1. Prepare trace metal solution A.

Boric acid ($B(OH)_3$)	0.1 g
Zinc sulfate heptahydrate ($ZnSO_4 \cdot 7H_2O$)	0.04 g
Ammonium molybdate	0.02 g
Manganese sulfate tetrahydrate ($MnSO_4 \cdot H_2O$)	0.04 g
Ferrous sulfate heptahydrate ($FeSO_4 \cdot 7H_2O$)	0.25 g
Distilled water	800 mL

   a. Once dissolved, dilute to volume (1000 mL).

2. Prepare basal medium B.

Glucose	50 g
Potassium dihydrogen phosphate (KH_2PO_4)	2.0 g
Magnesium sulfate heptahydrate ($MgSO_4 \cdot 7H_2O$)	1.0 g
Calcium chloride ($CaCl_2$) fused	0.2 g
Sodium chloride (NaCl)	0.1 g
Adenine	0.002 g
DL-Methionine	0.001 g
L-Histidine	0.001 g
DL-Tryptophan	0.001 g
Trace metal solution A	1.0 mL
Potassium lactate (50% w/w solution)	12.0 mL
Distilled water	800 mL

  a. Once dissolved, dilute to volume (1000 mL) and adjust the pH to 5.0 to 5.2 using additional lactic acid.
  b. Add 20 g agar.
3. Prepare lysine solution C.

Lysine	10 g
Distilled water	800 mL

  a. Once dissolved, dilute to volume (1000 mL).
4. Prepare solution D.

Inositol	2.0 g
Calcium pantothenate	0.2 g
Aneurine	0.04 g
Pyridoxine	0.04 g
p-Aminobenzoic acid	0.02 g
Nicotinic acid	0.04 g
Riboflavin	0.02 g
Biotin	0.0002 g
Folic acid	0.0001 g
Distilled water	800 mL

  a. Once dissolved, dilute to volume (1000 mL).
5. According to Morris and Eddy (1957), basal medium B, lysine solution C, and solution D are separately sterilized by steaming each for 30 minutes on three successive days. Alternatively, solutions C and D could be filtered through 0.45 μm membranes. Solution B can also be sterile filtered but the agar would have to be autoclaved separately from the other ingredients.
6. Warm medium B and solutions C and D to 45°C/113°F to 50°C/122°F in order to avoid solidification prior to mixing. Aseptically mix basal medium B (89 parts), lysine solution C (10 parts), and solution D (1 part).

13.5.9 *Zygosaccharomyces* Medium

1. Mix and dissolve the following ingredients.

Glucose	100 g
Yeast extract	10 g
Tryptone	10 g
Distilled water	800 mL
Agar (if desired)	25 g

2. Once dissolved, dilute to volume with distilled water (1000 mL) and autoclave at 121°C/250°F for 15 min.
3. After sterilization, place the flask containing the medium in a 45°C/113°F to 50°C/122°F water bath and allow the temperatures to equilibrate.

4. Aseptically add 10 mL glacial acetic acid, thoroughly mix, and pour plates (20 to 25 mL per plate).

13.6 MEDIA FOR LACTIC ACID BACTERIA

Lactic acid bacteria are nutritionally fastidious, microaerophilic microorganisms, which grow best on media enriched with fruit or vegetable juices and under conditions of low oxygen tension. Although specially designed incubators capable of maintaining controlled gaseous environments are available, their cost is high and, in most cases, unjustified. Many wineries have been successful cultivating lactic acid bacteria using large-mouth glass jars with candles or GasPak® systems (Section 18.5). When incubated anaerobically at 22°C/72°F to 25°C/77°F, 5 to 10 days are required to visualize colonies depending on the species (shorter times for many lactobacilli but longer times for oenococci and pediococci).

The media outlined for the cultivation of lactic acid bacteria are the apple juice Rogosa medium (King and Beelman, 1986) and the tomato juice–glucose–fructose–malate medium (Izuagbe et al., 1985). These media use either apple juice or tomato juice serum to provide the so-called tomato juice factor (Section 2.3). Liver extract or concentrate has a number of vitamins that improves bacterial growth and is available from Sigma Chemical Company. Both media can be made selective against *Saccharomyces* by the addition of cycloheximide.

The heterofermentation-arginine medium (Section 13.6.3) was developed by Pilone et al. (1991) for use in characterizing different physiological traits of lactic acid bacteria (Sections 15.4.1 and 15.4.5.2).

13.6.1 Apple Juice Rogosa Medium

1. Mix 1 g liver extract or concentrate with 100 mL distilled water for at least 30 min. Filter through Whatman no.1 filter paper.
2. Mix and dissolve the following ingredients.

Tryptone	20 g
Peptone	5 g
Yeast extract	5 g
Glucose	5 g
Apple juice	200 mL
Tween 80 (5% w/w solution)	1 mL
Distilled water	700 mL

3. Add the filtered liver extract to the ingredients specified in step 2 and adjust to pH 4.5 using 50% v/v H_3PO_4 and/or 6 M KOH.

4. To prepare a solidified medium, add 20 g agar prior to autoclaving at 121°C/250°F for 15 min.
5. If the medium is to be selective against *Saccharomyces*, mix 100 mg cycloheximide with 10 mL distilled water and filter sterilize through an 0.45 μm filter. Add this solution to the cooled media just after autoclaving.

13.6.2 Tomato Juice–Glucose–Fructose–Malate Medium

1. Mix 1 g liver extract with 100 mL water for at least 30 min. Filter through Whatman no.1 filter paper prior to adding to the medium.
2. Select a commercially processed tomato juice that contains only juice and salt (no preservatives and not from concentrate) or prepare juice from fresh tomatoes. Centrifuge 250 mL of the juice at 3000 × *g* for 30 min and decant, saving the supernatant (serum). Serum can be stored in small volumes at −20°C/−4°F until used.
3. Mix and dissolve the following ingredients.

Tryptone	20 g
Peptone	5 g
Yeast extract	5 g
Glucose	5 g
Fructose	3 g
Tomato juice serum	200 mL
Tween 80 (5% w/w solution)	1 mL
Distilled water	700 mL

4. Add the filtered liver extract from step 1 to the ingredients specified in step 3 and adjust to pH 5.5 using 50% v/v H_3PO_4 and/or 6 M KOH.
5. To prepare a solidified medium, add 20 g agar prior to autoclaving at 121°C/250°F for 15 min.
6. If the medium is to be selective against *Saccharomyces*, mix 100 mg cycloheximide with 10 mL distilled water and filter sterilize (0.45 μm). Add this solution to the cooled autoclaved media.

13.6.3 Heterofermentation-Arginine Broth

1. Prepare vegetable juice serum by centrifuging Campbell's V-8 Vegetable Juice® at 3000 × *g* for 30 min. Decant off from the solids, saving the supernatant (serum).

2. Mix and dissolve the following ingredients.

Tryptone	5 g
Yeast extract	5 g
Peptone	5 g
Fructose	20 g
Glucose	5 g
L-Arginine	6 g
Tween 80 (5% w/w solution)	1 mL
Vegetable juice serum	200 mL
Distilled water	800 mL

3. Adjust pH to 5.5 with 50% v/v H_3PO_4 and/or 6 M KOH.
4. Autoclave the medium at 121°C/250°F for 15 min.

13.7 MEDIA FOR ACETIC ACID BACTERIA

The glucose–yeast extract–carbonate medium (GYCM) described by Swings (1992) detects the presence of acid-producing microorganisms and is regarded as "standard growth medium" for acetic acid bacteria (De Ley et al., 1984). Owing to the buffering capacity of the carbonate, acetic acid bacteria can be maintained on GYCM at 4°C with once a month transfers. Mannitol–yeast extract–peptone medium (MYPM), nonselective WL medium, and the yeast extract–peptone–ethanol medium (YPE), the latter described by Du Toit and Pretorius (2002), can also be used to cultivate acetic acid bacteria.

Gluconobacter growing on GYCM over time (3 to 5 weeks) produces water-soluble brown pigments, which are not seen in the case of any similarly cultivated *Acetobacter* species. Grown on this medium, *Acetobacter* will produce clear zones or halos around colonies because the acid being produced will neutralize the $CaCO_3$. Unlike the lactic acid bacteria, acetic acid bacteria are obligate aerobes and so it is necessary to use spread plates.

13.7.1 Glucose–Yeast Extract–Carbonate Medium

1. Mix and dissolve the following ingredients.

Glucose	50 g
Yeast extract	10 g

Calcium carbonate ($CaCO_3$)	30 g
Distilled water	800 mL

2. Once ingredients are dissolved, dilute medium to volume (1000 mL).
3. Add 25 g agar prior to autoclaving at 121°C/250°F for 15 min.

13.7.2 Mannitol–Yeast Extract–Peptone Medium

1. Mix and dissolve the following ingredients.

Mannitol	25 g
Yeast extract	5 g
Peptone	3 g
Distilled water	800 mL

2. Once ingredients are dissolved, dilute medium to volume (1000 mL).
3. Add 25 g agar prior to autoclaving at 121°C/250°F for 15 min.

13.7.3 Yeast Extract–Peptone–Ethanol Medium

1. Mix and dissolve the following ingredients.

Yeast extract	10 g
Peptone	5 g
Distilled water	800 mL

2. Once ingredients are dissolved, adjust pH to 5.5.
3. Add 15 g agar prior to autoclaving at 121°C/250°F for 15 min.
4. Dissolve 20 mL ethanol in 200 mL distilled water and sterile filter through an 0.45 μm membrane. Add this solution to the tempered autoclaved media prepared in step 3 above and prior to pouring plates.

13.8 ASEPTIC TRANSFER TECHNIQUES

The ability to transfer microorganisms from one container to another without contamination is crucial to success in the microbiology laboratory. These techniques serve as the basis for subsequent work such as starter culture preparation or maintaining viable cultures in long-term storage. Transfer loops are normally used to transfer to the surface of agar (Petri plates and slants), whereas transfer needles are used to prepare stab cultures. Both implements are sterilized by heating in an open flame until red hot (Fig. 13.1).

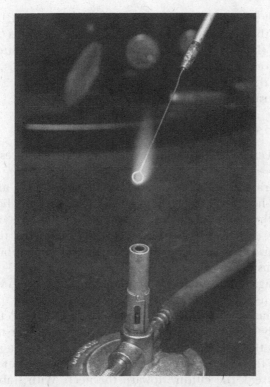

Figure 13.1.　Sterilizing transfer loop by heating in an open flame until red hot.
Photograph provided with the kind permission of WineBugs LLC.

13.8.1 Transfers from Solid to Solid Media

13.8.1.1 From Slant or Stab Cultures

1. Hold the test tube containing the culture of interest in one hand.
2. Hold the transfer loop or needle in the other hand, flame the loop/ needle to red hot in an open flame, and cool by carefully touching the surface of sterile agar.
3. Remove the test tube cap using the hand holding the loop or needle, flame the tube just inward of the neck, and carefully collect a small portion of growth with the loop or needle on the agar surface or along the stab line.
4. Withdraw the loop or needle, flame the tube just inward of the neck, and replace the cap.

5. Immediately transfer the microorganism to fresh agar by moving the loop across the surface of the agar in Petri plates or slants/stabs (Section 13.10.1). In the case of slants, place the loop at the bottom of the test tube and streak the culture toward the neck. With stabs, the needle is thrust into the agar and then withdrawn.
6. Flame the transfer loop or needle.

13.8.1.2 From Petri Dishes

1. Place the Petri dish containing the culture of interest within easy reach.
2. Hold the transfer loop or needle in the one hand and flame the loop to red-hot. Cool the loop by carefully raising the lid on the Petri dish with the other hand and touching the surface of agar where no colonies can be seen. Avoid completely removing the Petri plate cover because this can lead to microbial contamination by airborne microorganisms.
3. Collect a single colony in a sweeping or scooping motion with the transfer loop or needle.
4. Withdraw the loop/needle and replace the lid of the Petri plate.
5. Immediately transfer the microorganism to fresh agar by moving the loop across the surface of the agar in Petri plates or a slant (Section 13.10.1). In the case of slants, place the loop at the bottom of the test tube and streak the culture toward the neck. With stabs, a needle is thrust into the agar and then withdrawn.
6. Flame the transfer loop or needle.

13.8.2 Transfers from Solid to Liquid Media

1. Flame the loop to red-hot in an open flame. Cool the loop by carefully touching the surface of agar where no colonies can be seen.
2. Select a colony from the Petri plate, slant, or stab culture.
3. Pick up the test tube containing the sterile liquid medium with the other hand.
4. Remove the test tube cap with the same hand that is holding the loop. Some microbiologists remove and keep caps between their "ring" and "pinkie" fingers on the backside of their hands so that the thumb and hand can hold the loop.
5. Flame the test tube and immerse the loop into the liquid, rubbing the loop against the glass to assist in cell removal.
6. Remove loop, immediately re-flame the tube, and close. Re-flame transfer loop.

13.8.3 Transfers from Liquid to Solid Media

This type of transfer is slightly more complicated than the transfer of a colony from solid to liquid media but is as important as these techniques are used to enumerate microorganisms. To perform these procedures, graduated pipettes are used to transfer specific volumes to dilution blanks or to pour/spread plates (Section 14.3). Where reusable glass pipettes are used, these should be washed and placed into a metal canister for autoclaving and subsequent drying in an oven prior to use. Alternatively, sterilized disposable pipettes can be used for transfers (Section 18.7).

To limit microbial contamination of liquid media, caps are removed and the neck of the flask or test tube is placed in an open flame (Fig. 13.2). This action heats the air within the container, thereby creating a positive pressure that limits contaminates from entering the flask.

Figure 13.2. Heating the neck of a test tube prior to transfer of liquid. Photograph provided with the kind permission of WineBugs LLC.

Once used, pipettes and any nonsterile utensils should be placed in disinfectant and autoclaved at the end of the day. Disposable plasticware, such as pipettes and Petri plates, should never be discarded without first autoclaving. Because this operation will cause the plastic to partially melt, such disposables should be placed in trays before autoclaving. Alternatively, one may purchase disposable autoclavable polypropylene bags for packaging.

One technique that should be avoided is the practice of mouth pipetting. Although common among earlier generations of microbiologists, this practice should be prohibited due to safety concerns. In place, a number of relatively inexpensive pipetting devices are now available.

1. Place the culture as well as sterile broth or dilution tubes adjacent to one another in a test tube rack. With experience, laboratory personnel should be able to handle both tubes at one time by placing the tubes between the index and middle fingers of one hand. The thumb is then positioned to support both.
2. Remove the sterile pipette from the canister or wrapper and insert suction end into the pipetting device. Do not allow the pipette to touch any surface.
3. With the other hand, agitate the flask (vortex), remove the culture tube cap, flame the tube just inward of the neck, insert the pipette, and withdraw a defined volume.
4. Re-flame the tube just inward of the neck, stopper, and replace on rack.
5. Immediately pick up tube to be inoculated, remove cap, and flame the tube just inward of the neck.
6. Insert pipette to just above liquid level and release the inoculum volume. The last drop in the pipette tip may be delivered by touching the tip to the glass, not the liquid. Where transferring to a Petri plate (pour or spread), delivery of the sample is best accomplished by holding the pipette at a 45° angle to the surface. Allow inoculum liquid to drain from the pipette. Many laboratories use cotton-plugged "blow-out" pipettes that require positive pressure to dispense the complete volume. This should be done carefully to avoid splattering the sample on the cover and sides of the Petri plate.
7. Remove pipette and immediately flame the tube just inward of the neck, then stopper the tube. Discard pipette into disinfectant.

13.9 ISOLATION OF MICROORGANISMS

When an aliquot of grape must or wine is placed on a suitable solidified medium, colonies of various morphologies will appear on the plate. In theory, a single viable cell in the original sample continuously multiplies to eventually yield one visible colony after a suitable incubation.

The presence of different microorganisms in the sample is evidenced by variety of colony size, color, and shape observed after incubation. In order to separate these microorganisms, the culture is "streaked" on general (nonselective) media. The objective of streaking is to inoculate fewer and fewer microorganisms on each of the three consecutive sectors thereby increasing the chance of producing well-separated isolated colonies. When working with yeasts that secrete a capsule, contamination with bacteria is common. In such cases, the culture may need to be restreaked several times to yield pure isolates.

1. Hold the transfer loop in the right hand and flame the loop to red-hot in an open flame. Cool the loop by carefully touching the surface of agar.
2. Using the same procedures as described in Section 13.8, remove a colony (or a mass of colonies) and place the loop at position 1 shown in Fig. 13.3.
3. Carefully move the loop across the agar surface following the 1 pattern ("zig-zag"). The best separation is effected when streaks are kept as tight as possible and the maximum surface area of the plate is utilized. Avoid digging into the agar with the loop.
4. When 1 pattern is completed, replace the Petri plate lid, and flame the loop to sterilize.
5. Rotate the plate one-third turn counterclockwise, lift the lid, and cool the loop by again touching the agar.
6. Starting at position 2, streak the second pattern as per step 3. Except for the initial pass-through of the previously streaked area, one should not allow the loop to reenter that area.
7. Starting at position 3, streak the third pattern as per step 3.
8. Invert plates and Incubate under conditions optimal for the desired microorganism(s). Well-separated colonies should be found in the third streaked area such as seen in Fig. 13.4.

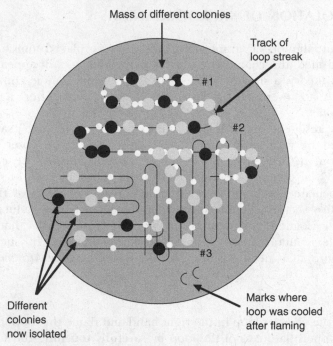

Figure 13.3.   Streaked sample showing isolated colonies of different hypothetical microorganisms of various morphologies.

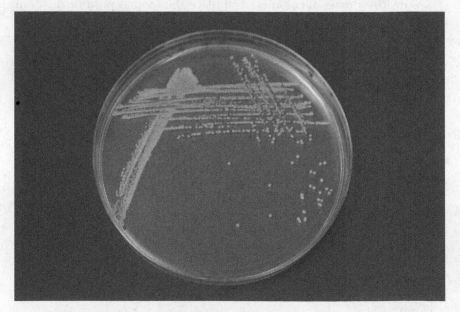

Figure 13.4.   Streaked sample showing isolated colonies of *Saccharomyces*. Photograph provided with the kind permission of WineBugs LLC.

13.10 MAINTENANCE AND STORAGE OF CULTURES

Microorganisms are commonly stored on the appropriate solidified medium, either as slants or stabs (weeks) or for longer periods (months) as liquid cultures suspended in glycerol kept at low temperatures ($-70°$C/ $-94°$F). Alternatively, microorganisms can be preserved for very long term storage (years) by freeze-drying. Method 13.10.3 has been used to freeze-dry yeasts (C.M.L. Joseph, personal communication, 2005), and method 13.10.4 has been used successfully for lactic acid bacteria (Duke, 1979).

13.10.1 Preparation of Agar Slants and Stabs

1. Prepare the suitable agar medium for the microorganism of choice.
2. Heat the medium to dissolve agar (solution will clear indicating the agar has dissolved).
3. While hot, distribute either 8 mL (slants) or 10 mL (stabs) of medium into each 15×125 mm test tube.
4. Put caps on test tubes loosely and autoclave at $121°$C/$250°$F for 15 min.
5. After removal from the autoclave, tighten the caps.
6. For slants, lay tubes to allow the agar to solidify at an angle. Laying the cap-end on a meter stick works well.
7. Transfer cultures to the slants or stabs using the procedures outline in Section 13.8.1.

13.10.2 Glycerol Suspensions

1. Prepare the appropriate broth for the microorganism being stored.
2. Inoculate the broth from a single colony off of an agar plate and incubate for the appropriate time and temperature.
3. Harvest the cells by centrifugation and aseptically add a minimal amount of broth to transfer the pellet to a small sterile test tube.
4. Re-centrifuge, discard the supernatant, and add a minimal volume of broth that contains 15% v/v glycerol. Place tubes in a low temperature freezer, ideally $-70°$C/$-94°$F.

13.10.3 Freeze-drying (Yeasts)

1. Sterilize the following supplies.
   a. Lyophylization vials (1 mL volume that can be sealed).
   b. Graduated pipettes (10 and 25 mL).
   c. Centrifuge bottles with screw caps (250 mL).

2. Dissolve 10 g skim milk powder (Difco) and 2 g sodium glutamate in 100 mL previously sterilized water in a sterile bottle. Autoclave medium at 110°C for 10 min.
3. Grow cells on an agar slant of appropriate medium for 3 days.
4. Transfer a loopful of the culture into a 1 mL solution of skim milk medium.
5. Distribute 200 µl of the 1 mL culture suspension into 1 mL lyophylization vials.
6. Quickly freeze the cultures and lyophilize under vacuum. Store cultures at −20°C/−4°F.

13.10.4 Freeze-drying (Lactic Acid Bacteria)

1. Sterilize the following supplies.
   a. Serum vials (30 mL, autoclaved with aluminum foil wrapped around the openings).
   b. Gray septa for serum vials (autoclave in a large beaker covered with aluminum foil).
   c. Graduated pipettes (10 and 25 mL).
   d. Centrifuge bottles with screw caps (250 mL).
2. Prepare the appropriate broth and distribute into test tubes (10 mL) and screw-capped Erlenmeyer flasks (1000 mL) prior to autoclaving at 121°C/250°F for 15 min.
3. Prepare buffer A by dissolving 1.11 g NaH_2PO_4 together with 11.27 g $Na_2HPO_4•7H_2O$ in 800 mL distilled water. Adjust the pH to 7.5 and dilute to volume (1000 mL).
4. Prepare buffer B by dissolving 2.22 g NaH_2PO_4 together with 22.54 g $Na_2HPO_4•7H_2O$ in 800 mL distilled water. Adjust the pH to 7.5 and dilute to volume (1000 mL).
5. Prepare the Naylor–Smith suspending medium.
   a. Mix 40 g dextrin, 10 g NH_4Cl, and 10 g thiourea with 800 mL distilled water, transfer into 1000 mL screw-capped bottles, and autoclave at 121°C/250°F for 15 min.
   b. Mix 10 g ascorbic acid in 200 mL distilled water. Sterile filter through a 0.45 µm membrane and aseptically add to the sterilized solution prepared in step 5a.
6. Inoculate a single colony from an agar plate into 10 mL of the appropriate broth. Incubate at the optimal temperature and time for the microorganism.
7. After incubating, inoculate 500 mL broth with all 10 mL of the culture and continue incubation under the same conditions.

8. Harvest the cells from the 500 mL broth using a centrifuge (3000 × *g* for 30 min).
9. Decant the medium and suspend the pellet in 25 mL buffer A prepared in step 3. Mix (vortex) the culture and re-centrifuge.
10. Decant buffer A and resuspend the pellet in 25 mL buffer B prepared in step 4. Mix (vortex) the culture and 25 mL Naylor–Smith suspending medium prepared in step 5.
11. Pipette 5 mL of the microbial suspension into each of 10 vials using a sterile 10 mL pipette. Aseptically cap with septa, making sure that they are not fully inserted in the bottle (side groove in the septa allows water to escape during freeze-drying).
12. Freeze the cultures at −20°C/−4°F or lower and begin the freeze-drying process as quickly as possible. Seal the vials under vacuum and store cultures at −20°C/−4°F.

CHAPTER 14

ESTIMATION OF
POPULATION DENSITY

14.1 INTRODUCTION

Estimating microbiological population density and diversity plays important and often pivotal roles at several junctures in the winemaking process. For instance, it is frequently necessary to determine changes in microbial populations during the preparation of starter cultures, growth and decline phases of malolactic fermentation, or monitoring potential *Brettanomyces* infections.

Population densities can be measured using many methods, but the three most important to enologists are microscopic counting (Section 14.4), direct plating, either pour or spread plates (Section 14.5), and membrane filtration (Section 14.5.3). Microscopic counting techniques are the most rapid but requires at least 10^4 cells per mL. As low population densities will not be detected microscopically, direct plating methods are normally used. These methods sometimes require dilution of the sample and time for incubation. Membrane filtration, followed by direct plating, is applied to those wines suspected of having a low viable population (<25 cells per mL). Here, cells are concentrated on the membrane before transfer to solid media. Bioluminescence (Section 14.6) is another rapid enu-meration method frequently used to monitor the effectiveness of cleaning and sanitizing programs. Finally, microbial populations can be estimated using a spectrophotometer or nephlometer (Section 14.7) but high popu-lations in the sample are required, limiting the effectiveness of these instrument, in routine applications.

14.2 SAMPLING

To estimate the population of viable microorganisms, it is important to obtain a homogeneous sample. As such, tanks and barrels should be agitated to uniformily mix the contents just prior to sampling. Alterna-tively, one can sample the surface, middle, and/or bottom of the tank or barrel without prior mixing. This can be advantageous because micro-organisms will stratify within a tank/barrel depending on metabolic requirements. As an example, *Acetobacter* is commonly found on the surface of wine due to its requirement for oxygen. Tank mixing prior to sampling would evenly distribute the entire population of *Acetobacter* within the tank or barrel. This action would reduce the number of cells per unit volume, potentially to undetectable levels depending on the bacterial population and the volume of the tank. In this case, removing a sample from the surface of the wine would allow maximum detection of the bacterium.

Whenever possible, samples for microbiological analysis (50 to 100mL) should be removed aseptically and placed into sterile containers to reduce the potential for secondary contamination due to non-wine microorganisms. Samples can be removed through sampling ports on tanks or by wine thiefs/pipettes. Sampling devices should be sterilized either by flame or 70% v/v ethanol prior to use. If a steriliant is used, the thief or pipette should be rinsed with sterile water before obtaining the sample.

In the case of sanitation monitoring, a traditional method for sampling equipment surfaces involves rubbing a sterile, cotton swab over a specified area and then enumerating the microorganisms that adhere to the swab. An adequately cleaned and sanitized area should have no more than 100 colonies per surface area sampled (Sveum et al., 1992), depending on the medium used to enumerate the microorganisms. Rather than use direct enumeration methods, faster results can be obtained using bioluminescence (Section 14.6). The following procedure is that of Sveum et al. (1992).

1. Open the sterile swab container, grasping the end of the stick. Do not touch any portion that might be inserted into the 18×150 mm test tube containing sterile 9 mL 0.1% w/v peptone.
2. Aseptically open the sterile diluent, moisten the swab head, and press out excess by rolling the swab on the inside of the diluent.
3. Hold the swab handle at a 30° angle and rub the head over a surface of approximately $50 \, cm^2/7.7 \, in^2$ three times, reversing direction between strokes.
4. Return the swab to the diluent vial, briefly rinse in the diluent and press out the excess to remove the microorganisms from the swab.
5. Swab four more $50 \, cm^2/7.7 \, in^2$ areas of the surface being sampled, repeating steps 3 and 4.
6. Once the areas have been swabbed, position the swab head in the test tube and break or cut the stick using sterile scissors so that only the head is left in the test tube.
7. Mix the test tube very well and enumerate the microorganisms present using the appropriate method and media (Section 14.5). If very low populations are suspected, the entire volume of the diluent should be passed through a sterile membrane filter and the pad incubated (Section 14.5.3).

14.3 SAMPLE DILUTION

Before enumeration, samples may have to be serially diluted because the number of microorganisms is too high for a valid measurement. The volume of diluent into which the sample is added is referred to as a "dilution blank." Although dilution blanks are most commonly made using sterile 0.1% w/v peptone, other media can be used such as 0.85% w/v saline or Butterfield's phosphate buffer (Swanson et al., 1992). Use of sterile distilled water for purposes of dilution should be avoided due to osmotic stress, which decreases viability. Normally, 18×150 mm autoclavable test tubes equipped with polypropylene or metal closures or caps are used for making dilution blanks rather than cotton stoppers.

In a dilution series, the minimum dilution is 10-fold ($1 \, mL + 9 \, mL$, $1:10$, or 10^{-1}). Practically, dilutions are performed by aseptically transferring 1 mL of sample into a tube containing 9 mL diluent. Dilutions of higher magnitude are then made by sequential $1:10$ dilutions of previous dilution as shown in Fig. 14.1. For example, a $1:100$ dilution of a wine can performed by two sequential $1:10$ dilutions. However, a $1:100$ dilution can also be made by adding 1 mL of the wine to 99 mL diluent (e.g., milk dilu-

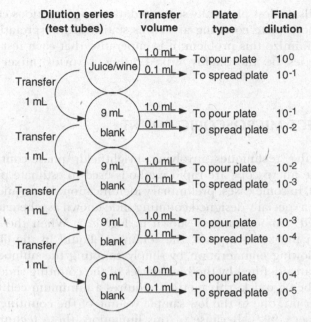

Dilution series (test tubes)	Transfer volume	Plate type	Final dilution

Figure 14.1. Preparation of a dilution series scheme to enumerate microorganisms.

tion bottles). Although carrying out several smaller dilutions rather than fewer larger dilutions (e.g., four sequential 1:10 dilutions vs. two sequential 1:100 dilutions) reduces experimental error, time constraints and extra costs of pipettes and media may make this procedure impractical. Because of slight volume losses during autoclaving, some laboratories will adjust the volume of dilution blanks prior to autoclaving to contain slightly more than 9 mL (9.1 or 9.2 mL) with the expectation that the final volume will be closer to 9 mL.

When preparing a dilution series, it is important to maintain aseptic conditions (Section 13.8). For example, dilution blanks must be flamed when opened to transfer sample (Fig. 13.2). Sterile 1.0 mL pipettes (0.1 mL graduations) or variable volume pipettors equipped with sterile plastic tips must be used to aseptically transfer liquid between dilution blanks. Sterile pipettes or pipette tips should be used only once, and then a new pipette or tip is used. Pipettors can be sterilized by wiping down their surfaces with 70% v/v ethanol and should be calibrated yearly.

One problem in preparing a dilution series is the potential agglomeration of cells. For instance, *Pediococcus* as well as some yeasts secrete a sticky capsule that allows cells to clump together in aggregations of a few to 20

or more cells. Unless physically separated, these aggregations cause difficulties in both plate counting as well as statistical interpretation of the
data. To minimize this problem, it is imperative that each test tube in a
dilution series be thoroughly mixed using a vortex mixer prior to
transfer.

14.4 DIRECT MICROSCOPIC COUNT

Cell counting techniques involving brightfield, phase-contrast, and
fluorescence microscopy are rapid methods used to estimate population
density and, in some cases, preliminary identification. Here, juice or wine
is placed in a specially designed counting slide known as a hemacytometer
(also spelled hemocytometer) shown in Fig. 14.2. When the cover slip
provided in the kit is applied, the volume of liquid in the chamber is
defined, allowing enumeration by simply counting the number of cells
within the area defined by the boundaries of the counting slide grid.

Despite being quick, this method requires a minimum cell density of
10^4 cells per mL due to the low sample volume of the counting chamber
(Splittstoesser, 1992). Because of this limitation, these techniques find

Figure 14.2.   A hemacytometer. Photograph provided with the kind permission of
WineBugs LLC.

most appropriate application in monitoring starter cultures or following the course of fermentation when populations are in excess of 10^6 cells per mL.

Another difficulty with the direct microscope count is the fact that the method views both viable and nonviable (dead) cells. Depending on stage in the growth cycle, as well as history of the sample, the ratio of viable to nonviable cells may vary considerably and makes comparing results to those of direct plating difficult. Because of this, plating normally provides lower estimates of viable populations than microscopy. To make the distinction between viable and nonviable cells, various stains and dyes can be used either singly or in combination (Section 14.4.2, 14.4.3, and 14.4.4).

14.4.1 Using a Microscope Counting Chamber

The specific grid pattern etched into a hemacytometer divides the space into nine large squares, each of which is $1\,\text{mm}^2$ (Fig. 14.3). The central counting area (square 5) contains 25 middle-sized squares and each of these has 16 smaller squares. Whereas some microbiologists only count the central large square, others will tally the corner squares as well (squares

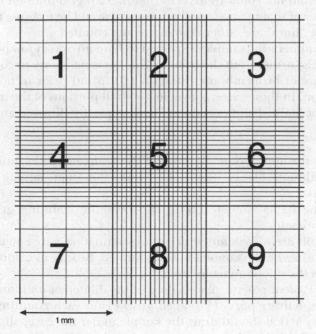

1 mm

Figure 14.3.   Typical counting grid for a hemacytometer as viewed using a microscope.

1, 3, 7, and 9). Because the cover glass rests exactly 0.1 mm above the bottom of the chamber, the volume of liquid under each of the large squares is $0.1 \, mm^3$ or $0.1 \, \mu L$.

Hemacytometers have two chamber wells, each containing a specific volume of liquid (Fig. 14.2). The wells are filled with juice or wine by placing a pipette in the V-shaped groove and allowing capillary action to draw the liquid into the chamber. After allowing the cells to settle, counting is initiated. For best results, the number of cells should be within the approximate range of 20 to 50 cells per large square (Fig. 14.3). Knowing the dilutions used in the original suspension and the volume of liquid microscopically examined, the original population can be estimated (cells/mL).

To improve accuracy, both chambers on a hemacytomer should be counted and a sufficient number of squares should be counted to give a total count of about 600 cells (Splittstoesser, 1992). Most importantly, it is necessary that each laboratory establish whether cells lying on grid boundaries or lines be included in the count or be rejected. Depending on the initial dilutions, inconsistency in this regard may have a large effect on estimating the actual population of microorganisms. One procedure would be to count those cells that overlap a grid line on the top or right of a square but not count those cells that overlap grid lines on the left or bottom lines of that square. Another protocol would be to agree that those cells on the "lines" are either counted or not counted.

When enumerating samples expected to have no or very low population densities (e.g., wines from the bottling line), it is necessary to concentrate known volumes by membrane filtration ($<0.45 \, \mu m$) prior to microscopic examination. In these cases, conveniently sized portions of the membrane may then be stained and examined microscopically (Kunkee and Neradt, 1974).

1. Prepare the appropriate dilutions of the sample to be examined using 9 mL dilution blanks (Fig. 14.1) with thorough mixing between dilutions.
2. Place the hemacytometer cover slip over the counting grid of the hemacytometer.
3. If the cells are to be stained prior to examination, mix equal portions of the stain with the samples to be examined. Be sure to record volumes and specific dilution factors used.
4. Using a Pasteur pipette, place a drop of the diluted or undiluted sample into the delivery ports (V-shaped groove) on each counting surface. Capillary action should draw the sample under the cover slip and uniformly fill the counting area. Alternatively, transfer a drop of the final

dilution directly to the counting surfaces and then place a cover slip on top. Although this technique works, care must be taken to avoid creating and trapping air bubbles within the counting area. It is important not to under or over fill the counting chamber.

5. Using 40× or 100× objectives, count a sufficient number of squares to give a total count of about 600 cells. The results between the two grids can then be averaged to provide a mean value.

6. Calculation:

cells/mL in original sample = (average number of cells) × (the volume of liquid examined) × (original sample dilution)

7. Sample calculation: Analyzing an undiluted wine sample, 135 cells were counted in 5 of the 25 middle-sized squares within a large square (square 5) for one chamber. The other chamber contained a total of 141 cells. To calculate the cell population in the wine, the average number of cells (138) is first multiplied by 5 to yield total estimated cells in square 5 (690) and then multiplied by 10,000 (reciprocal of $0.1\,\mu L$, or $0.0001\,mL$ sample volume). In this case, total viability in the wine sample = 6.9×10^6 cells/mL.

14.4.2 Methylene Blue

Methylene blue exists in two redox-dependent states; reduced and oxidized. The reduced state (*leuco*-form) is colorless whereas the oxidized form is blue. If a viable yeast cell takes up the dye, methylene blue is reduced to the colorless form so that the cell appears colorless (or white) against the blue background. Thus, cells that reduce the dye to the colorless form are assumed to be viable, whereas dead yeast do not reduce the dye and appear blue or black.

Although traditionally used in brewing and winemaking industries, some researchers have argued against the application of methylene blue as a means to determine yeast viability. For example, O'Connor-Cox et al. (1997) noted that the dye does not stain all dead cells in a sample, a fact that can lead to overestimating yeast viability.

Although several forms of methylene blue are available, the preparation used for biological work is methylene blue chloride. Methylene blue thiocyanate, a salt used as a redox indicator in milk testing, should not be used as a biological stain. Because methylene blue rapidly becomes toxic to microorganisms, preparations should be microscopically examined within 10 min.

1. Prepare a citrate buffer by dissolving 2.4 g disodium hydrogen citrate ($Na_2HC_6H_5O_7$) and 2.1 g sodium dihydrogen citrate ($NaH_2C_6H_5O_7$) in a minimal amount of distilled water and dilute to 100 mL. Adjust pH to 4.6 if necessary.
2. Dissolve 0.3 g methylene blue chloride in 30 mL of 95% v/v ethanol.
3. Add 100 mL of the citrate buffer to the methylene blue solution.

14.4.3 Ponceau-S

Ponceau-S is a general stain for cytoplasmic protein and does not distinguish between living, dead, or dying cells that have not undergone autolysis. Because of this problem, Kunkee and Neradt (1974) suggested staining previously examined methylene blue–stained preparations with Ponceau-S because the latter will stain both viable and dead yeast cells.

1. Dissolve 0.9 g Ponceau-S, 13.4 g sulfosalicylic acid, and 13.4 g trichloroacetic acid in 72.3 g distilled water.

14.4.4 Wolford's Stain

Wolford's stain is used to monitor viable yeast by preparing a wet mount (Section 12.5) that is examined using a microscope (McDonald, 1963). Viable cells reject the stain and will appear clear, whereas dead cells absorb Wolford's stain and appear deep red or reddish-purple.

1. Prepare North's stain.
   a. Add enough methylene blue to 95% v/v ethanol to make a saturated solution (at least 1.5 g methylene blue per 100 mL).
   b. Dissolve 3 mL aniline oil in 10 mL 95% v/v ethanol in a 100 mL volumetric flask.
   c. Carefully, add 1.5 mL concentrated 12 M HCl and, with stirring, add 30 mL of the saturated solution of methylene blue.
   d. Dilute to volume (100 mL) with distilled water.
2. Prepare basic fuchsin solution.
   a. Dissolve 1 g basic fuchsin in a 100 mL volumetric flask containing 95% v/v ethanol.
   b. Dilute to volume (100 mL) with 95% v/v ethanol.
3. Transfer 10 mL North's stain and 2 mL basic fuchsin solution to a 100 mL volumetric flask. Dilute to volume with distilled water.
4. Filter the stain and store under refrigeration until used (maximum storage time is 3 to 4 weeks).

14.5 DIRECT PLATING

In direct plating, a known volume of grape juice or wine (0.1 to 1.0 mL) is placed into either a sterile Petri dish and mixed with a tempered agar medium (pour plate method) or transferred onto a prepoured, solidified agar medium in a sterile Petri dish (spread plate method). After incubation for a specific period of time at a specified temperature, each cell present in the original sample will theoretically grow using the nutrients present in the agar. Eventually, the colony will be visible to the naked eye and can be counted. Based on the number of colonies on a given plate and its dilution, the number of viable microorganisms per milliliter of sample can be calculated. This value is normally expressed as "colony forming units per milliliter," or CFU/mL, rather than cells/mL. The term "CFU" is used because many wine microorganisms do not exist solely as single cells. Rather, many occur as pairs, tetrads, or short chains mak-ing the term CFU a more accurate reflection of how the cells were enumerated.

By convention, the cell population on a countable individual agar plate must be between 25 and 250 individual colonies (Swanson et al., 1992), although some microbiologists count plates that contain between 30 and 300 colonies. Thus, a wine sample that contains 10,000 cells per mL must be diluted 1 : 100 to achieve a population of approximately 100 colonies in 1 mL of diluted wine in an agar plate. The number of dilutions to be prepared can be estimated using an approximate microscopic count (Table 14.1).

Both spread and pour plate methods are commonly used for microbial enumeration of juice and wine samples. In fact, commercial kits can be purchased with pre-sterilized and pre-poured agar as well as other reagents and materials so that even the smallest of wineries will have the ability to perform these tests. However, both methods also suffer from logistical and interpretational difficulties. For instance, it is generally necessary to plate

Table 14.1. Estimating sample dilutions needed when enumerating microorganisms using plating methods.

Number of cells in a microscopic field (1000×)	Potential population (cells/mL)	Estimated dilutions to yield countable plates
1 to 10	10^4 to 10^5	10^{-2} to 10^{-3}
10 to 100	10^5 to 10^6	10^{-3} to 10^{-4}
>100	$>10^6$	10^{-4} to 10^{-7}

multiple dilutions (in duplicate) of the same sample in order to arrive at plates that are countable and statistically valid. In the case of pour plates, embedded colonies may be difficult to recover and transfer.

14.5.1 Pour Plates

Preparation of pour plates is a relatively simple procedure that generally yields well separated colonies compared with the spread plate method. Although some microorganisms grow better under reduced oxygen as would be present in colonies growing under agar (pour plates), obligate aerobes (e.g., *Acetobacter*) grow better using spread plates.

Success in using this technique requires that the agar be sufficiently liquid to permit introduction of the sample, dispersion of cells, and plating without cooling to the point of solidifying. As such, agar must be "tempered." Here, the agar is held at a temperature just above the solidification point prior to pouring. Agar solidifies at temperatures slightly below 40°C/104°F and cannot re-melt until 90°C/194°F to 100°C/212°F. To maintain the liquid state for a sufficient time to carry out the steps involved in plating, liquefied agar must be maintained at 45°C/113°F to 50°C/122°F prior to pouring, commonly using a water bath. Although media could be poured at a higher temperature, there is the potential for thermal shock and subsequent death of the microorganisms present.

1. See Fig. 14.1 for setting up a dilution series using 9 mL dilution blanks. Serially dilute the juice or wine, thoroughly vortexing each dilution blank before transferring 1 mL into the next blank.
2. Place 0.1 or 1.0 mL from each dilution blank into a sterile Petri dish.
3. Immediately after autoclaving, place the appropriate agar medium in a water bath at 45°C/113°F to 50°C/122°F. Once the temperature has equilibrated, aseptically add approximately 25 mL of the agar to each plate (smaller Petri plates will use less agar medium).
4. Carefully mix each plate using a "figure eight" pattern. Allow the plates to cool and solidify undisturbed, usually for at least an hour.
5. Invert the plates (upside down) and incubate at the appropriate temperature and time.
6. After incubation, find the plate (dilution) that has between 25 and 250 visible colonies and count all colonies.
7. Calculation: Multiply the count obtained in step 6 by the reciprocal of the overall dilution (dilution and the volume of sample on the plate).
8. Sample calculation: If the 10^{-3} dilution plate has 104 visual colonies, the estimated count of viable microorganisms in the original juice or wine sample would be reported as 104×10^3, or 1.04×10^5 CFU/mL.

14.5.2 Spread Plates

Spread plates are used to enumerate aerobic or heat-sensitive microorganisms and can be prepared a few days prior to use. This procedure utilizes bent glass rods ("hockey sticks") to distribute a defined volume of liquid evenly over the surface of the solidified agar medium (Fig. 14.4). Normally, the maximum volume that can be placed on a spread plate is 0.1 mL, unlike pour plates, which can receive up to 1 mL. The method suffers from difficulties in complete transfer and separation of individual cells needed to yield separate countable colonies. Excess moisture present on the agar surface can also result in unexpected colony spread and uncountable plates.

1. See Fig. 14.1 for setting up a dilution series using 9 mL dilution blanks. Serially dilute the juice or wine, thoroughly mixing each dilution blank before transferring 1 mL into the next blank.

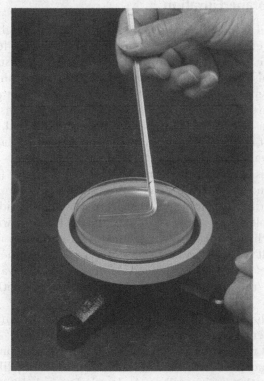

Figure 14.4. A glass "hockey stick" being used to spread a liquid sample over the surface of solidified agar. Photograph provided with the kind permission of WineBugs LLC.

2. Place 0.1 mL from each dilution blank onto a sterile Petri dish containing the appropriate solidified medium.
3. Holding the handle, dip a glass hockey stick into 70% v/v ethanol, flame, and allow flame to burn itself out and cool. Spread the 0.1 mL around on the plate using the hockey stick.
4. Invert the plates (upside down) and incubate at the appropriate temperature and time.
5. After incubation, find the plate (dilution) that has between 25 and 250 visible colonies and count all colonies.
6. Calculation: To calculate the original population, multiply the count by the reciprocal of the overall dilution (dilution of sample × volume on plate).
7. Sample calculation: If the 10^{-5} plate (dilution) has 86 visual colonies, the estimated count of viable microorganisms in the original juice or wine sample would be reported as 86×10^5, or 8.6×10^6 CFU/mL.

14.5.3 Membrane Filtration

For this method, a specific volume of wine is passed through a sterile filter membrane, often with an absolute pore size of 0.45 μm or less. This membrane, with the trapped microorganisms on the upper surface, is then aseptically transferred face-up onto a Petri dish containing a solidified medium. This method is useful for wines that should not contain or are expected to have low viable populations of microorganisms (e.g., bottling line) because a large volume of the sample can be tested. However, this method cannot be used if particulate matter present in the sample plugs or clogs the membrane.

Many choices of membrane filtration units are available from suppliers. Some units can be reused after cleaning and sterilization (autoclaving) and use presterilized membranes that are packaged individually. Alternatively, disposable units are easy to use and convenient with some even containing both the filtration apparatus and a Petri dish attached. With both systems, it is necessary to use aseptic techniques when transferring the membrane to the agar surface. Membranes are available with or without grids, the former of which can be helpful with microscopic counting.

When sampling from the bottling line where microorganisms are either absent or present at low populations (<5 CFU/1000 mL), the volume of sample needed to provide statistically reliable results becomes important. If, by experience, recovery of >10 cells/L at bottling likely results in instability, detection limits can be calculated when only 100 mL of a 750 mL bottle is membrane filtered (Cases 1 and 2).

Case 1:

$$\frac{10\ \text{CFU}}{1000\ \text{mL}} \times 100\ \text{mL} = 1\ \text{CFU}$$

Case 2:

$$\frac{20\ \text{CFU}}{1000\ \text{mL}} \times 100\ \text{mL} = 2\ \text{CFU}$$

Comparing cases 1 and 2, the difference between "stability" and "potential refermentation" is less than 1 CFU per plate, a population that would probably not be recovered. Using the same acceptance levels, detection limits are far better when the entire bottle contents (750 mL) are membrane filtered (cases 3 and 4).

Case 3:

$$\frac{10\ \text{CFU}}{1000\ \text{mL}} \times 750\ \text{mL} = 7.5\ \text{CFU}$$

Case 4:

$$\frac{20\ \text{CFU}}{1000\ \text{mL}} \times 750\ \text{mL} = 15\ \text{CFU}$$

Although the theoretical ability to detect microorganisms has improved with an increase in sample volume, these results are still well below the minimal CFU per plate requirement of 25 colonies (Section 14.5). To develop 25 colonies per plate, a minimum of 2500 mL would have to be filtered. For this reason, bottling lines often utilize in-line samplers that continuously sample the wine. These are equipped with a membrane filter that can easily be disassembled and a new one replaced at regular intervals. Alternatively, wineries should consider membrane filtering several bottles from a given sampling. When microbiologically sampling wines that have just been bottled, some wineries prefer to wait 2 to 3 days before plating in order to allow (a) time for any antimicrobials (e.g., SO_2) added at bottling to act and (b) death of any non-wine microorganisms that may be present.

1. Aseptically obtain a sample from either a bottle or a tank.
   a. Sampling from bottles: Swab the neck of the bottle and dip the corkscrew in 70% v/v ethanol. Ignite the ethanol but keep the

corkscrew and the bottle pointed downward to avoid burning ethanol on hands or clothing.

b. Sampling from tanks: Agitate the tank before removing a sample because microorganisms will either be at the top (e.g., some non-*Saccharomyces* yeasts) or bottom (e.g., lactic acid bacteria) of the tank. Ideally, flame the sampling instrument using an open flame and place the sample into a sterile container.

2. Flame the open neck of the bottle or sample container for a couple of seconds and immediately pour at least 250 mL into the top of a membrane filter unit.

3. Apply vacuum until all of the wine has been filtered into the lower chamber. Be sure to have a trap flask located between the vacuum source and the sterile filter unit to catch additional wine.

4. Flame forceps and aseptically remove the membrane that now contains microorganisms.

5. Roll the membrane (grid up) onto a sterile Petri dish containing the agar medium, making sure of complete contact between the membrane and the agar. If the membrane is not in direct contact with the agar, the microorganisms will not be able to grow. Any air bubbles can be removed from underneath the membrane by pushing lightly with sterile forceps.

6. Incubate and watch for the appearance of colonies.

14.6 BIOLUMINESCENCE

Bioluminescence is the process by which a molecule in the excited state emits light, which is then measured (Hartman et al., 1992). This process can be used to measure the amount of ATP produced by microorganisms as part of their metabolism. In theory, measurement of this compound should provide an estimate of viable cell numbers because higher populations of microorganisms produce more ATP. Because a single yeast cell will have generally more ATP than a bacterial cell, the detection limit for yeast could be as low as 10 cells (Hartman et al., 1992). These authors further suggested that a practical limit for bacteria is closer to 1000 to 10,000 cells.

The luciferin–luciferase assay commonly used to measure ATP is as follows:

Luciferin + enzyme + ATP + Mg^{2+} → Luciferin–enzyme–AMP + pyrophosphate

Luciferin–enzyme–AMP + O_2 → Oxyluciferin + enzyme + AMP + CO_2 + light

Because this assay can be completed in just a few minutes, the luciferin–luciferase bioluminescence technique is increasingly being used *in lieu* of traditional swab and plate methods to monitor sanitation programs (Section 14.2). Currently, several luminometers and test kits are commercially available. Once the monitor is purchased, the cost per test including sample collection container, swab, and reagents can be inexpensive. Although this method has been used in the alcoholic and nonalcoholic beverage industries (Thompson, 2000), a major drawback lies in the translation of ATP measurements to viable cell counts (Hartman et al., 1992). ATP production will vary between microorganisms, as well as their physiological state (injured, starved, etc.). In addition, this assay cannot distinguish between ATP produced by yeast or other microorganisms, making interpretation of the results difficult.

14.7 NEPHELOMETRY AND OPTICAL DENSITY

Spectrophotometers and nephelometers have a long history of use by microbiologists as tools to rapidly estimate microbial population density as well as concentrations of specific compounds (e.g., color, malic acid, etc.) by measuring optical density (absorbance). Whereas a spectrophotometer measures how much light being passed through a liquid sample is absorbed or transmitted, nephelometers measure light scatter. As illustrated in Fig. 14.5, spectrophotometric measurements of transmitted light (420 or 650 nm) are at made in-line with the light source, whereas nephelometric measurements are taken at 90° to the source. Standard curves plotting absorbance against populations of microorganisms or other analytes must be prepared to determine the concentration in an unknown sample. As such, successful applications require calibration of the absorbance value to other enumeration methods, most commonly direct plating (Section 14.5).

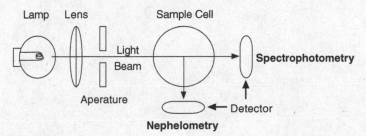

Figure 14.5. Diagram of spectrophotometic and nephelometic instruments. Spectrophotometric assays for optical density utilize wavelengths of either 420 or 650 nm.

The accuracy of using a spectrophotometer or nephelometer to estimate microbial populations depends on several factors. Because larger cells such as yeast will absorb more light than smaller cells such as bacteria, individual calibration curves for specific microorganisms must be generated. In addition, yeasts tend to aggregate, particularly toward the end of fermentation, which will also influence absorbance values. In terms of sensitivity, spectrophotometric estimations require relatively dense suspensions compared with nephelometric measurements. For instance, the minimal detection limit for spectrophotometers can be quite high, approaching 10^6 to 10^7 CFU/mL (Parish and Davidson, 1993). Although nephelometry is ideal for measuring light scattering associated with suspended microorganisms, these instruments are not commonly used.

CHAPTER 15

IDENTIFICATION OF WINE MICROORGANISMS

15.1 INTRODUCTION

Biologists use a system of identification based on the degree of physical and physiological similarity between organisms known as taxonomy. Early attempts at classification involved comparison of observable phenotypic features such as cell shape (morphology), as well as use of a variety of often crudely prepared sugar and nitrogen substrates. From these evolved extensive diagnostic schemes for identification, which still initially rely on cell and colony morphology. Today, other phenotypic markers include specific physiological and biochemical characteristics, sequences of genetic material, and immunochemical responses (Chapter 16).

The fundamental unit or taxon of biological classification is the species. Classically, general biology texts describe the species as being a reproductively isolated population; that is, it is different from all other similar organisms to the extent that it can no longer interbreed. Whereas this definition is adequate for organisms with an established and regular sexual phase in the life cycle, it is less clear how it may be applied in the case of those that normally reproduce by asexual means. Thus, the concept of

241

species among yeasts and bacteria is difficult to interpret. Although bacteriologists retain use of the concept of species, it has been extended to encompass (potentially) many very similar strains. In theory, a strain consists of progeny of a single cell, analogous to the propagation of grape clones. Thus, the use of strain includes a collection of similar-appearing microorganisms that differ only in terms of a few "minor" physiological properties. A good example of this is the bacterium *Oenococcus oeni* of which there are many known strains including the familiar ML-34, PSU-1, Viniflora, and EQ-54.

With the influx of information regarding wine microorganisms and the use of genetic sequencing, there is significant debate as to the extent of differences between microorganisms and whether these constitute different species or strains. As taxonomists increasingly rely on similarities at the genomic level, the relationships between microorganisms will both clarify as well as become confused. The taxonomy of microorganisms will continue to evolve as new relationships are determined. For general reference, the most current editions of taxonomic guides for yeasts, bacteria, and molds, including *The Yeasts*, *Bergey's Manual of Systematic Bacteriology*, and *The Procaryotes*, should be consulted.

15.2 IDENTIFYING MICROORGANISMS

Classic microbiology relied on the information gained about an unknown microorganism from various biochemical tests to identify genus and species. Because a great number of these tests are normally required, identification is commonly a very slow and expensive process. However, experienced wine laboratory personnel can generally shortcut classic protocols for rapid, but tentative, identification.

One of the first microbiological tests performed in a winery is to examine the microorganism under a microscope. Here, a juice or wine sample can be prepared as a wet mount (Section 12.5) and then examined using phase-contrast microscopy (Section 12.2.3). This examination will yield information related to the shape (cocci, rods, pointed ends, bowling pin, egg, ogival, elongated, lemon, needle-like, etc.), size (dimensions), and arrangement (single, pairs, tetrads, groups, or chains) of the cells. For instance, detection of small, lemon-shaped yeasts early in alcoholic fermentation could indicate the presence of *Kloeckera* or *Hanseniaspora*, whereas observation of rod-shaped bacteria suggests a potential spoilage due to *Lactobacillus* spp. Although microscopy can be used to quickly examine a must or wine, it must be noted that a number of microscopic fields should be examined because a high population is needed for detection (Section 14.4). Because of this, concentration of cells by centrifuge is sometimes required. A comprehensive reference with photographs of wine microorganisms as these appear as colonies and microscopically can be found in Edwards (2005).

Although very quick, microscopic evaluation of a wine can lead to errant results. First, the appearance of yeasts varies depending on age and culture conditions. For example, a culture grown on malt agar for 72 h may appear distinctly different from cells isolated from a wine toward the completion of alcoholic fermentation. Second, yeasts exhibit substantial variation in size and shape, even when in pure culture, reflecting the fact that asexual reproduction results from budding. For example, young daughter cells of the apiculate yeasts *Kloeckera* and *Hanseniaspora* are nearly spherical in shape but upon repeated budding, older mother cells become distinctly lemon-shaped. Because lemon- and spherical-shaped cells would be observed in a 1 week old culture of this microorganism, this observation could be misinterpreted as a culture contamination problem.

If identity confirmation is required, then the microorganism must be isolated from the wine (Section 13.9) prior to any characterization. It will probably be necessary to re-streak the isolate several times in order to obtain a pure isolate completely free from contamination. Unless a specific genus of microorganism is suspected, different media should be used initially in order to isolate the various microorganisms present in the sample.

When examining microorganisms growing on agar, it is important to make note of colony characteristics as illustrated in Fig. 15.1. Such characteristics include shape (circular, irregular, or rhizoid), size (dimensions, normally expressed in millimeters), topography (flat, raised, convex, concave, or umbonate), presence of pigments, opacity (transparent,

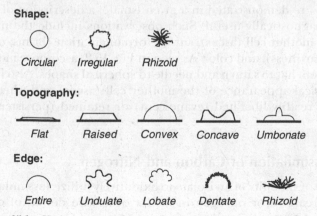

Shape:

Circular   Irregular   Rhizoid

Topography:

Flat   Raised   Convex   Concave   Umbonate

Edge:

Entire   Undulate   Lobate   Dentate   Rhizoid

Figure 15.1.   Various morphologies of colonies growing on solidifed media.

translucent, or opaque), surface (smooth, rough, dull, or glistening), edge (entire, undulate, lobate, or dentate, or rhizoid) or any changes to the agar (color or opacity changes due to pH or $CaCO_3$ indicators, respectively).

Once isolated, the unknown microorganism(s) can be better characterized using the methods detailed in the following sections for identification. This chapter describes those classic methods used to characterize and identify yeasts and molds (Section 15.3) as well as bacteria (Section 15.4) found in grape juices and wines.

15.3 YEASTS AND MOLDS

As yeast cellular morphology may vary with cultivation techniques as well as the stage in the growth cycle, this step in identification must be standardized. Ideally, recently isolated colonies are transferred to a nutrient broth medium and incubated at 25°C/77°F for 72 h or until colonies develop and then examined microscopically. Generally, most yeasts require only 2 to 4 days for growth while longer times (10 to 14 days) are needed for those isolates suspected of being *Dekkera/Brettanomyces*. When growing on agar substrate, yeast isolates typically appear as larger opaque (creamy) colonies compared with smaller (pinpoint) transparent bacterial colonies. Additionally, color development (chromogenesis) provides immediate diagnostic information. For example, *Rhodotorula* produces a salmon-pink-colored colony on laboratory media.

As discussed previously, the absence or presence of ascospores provides important information regarding identification (Section 1.2.1). Commonly used sporulation media are presented in Table 15.1. Because some of these media are also used for general isolation and growth, it is not uncommon to observe ascospore formation in recently isolated yeasts. If ascospores are demonstrated in a given isolate, a description of the ascospore is diagnostically useful. Such observations include the number of spores per mother cell (ascus), surface ornamentation (brims, equatorial rings, or wartiness), and color. As seen in Fig. 15.2, ascospore morphology can vary from hat to saturn and needle to spherical shapes. Also important is the physical appearance of the mother cell (ascus) and whether ascospores are readily liberated (evanescent) or retained (persistent) within the ascus.

15.3.1 Assimilation of Carbon and Nitrogen

The ability or inability of an isolate to oxidatively utilize (assimilate) single sources of carbon or nitrogen in media otherwise devoid of carbon or

Table 15.1.   Suggested sporulation media for yeast isolates.

Yeast	Sporulation medium
Dekkera	Gorodkowa, enriched yeast extract–malt extract[b]
Hanseniaspora	Malt extract,[b] potato dextrose agar[b]
Hansenula[a]	Yeast extract–malt extract,[b] malt extract,[b] V-8
Metschnikowia pulcherrima	Yeast extract–malt extract,[b] malt extract,[b] V-8
Pichia[a]	V-8, acetate, malt extract,[b] Gorodkowa
Saccharomyces	Acetate agar[b]
Saccharomycodes ludwigii	Gorodkowa, malt extract[b]
Schizosaccharomyces pombe	Yeast extract–malt extract,[b] malt extract,[b] V-8
Torulaspora delbrueckii	Yeast extract–malt extract[b]
Zygosaccharomyces bailii	Wort,[b] yeast extract–malt extract[b]

[a]Sporulation on any given medium is species dependent.
[b]Although these media can be made in the laboratory, it may be easier to purchase commercially available preparations.

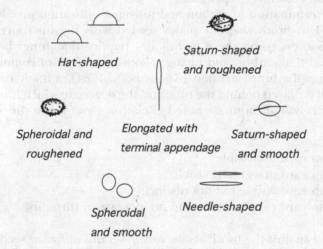

Hat-shaped

Saturn-shaped and roughened

Spheroidal and roughened

Elongated with terminal appendage

Saturn-shaped and smooth

Spheroidal and smooth

Needle-shaped

Figure 15.2.   Representative shapes of ascospores found in yeasts. Adapted from Yarrow (1998) with the kind permission of Elsevier Ltd.

nitrogen sources is a widely used diagnostic tool for identification of yeasts and molds. For instance, *Pichia* spp. are distinguished from *Hansenula* by their inability to grow on nitrate as sole source of nitrogen (Yarrow, 1998).

Only young, vigorously growing cultures should be used for the evaluation of nitrogen and carbon utilization profiles. Furthermore, cultures should be previously transferred from yeast extract–malt extract medium because nutrient carry-over by the microorganism is also possible. To minimize this problem, appropriate controls containing only basal medium and yeast (either no carbon or nitrogen source) should be incorporated in these assays.

Assimilation tests for carbon compounds use yeast nitrogen base (YNB) without carbon sources, and the assimilation test for nitrogen compounds use yeast carbon base (YCB) without assimilable nitrogen sources. For the tests described, YNB without amine acids and ammonium sulfate is used (Anonymous, 1984). Carbon sources normally examined include a number of pentoses, hexoses, disaccharides, trisaccharides, polysaccharides, alcohols, organic acids, and glycosides as specified by Yarrow (1998). Nitrogen sources commonly tested include nitrate, nitrite, ethylamine hydrochloride, cada-verine dihydrochloride, L-lysine, imidazole, glucosamine, creatine, and creatinine. When nitrite is used as a test compound, it is necessary to adjust the pH of the medium to 6.5 because toxic nitrous acid is formed at pH <6.0.

The determination of carbon and nitrogen utilization profiles can be performed in broth or pour plates seeded with the microorganism of interest. Success using these techniques require that fungi be in pure culture and that carbon and nitrogen sources be free of contamination. To interpret the broth method (Anonymous, 1984), a finely ruled white index card is placed behind the tube and the degree to which haze/turbidity hampers visualization is noted. Relative growth may be scored as follows:

Lines on card not visible	(+++)
Lines on card marginally visible	(++)
Lines on card visible but not distinct	(+)
Lines on card clearly visible and no increase in turbidity	(−)

Contrasted against the broth assays, auxanograms utilize a "seeded" background of the yeast isolate onto which various carbon or nitrogen sources are added. Growth on the carbon or nitrogen source is visualized as haze or turbidity around the source.

One difficulty with auxanograms is that excess moisture on the agar surface may cause test compounds to flow, potentially causing contamination. Although some recommend drying moist agar in an oven at 37°C/99°F for 90 min, this technique not only risks microbial contamination but also thermal injury or death of sensitive yeasts. In those situations where tubes are scored as "positive" in the nitrogen utilization assay, results should be reconfirmed by transferring a loop of medium to a fresh tube containing the same nitrogen source and reevaluated after 1 week.

While the choice whether to use broth or auxanograms for nitrogen and carbon utilization depends on individual preference, auxanograms can be easier to perform.

15.3.1.1 Carbon Utilization by Broth Culture

1. Prepare a 10× stock solution by dissolving 6.7 g YNB (without amino acids and ammonium sulfate) in approximately 90 mL distilled water. Once dissolved, bring to volume (100 mL) and filter sterilize through an 0.45 μm membrane. Keep in the refrigerator for use as needed.
2. Prepare the yeast inoculum.
   a. Combine 1 mL of 10× stock solution of YNB with 9 mL distilled water.
   b. Dissolve 10 mg glucose, 50 mg ammonium sulfate, 10 mg histidine, 20 mg methionine, and 20 mg tryptophan in the medium and sterile filter through an 0.45 μm membrane.
   c. Inoculate with a single colony and incubate for 48 h at 25°C/77°F.
   d. Using sterile water, dilute culture to a "slightly hazy" just prior to inoculation.
   e. Repeated transfers to/from this medium can improve cell adaptation and reduce carbohydrate reserves which can compromise results.
3. Aseptically dispense 0.5 mL aliquots of the 10× stock YNB into sterile capped or cotton-plugged 16 × 125 mm test tubes.
4. Prepare a series of carbon sources (sugars, sugar alcohols, and organic acids) at concentrations of 0.5% w/v in distilled water. In the case of raffinose, prepare a 1% w/v solution. Filter sterilize all solutions through on a 0.45 μm membrane.
5. Aseptically transfer 4.5 mL of a given carbon source to a test tube containing 0.5 mL 10× stock YNB. Control treatments contain 0.5 mL 10× stock YNB and 4.5 mL sterile distilled water.
6. Using a sterile pipette, aseptically transfer one to two drops of the yeast inoculum to all test tubes.
7. Incubate the test tubes at 25°C/77°F for 1 week. To ensure oxidative conditions throughout the incubation period, some laboratories prefer to store cultures on a slant rather than maintain the tubes upright.

15.3.1.2 Carbon Utilization by Auxanogram

1. Prepare a 10× stock solution by dissolving 6.7 g YNB in approximately 90 mL distilled water. Once dissolved, bring to volume (100 mL) and filter sterilize through a 0.45 μm membrane.

2. Prepare the yeast inoculum.
   a. Add 1 mL of 10× stock solution of YNB to 9 mL distilled water.
   b. Dissolve 10 mg glucose, 50 mg ammonium sulfate, 10 mg histidine, 20 mg methionine, and 20 mg tryptophan in the medium and sterile filter through an 0.45 μm membrane.
   c. Inoculate with a single colony and incubate for 48 h at 25°C/77°F.
   d. Repeated transfers to/from this medium can improve cell adaptation and reduce carbohydrate reserves, which can compromise results.
3. Mix 0.67% w/v YNB and 2% w/v agar and dissolve in a boiling water bath. Dispense 20 mL into capped or plugged 20 × 150 mm test tubes and autoclave at 121°C/250°F for 15 min.
4. Once removed from the autoclave, temper the media in a water bath at 45°C/113°F to 50°C/122°F.
5. Prepare a pour plate by transferring 0.1 mL of the yeast inoculum to center of Petri plate. Pour the tempered agar into the Petri plate and using a "figure eight motion," disperse the yeast throughout the agar. Allow agar to solidify completely and cool before proceeding.
6. Using a laboratory marker, write the compounds of interest on the bottom portion of the Petri dish. A maximum of three compounds per plate can be evaluated.
7. Transfer approximately 5 mg of each carbon compound to the surface of the agar, placing the compounds 1 to 1.5 cm from edge of the plate. If the same spatula is used for transfer, thoroughly clean and sterilize between each compound. Once the first compound has been placed on agar substrate, maintain the plate upright.
8. Incubate in upright position (agar in bottom, not in lid) at 25°C/77°F for 3 to 7 days prior to evaluation.

15.3.1.3 Nitrogen Utilization by Broth Culture

1. Prepare a 10× stock solution by dissolving 11.7 g YCB in approximately 90 mL distilled water. Once dissolved, bring to volume (100 mL) and filter sterilize through an 0.45 μm membrane. Keep in the refrigerator for use as needed.
2. Prepare the yeast inoculum.
   a. Add 1 mL of 10× stock solution of YNB to 9 mL distilled water. It may be necessary to warm the solution to improve solubilization.
   b. Dissolve 10 mg glucose, 50 mg ammonium sulfate, 10 mg histidine, 20 mg methionine, and 20 mg tryptophan in the medium and sterile filter through an 0.45 μm membrane.

c. Inoculate with a single colony and incubate for 48 h at 25°C/77°F.

d. Repeated transfers to/from this medium will reduce cellular nitrogen reserves which can compromise results.

e. Using sterile water, dilute culture to a "slightly hazy" just prior to inoculation.

3. Aseptically dispense 0.5 mL aliquots of 10× stock YCB into sterile capped or cotton-plugged 16 × 125 mm test tubes.

4. Separately dissolve 0.078 g potassium nitrate (KNO_3), 0.026 g sodium nitrite ($NaNO_2$), 0.064 g ethylamine hydrochloride, and 0.080 g L-lysine in 10 mL distilled water and filter sterilize through an 0.45 μm membrane.

5. Aseptically transfer 4.5 mL of a given nitrogen source to a test tube containing 0.5 mL 10× stock YCB. Control treatments contain 0.5 mL 10× stock YCB and 4.5 mL sterile distilled water.

6. Using a sterile pipette, aseptically transfer one to two drops of the inoculum to all test tubes.

7. Incubate at 25°C/77°F for 1 week. To ensure oxidative conditions throughout the incubation period, some laboratories prefer to store cultures on a slant rather than maintain the tubes upright.

15.3.1.4 Nitrogen Utilization by Auxanogram

1. Prepare a 10× stock solution by dissolving 11.7 g YCB in approximately 90 mL distilled water. Once dissolved, bring to volume (100 mL) and filter sterilize through an 0.45 μm membrane. Keep in the refrigerator for use as needed.

2. Prepare the yeast inoculum.

   a. Add 1 mL of 10× stock solution of YCB to 9 mL distilled water.

   b. Dissolve 10 mg glucose, 50 mg ammonium sulfate, 10 mg histidine, 20 mg methionine, and 20 mg tryptophan in the medium and sterile filter through an 0.45 μm membrane.

   c. Inoculate with a single colony and incubate for 48 h at 25°C/77°F.

   d. Repeated transfers to/from this medium will reduce cellular nitrogen reserves, which can compromise results.

3. Mix 1.17% w/v YCB and 2% w/v agar and dissolve in a boiling water bath. Dispense 20 mL into capped or plugged 20 × 150 mm test tubes and autoclave at 121°C/250°F for 15 min.

4. Once removed from the autoclave, temper the media in a water bath at 45°C/113°F to 50°C/122°F.

5. Prepare a pour plate by transferring 0.1 mL of the yeast inoculum to center of Petri plate. Pour the tempered agar into the Petri plate and

using a "figure eight motion," disperse the yeast throughout the agar. Allow agar to solidify completely and cool before proceeding.

6. Using a laboratory marker, write the compounds of interest on the bottom portion of the Petri dish. A maximum of three compounds per plate can be evaluated.

7. Transfer approximately 5 mg of each carbon compound to the surface of the agar, placing the compounds 1 to 1.5 cm from edge of the plate. If the same spatula is used for transfer, thoroughly clean and sterilize between each compound. Once the first compound has been placed on agar substrate, maintain the plate upright.

8. Incubate in upright position (agar in bottom, not in lid) at 25°C/77°F for 3 to 7 days and evaluate each zone for growth.

15.3.2 Demonstration of Ascospores

Upon isolation of the yeast, details of cell morphology, budding type, and relative size should be noted. Recently isolated yeasts may already be sporulating and further effort in this regard may not be necessary. Assuming this is not the case, transfer the isolate to an appropriate medium as specified in Table 15.1 and incubate at 25°C/77°F, examining the culture after 2 to 3 days and then once a week for 6 weeks. Examination requires preparation of a wet mount (Section 12.5) and viewing under phase-contrast microscopy. Alternatively, the preparation can be stained using malachite green (Section 15.3.2.5) and viewing using brightfield microscopy. If sporulation is not seen, transfer the culture to an alternative medium and continue to monitor for an additional 6 weeks.

Acetate Agar 2, Gorodkowa, and the V-8 media are described by Yarrow (1998), and the enriched yeast extract–malt extract medium was originally described by van der Walt and van Kerken (1961). The staining procedure for spore cultures using malachite green is from Yarrow (1998).

15.3.2.1 Acetate Agar 2

1. Mix and dissolve the following ingredients.

Glucose	1 g
Potassium chloride (KCl)	1.8 g
Sodium acetate trihydrate	8.2 g
Yeast extract	2.5 g
Agar	15 g
Distilled water	900 mL

2. Place the flask in boiling water to dissolve the ingredients. Once dissolved, add distilled water to dilute to volume (1000 mL) and autoclave at 121°C/250°F for 15 min.

3. Once removed from the autoclave, temper the media in a water bath at 45°C/113°F to 50°C/122°F.
4. Prepare pour plates and allow agar to solidify.
5. Using a flamed and cooled loop, transfer a young culture (incubation time of 24 to 48 h) to the medium and incubate at 25°C/77°F for 2 to 3 days prior to examination with a microscope.

15.3.2.2 Gorodkowa Agar

1. Mix and dissolve the following ingredients.

Glucose	1 g
·Peptone	10 g
Sodium chloride (NaCl)	5 g
Agar	20 g
Distilled water	900 mL

2. Place the flask in boiling water to dissolve the ingredients. Once dissolved, add distilled water and dilute to volume (1000 mL). Autoclave at 121°C/250°F for 15 min.
3. Once removed from the autoclave, temper the media in a water bath at 45°C/113°F to 50°C/122°F.
4. Prepare pour plates and allow agar to solidify.
5. Using a flamed and cooled loop, transfer a young culture (incubation time of 24 to 48 h) to the medium and incubate at 25°C/77°F for 2 to 3 days prior to examination with a microscope.

15.3.2.3 V-8 Agar

1. In a 1000 mL flask suspend 5 g compressed baker's yeast in 10 mL distilled water and add to 350 mL Campbell's V-8 Vegetable Juice® (Campbell Soup Company, Camden, NJ, USA).
2. Adjust to pH 6.8 using 50% NaOH and heat in a boiling water bath for approximately 10 min. Recheck pH and adjust to 6.8 if necessary.
3. In a second flask, add 14 g agar to 340 mL distilled water and heat to dissolve.
4. Combine contents of the two flasks and bring to 1 L final volume with distilled water. Autoclave at 121°C/250°F for 15 min.
5. Once removed from the autoclave, temper the media in a water bath at 45°C/113°F to 50°C/122°F.
6. Prepare pour plates of the medium and allow agar to solidity.
7. Using a flamed and cooled loop, transfer a young culture (incubation time of 24 to 48 h) onto the medium and incubate at 25°C/77°F for 2 to 3 days prior to examination with a microscope.

15.3.2.4 Enriched Yeast Extract–Malt Extract Agar

1. Prepare the 10× vitamin stock solution by mixing and dissolve the following ingredients.

Biotin	0.0010 g
Folic acid	0.0010 g
Calcium pantothenate	0.20 g
Inositol	1.0 g
Niacin	0.20 g
p-Aminobenzoic acid	0.10 g
Pyridoxine hydrochloride	0.20 g
Riboflavin	0.10 g
Thiamin	0.5 g
Distilled water	400 mL

2. Once dissolved, dilute to volume (500 mL) with distilled water. Sterile filter the stock solution through on 0.45 μm membrane and store at −15°C/5°F to −10°C/14°F until used. Solution must be diluted 1 : 10 prior to use.
3. With thorough mixing, add 1% v/v of the single strength vitamin solution to recently autoclaved and cooled (50°C/122°F) yeast extract–malt extract medium.
4. Using a flamed and cooled loop, transfer a young (incubation time of 24 to 48 h) culture to the medium and incubate at 25°C/77°F for 2 to 3 days prior to examination with a microscope.

15.3.2.5 Staining Spore Cultures

1. Heat-fix the culture onto a microscopic slide.
2. Flood the slide with a solution of 0.5% w/v malachite green and 0.05% w/v basic fuchsin made in distilled water.
3. Heat the slide to steaming over a flame for 1 min and then wash thoroughly in flowing water. Blot the slide dry and examine using a bright-field microscope.

15.3.3 Demonstration of Mycelia/Pseudomycelia

The presence or absence of mycelium or pseudomycelium is demonstrated by growing the fungi as a slide culture (Section 12.6). However, microscopic evaluation requires removal of the slide culture from the Petri plate, thereby increasing the potential for microbial contamination. To minimize contamination, the number of examinations should be limited such that results can be obtained after only one to two observations.

1. Prepare the slide culture as illustrated in Fig. 12.3.
2. Prepare and autoclave the appropriate agar medium commonly potato dextrose agar or corn med agar (Anonymous, 1984). Pour into sterile Petri plates and allow to solidify.
3. Using a sterile spatula, cut the agar into cubes of a suitable size to fit within the dimensions of the microscope slide cover slip.
4. After resterilizing the spatula (dipping in 70% v/v ethanol and then flaming), carefully scoop out an agar cube and transfer to center of a sterile microscope slide inside the Petri plate (this step may take practice!).
5. Using a sterile loop, transfer a yeast colony onto the agar cube. With sterile forceps, carefully place a sterile cover slip over inoculated agar cube, taking care not to leave an exposed agar surface.
6. Transfer sufficient sterile distilled water to the Petri plate to maintain high humidity.
7. Incubate at 25°C/77°F for 4 to 7 days and examine microscopically.

15.3.4 Fermentation of Carbohydrates

The ability to utilize carbon and nitrogen oxidatively (with oxygen) does not always correlate with the ability of the yeast to use these sugars fermentatively. Thus, fermentation profiles provide further information necessary for identifying species of yeasts. Here, the ability to fermentatively utilize a specific sugar is detected by trapping CO_2 inside the Durham tube placed upside down in the liquid medium. Sugars most commonly tested are glucose, galactose, sucrose, maltose, lactose, raffinose, and trehalose. Other sugars (inulin, starch, melibiose, cellobiose, and D-xylose) are sometimes tested as well. Commercially-prepared carbohydrate testing kits are available (Section 15.4.4).

1. Prepare yeast extract broth by dissolving 0.5% w/v yeast extract in distilled water.
2. For sugars that can be autoclaved (e.g., glucose and galactose), prepare as 2% w/v solutions in the yeast extract broth and transfer 10 mL to capped 18 × 150 mm test tubes. Insert a Durham tube open-side down into each tube, loosely replace the cap, and autoclave at 121°C/250°F for 15 min.
3. Because many sugars should not be autoclaved, prepare as 2% w/v solutions in the yeast extract broth and individually filter sterilize the solutions through 0.45 μm membranes. For raffinose, prepare a 4% w/v solution because some strains only use part of the molecule (Yarrow, 1998). Transfer 10 mL aliquots of yeast extract broth to sterile 18 ×

150 mm test tubes, aseptically insert sterile Durham tubes open-side down into the test tubes, and cap. Note the volume of the air bubble prior to inoculation.

4. Inoculate each tube with yeast to yield a slightly turbid suspension and incubate at 25°C/77°F, examining for gas formation at intervals of 2 to 3 days for up to one week.

15.4 BACTERIA

In many situations, wine microbiologists can tentatively identify wine bacteria by genus based on cellular morphology. Due to morphological variation, microscopic identification should be coupled with selected physiological tests in order to confirm identification (Sections 2.2 and 3.2). For example, lactic acid bacteria can be tentatively identified by Gram stain and fermentation of carbohydrates. In fact, *Oenococcus, Pediococcus,* and *Lactobacillus* have all been classically identified to species level using carbohydrate fermentation patterns (Garvie, 1967a; 1986a; 1986b; Kandler and Weiss, 1986; Edwards et al., 1991; 1993; Edwards and Jensen, 1992).

One difficulty with some of these tests is the fact that many lactic acid bacteria isolated from grape musts or wines require the so-called "tomato juice factor" (Section 2.3), a compound present in fruit and vegetable juices and serums. These juices and serums are commonly added to media to improve bacterial growth, but they also contain sugars that can interfere with carbohydrate utilization tests. To exclude juices or serums but improve bacterial growth, Garvie (1984) suggested the incorporation of additional pantothenic acid or use of a heavy inoculation. Besides fruit or vegetable juices or serums, yeast extract (0.5% w/v) is frequently added as a nitrogen supplement along with Tween 80 as a source of fatty acids (Kunkee, 1974, Champagne et al., 1989).

De Ley et al. (1984), De Ley and Swings (1984), and Swings (1992) identified several tests that distinguish *Acetobacter* from *Gluconobacter* with the easiest being their relative ability to utilize ethanol as a carbon source. Here, *Acetobacter* oxidizes the alcohol completely to carbon dioxide and water while *Gluconobacter* can oxidize ethanol only to acetic acid. Similarly, *Acetobacter* will oxidize lactate to CO_2, and H_2O whereas *Gluconobacter* cannot. Presumably, *Gluconacetobacter* behaves similarly to *Acetobacter* as the former genus was originally classified by De Ley and Swings (1984) and De Ley et al. (1984) as *Acetobacter*. Differentiation of *A. aceti* and *A. pasteurianus* requires the ability to detect oxidation of glycerol to dihydroxyacetone, a property called ketogenesis. One difficulty in defining phenotypic characteristics of acetic acid bacteria is that the group undergoes spontaneous mutations rather frequently (Swings, 1992).

15.4.1 Ammonia from Arginine

Microscopically, it may be difficult to separate *O. oeni* from some species *Lactobacillus* due to morphological similarity. In these cases, Garvie (1984) suggested that separation could be made using ammonia formation from arginine (Section 2.4.2). The described method relies on the heterofermentation-arginine medium (Section 13.6.3) used by Pilone et al. (1991) but without inclusion of fructose.

Nessler's reagent contains mercury (II) iodide, which is dissolved in an aqueous solution of potassium iodide and potassium hydroxide to yield K_2HgI_4. The reagent is used to detect the presence of ammonia as per the following reaction:

$$NH_4^+ + 2[HgI_4]^{2-} + 4OH^- \rightarrow HgO \bullet Hg(NH_2)I + 7I^- + 3H_2O$$

The formation of a brick-orange color indicates the présence of ammonia. Nessler's reagent (CAS number 7783-33-7) is available from commercial sources and is not normally prepared in the laboratory.

1. Mix and dissolve the following ingredients.

Tryptone	5 g
Yeast extract	5 g
Potassium dibasic hydrogen phosphate (K_2HPO_4)	2 g
Glucose	0.5 g
Distilled water	800 mL

   a. Once ingredients are dissolved, dilute medium to volume (1000 mL) and then divide into two 500 mL aliquots.
   b. To one aliquot, add 1.5 g arginine-HCl.
   c. Adjust the pH of both media to pH 5.5 with 50% v/v H_3PO_4 or 6 M KOH.
   d. Dispense into 18×150 mm test tubes, add the caps, and sterilize by autoclaving at 121°C/250°F for 15 min.
2. Prepare the following basal broth medium.
   a. Mix and dissolve the following ingredients.

Peptone	10 g
Yeast extract	2.5 g
Tween 80 (5% w/w solution)	2 mL
Distilled water	800 mL

    b. Once ingredients are dissolved, dilute to volume (1000 mL) and adjust pH to 5.2 with 50% v/v H_3PO_4 or 6 M KOH.

    c. Distribute into bottles and sterilize by autoclaving at 121°C/250°F for 15 minutes.

3. Prepare the bacterial inoculum.

    a. Inoculate a colony into 10 mL apple juice Rogosa medium (Section 13.6.1) and incubate for 7 days.

    b. Harvest by centrifugation and suspend in a minimal amount of sterile basal broth medium.

    c. Inoculate test tubes at a rate of 0.1 to 0.2 mL inoculum per tube. Be sure to add uninoculated basal broth to the control test tubes.

4. Detection of ammonia.

    a. After incubation for up to 3 weeks, centrifuge 2 mL from inoculated broth (with and without arginine).

    b. Place 1 mL of each broth onto a white tile spot-plate and add 1 or 2 drops of Nessler's reagent.

15.4.2 Catalase

Hydrogen peroxide (H_2O_2) is a highly reactive by-product of the growth of some strains of microorganisms. Because H_2O_2 is toxic, these microorganisms contain the enzyme catalase to remove hydrogen peroxide as it is formed.

$$O_2 + 2H^+ + 2e^- \rightarrow H_2O_2 \text{ (oxidase)}$$

$$2H_2O_2 \rightarrow 2H_2O + O_2 \text{ (catalase)}$$

Lactic acid bacteria are generally catalase-negative. Hence, treatment with hydrogen peroxide does not elicit the bubbling reaction as noted with acetic acid bacteria. However, some strains of *Lactobacillus* have a pseudo-catalase, which will produce a positive reaction upon exposure to H_2O_2 (Edwards et al., 1993). Where weak catalase activity is suspected, carefully place a cover slip over the preparation and examine under a dissecting scope or low magnification of the compound microscope for bubbles. As *Saccharomyces* exhibits a strong catalase reaction, this microorganism can be used as a positive control in this test.

The protocol described originates from Whittenbury (1964) and Pilone and Kunkee (1972). Stock solutions of 30% H_2O_2 should be stored at 4°C/39°F, but the reagent will decompose over time leading to false negative responses. It is recommended to prepare fresh H_2O_2 when performing this test.

1. Transfer 0.15 g haematin into 30 mL distilled water. Add enough drops of 0.1 N NaOH to dissolve the haematin.
2. Sterile filter the solution through an 0.45 µm membrane into a previously sterilized bottle.
3. Prepare 1000 mL apple juice Rogosa agar (Section 13.6.1) but only add 0.5% w/v glucose and adjust the pH to 5.5.
4. After autoclaving the agar, aseptically add 10 mL of the haematin solution to the tempered medium just prior to pouring plates.
5. Streak the test microorganisms onto the solidified agar.
6. When good growth is apparent on the plates (7 to 10 days), transfer several drops of freshly prepared 3% v/v H_2O_2 onto some of the colonies.

15.4.3 Dextran from Sucrose

Some lactic acid bacteria have the ability to metabolize sucrose to produce polymers known as dextrans (viscous, slimy). While most strains of *Oeno-coccus oeni* are negative for this attribute (Edwards et al., 1991), some strains of *Lactobacillus* (Edwards et al., 1993) and *Leuconostoc mesenteroides* (Garvie, 1984) are positive. The outlined method originates from Pilone and Kunkee (1972). To test for dextran formation, touch a sterile inoculating needle straight down into an isolated colony and move the needle straight back up. A positive reaction is one where threads or strands are drawn up on the needle.

1. Prepare 1000 mL apple juice Rogosa agar (Section 13.6.1).
2. Adjust pH to 4.5 with 50% v/v H_3PO_4 or 6 M KOH, add agar (20 g/L), and autoclave at 121°C/250°F for 15 min. After autoclaving, temper the agar in a 45°C/113°F to 50°C/122°F water bath.
3. Add 50 g sucrose to 100 mL distilled water and dissolve.
4. Filter sterilize the sucrose solution through a 0.45 µm membrane and aseptically add to the tempered agar medium.
5. Add approximately 25 mL medium per Petri plate.
6. Once solidified, streak each test strain on the agar medium and incubate at 25°C/77°F for approximately 2 weeks before examining the culture for dextrans using a sterile needle.

15.4.4 Fermentation of Carbohydrates

Fermentation of specific carbohydrates by lactic acid bacteria with production of acid and/or gas (CO_2) has been used to identify lactic acid bacteria. Traditionally, fermentation patterns are determined by growing the microorganism in test tubes containing a medium with various carbohy-drates and a pH indicator (Kandler and Weiss, 1986). Carbohydrates commonly tested include amygdalin, L-arabinose, arbutin, D-cellobiose, esculin, fructose, D-galactose, β-gentibiose, glucose, lactose, maltose, mannose, melezitose, melibiose, α-methyl-D-glucoside, raffinose, ribose, salicin, sucrose, trehalose, D-turanose, and D-xylose. The presence of acid (and therefore bacterial growth) is indicated by the medium changing color from a blue-green to yellow while gas (CO_2) formation is seen as trapped bubbles.

In recent years, simple and convenient miniaturized multitest kits have been developed. One method, the API system, is produced for the identi-fication of *Lactobacillus* spp. but has also been applied to other lactic acid bacteria. A product of bioMerieux (Marcy l'Etoile, France), these prepack-aged kits include a gallery of biochemical tests selected to maximize the efficiency of identification of an unknown microorganism. Although Pardo et al. (1988) concluded that the system yielded inaccurate results for characterizing strains of *O. oeni*, Jensen and Edwards (1991) success-fully applied a modified medium for screening strains isolated from wines. Positive reactions (fermentation of the test carbohydrate) are those where a distinctive color change can be observed (e.g., from blue-green to yellow). Small bubbles under the mineral oil indicating gas formation can some-times be observed.

One important element with performing these tests is to use a specific inoculation rate. Rather than use direct plating to enumerate the bacteria in the inoculum (Section 14.5), the McFarland scale can be used to esti-mate populations based on turbidities of barium chloride ($BaCl_2$). In summary, different concentrations of $BaCl_2$ in the presence of H_2SO_4 cor-respond to the turbidity of dense bacterial populations present in a saline suspension (Gradwohl, 1948). $BaCl_2$ suspensions are prepared in small test tubes (Table 15.2) and the bacterial population estimated by comparing turbidities either visually or using a nephelometer (Section 14.7). To deter-mine the approximate bacterial density in broth cultures, dissolve sulfuric acid and $BaCl_2$ in the same sterile broth used to suspend the bacteria.

15.4.4.1 Test Tube Method

1. Centrifuge 10 mL of a bacterial culture grown for 7 days in apple juice Rogosa broth (Section 13.6.1), aseptically remove the broth, suspend the pellet in a minimal amount of 0.1% w/v peptone, and inoculate 0.1 mL into sterile 18 × 150 mm test tubes.
2. Mix and dissolve the following ingredients.

Peptone	10 g
Yeast extract	5 g

Potassium dibasic phosphate (K_2HPO_4)	5 g
Diammonium citrate	2 g
Sodium acetate ($NaC_2O_2H_3$)	5 g
Magnesium sulfate ($MgSO_4 \bullet 7H_2O$)	0.5 g
Manganese sulfate ($MnSO_4 \bullet 4H_2O$)	0.2 g
Tween 80	1.0 g
Distilled H_2O	800 mL

3. Once ingredients are dissolved, dilute medium to volume (900 mL). Add chlorophenol red (0.5 g/L) and adjust pH to 6.2 to 6.4 with either 50% v/v H_3PO_4 or 6 M KOH.
4. Add agar (15 g) and autoclave the medium at 121°C/250°F for 15 min.
5. Dissolve 10 g of each test carbohydrate into 100 mL distilled water, sterile filter through an 0.45 μm membrane, and add to the tempered medium. Be sure to check water solubilities of the various carbohydrates prior to testing (e.g., esculin and salicin are not as soluble as other carbohydrates).
6. As rapidly as possible, add 10 mL of the medium containing the test carbohydrate to a 18 × 150 mm test tube containing the bacterium. Thoroughly mix by vortexing and set aside. Once solidified, incubate the test tubes for 7 to 10 days. If the bacterial strain does not grow well, tryptone and/or calcium pantothenate can be added to the medium at concentrations of 20 g/L and 0.2 mg/L, respectively.

15.4.4.2 API Method

1. Centrifuge 10 mL of a bacterial culture grown in apple juice Rogosa broth (Section 13.6.1).
2. Aseptically decant the broth, and suspend the pellet in a minimal amount of a sterile basal broth that contains 1% w/v peptone, 0.25% w/v yeast extract, 0.01% w/v Tween 80, and 0.2 g/L calcium pantothenate.
3. Add the bacterial suspension to addition basal broth that contains a pH indicator (0.016% w/v bromcresol green) to obtain an optical density of McFarland Scale of 2 (Table 15.2).
4. Inoculate the bacterial suspensions into individual wells (galleries) of the API Rapid CH system (API Analytab Products, Plainview, NY). Overlay the wells with sterile mineral oil to maintain an anaerobic environment.
5. Incubate the API galleries at 25°C/77°F for up to 28 days and score each well as per the manufacturer's directions.

Table 15.2.   McFarland scale standards used to estimate bacterial populations.

McFarland scale	1% w/v H$_2$SO$_4$ (mL)	1% w/v BaCl$_2$ (mL)	Estimated population ($\times 10^6$ CFU/mL)
1	9.9	0.1	300
2	9.8	0.2	600
3	9.7	0.3	900
4	9.6	0.4	1200
5	9.5	0.5	1500
6	9.4	0.6	1800
7	9.3	0.7	2100
8	9.2	0.8	2400
9	9.1	0.9	2700
10	9.0	1.0	3000

15.4.4.3 McFarland Scale

1. Prepare the McFarland scale by mixing 1% w/v sulfuric acid and 1% w/v barium chloride (BaCl$_2$) according to Table 15.2.
2. If the scale is used with the API Rapid CH system to prepare the bacterial inoculum, it is best to make McFarland Scale 1 to 4 in order to compare densities of the inoculum and determine the specific dilutions required.

15.4.5 Gas from Glucose

Formation of carbon dioxide from glucose is characteristic of heterofermentative species (*O. oeni* and some *Lactobacillus*). Conversely, failure to produce CO$_2$ from glucose indicates homofermentative utilization of the sugar and is suggestive of *Pediococcus* or some species of *Lactobacillus*. It must be noted that some *Lactobacillus* spp. are classified as being facultatively heterofermentative but hexoses (e.g., glucose) are utilized homofermentatively with no gas formation (Section 2.4.1).

Method A is from Edwards et al. (1991), a protocol originally modified from Gibson and Abdel-Malek (1945) and Garvie (1984) while method B was described by Pilone et al. (1991). The presence of gas is indicated by the elevation of the agar plug inside of the tube (method A) or by gas trapped in the Durham tube (method B). If no gas is apparently produced using method A, it may be necessary to sonicate the tubes to force gas to coalesce into visible bubbles.

If poor growth of the bacterium is observed, 20% v/v tomato juice serum can be added. However, tomato juice serum will have some glucose and other sugars present and so may interfere with this assay. As such, it is very important to have three control tubes for either procedure:

Control 1: Medium + no glucose + no bacteria
Control 2: Medium + glucose + no bacteria
Control 3: Medium + no glucose + bacteria

Gas or acid production in control 1 or 2 indicates microbial contamination of the medium indicating all tubes were contaminated. Gas/acid production in control 3 suggests high amounts of sugar were present in the serum used to prepare the medium. If this is the case, 0.2 g/L calcium pantothenate may be substituted in lieu of tomato juice serum.

15.4.5.1 Method A

1. Select a commercially processed tomato juice that contains only juice and salt (no preservatives and not from concentrate) or prepare juice from fresh tomatoes. Centrifuge 250 mL of the juice at $3000 \times g$ for 30 min and decant, saving the supernatant (serum). Serum can be stored in small volumes at $-20°C/-4°F$ until used.
2. Mix and dissolved the following ingredients.

Peptone	5 g
Yeast extract	1.25 g
Tween 80 (5% w/w solution)	1 mL
Bromcresol green (1% w/w solution)	2 mL
Tomato juice serum	50 mL
Distilled water	200 mL

3. Adjust the medium to pH 5.2 with 50% v/v H_3PO_4.
4. Divide the medium into two 125 mL aliquots and add 1.25 g glucose to one, stirring to dissolve the sugar. The final glucose concentration is approximately 0.5% w/v, excluding the impact of tomato juice serum. Place these media into a $45°C/113°F$ to $50°C/122°F$ water bath.
5. Dissolve 2.5 g agar in 250 mL distilled water by steaming in the autoclave at $110°C/230°F$ for 10 min.
6. Stir 125 mL molten agar into each portion of broth (with and without glucose) from step 3.
7. Dispense media into 18×150 mm test tubes, cap, and sterilize by autoclaving at $121°C/250°F$ for 15 min.
8. Prepare the agar overlays.
   a. Mix 5 g agar in 250 mL water and dissolve by steaming in the autoclave at $110°C/230°F$ for 10 min.

    b. Distribute 5 mL per screw-capped 16 × 125 mm tube and sterilize by autoclaving at 121°C/250°F for 15 min.
9. Mix and dissolve the following basal broth.

Peptone	10 g
Yeast extract	2.5 g
Tween 80 (5% w/w solution)	2 mL
Distilled water	800 mL

    a. Once dissolved, dilute to volume (1000 mL) and adjust pH to 5.2 with 50% v/v H_3PO_4, distribute into 100 mL milk dilution bottles, and sterilize by autoclaving 121°C/250°F for 15 min.
10. Inoculate test tubes with bacterial cultures prepared as follows.
    a. Inoculate cultures from plates into 10 mL apple juice Rogosa broth (Section 13.6.1).
    b. Harvest cells by centrifugation (3000 × g) and suspend the pellet in a minimal amount of basal broth (step 8).
    c. Melt media (step 6) and overlays (step 7) by steaming in the auto-clave at 110°C/230°F for 10 min. Place the media and the overlays in a 45°C/113°F to 50°C/122°F water bath.
    d. After adding 0.1 to 0.5 mL inoculum to the test tubes (step 8b), vortex to mix, and then chill the tube immediately in an ice-water bath.
    e. When these media are completely solidified, overlay each with an agar overlay (plug should be 5 cm in height) and incubate at 25°C/77°F for 2 to 3 weeks.

15.4.5.2 Method B

1. Prepare heterofermentation-arginine broth (HFA) (Section 13.6.3).
2. Place 9 mL into sterilized 16 × 150 mm test tubes containing an inverted Durham tube, cap, and autoclave at 121°C/250°F for 15 min.
3. Inoculate the test tubes with 0.1 mL of an actively growing culture and incubate at 22°C/72°F to 30°C/86°F for up to 3 weeks.
4. As an alternative to using Durham tubes, the inoculated HFA broth can be overlaid with approximately 1 cm of molten vaspar (mixture of 1 part petroleum jelly + 6 parts paraffin) to assist in visualizing gas formation.

15.4.6 Gram Stain

One of the most commonly used procedures to assist in identification of bacteria, the Gram stain, differentiates species based on cell wall characteristics. The procedure calls for initial staining of cells with a blue-colored dye, crystal violet, removal of some or all of the dye from cells using a solvent ("decolorization"), and then re-staining ("counterstaining") with safranin, a red-colored dye. The preparation is then microscopically examined without a cover slip using an oil immersion objective under brightfield illumination. Depending on how much crystal violet is retained by the cells, counterstaining with a red dye (safranin) will yield either Gram-positive reaction (cells appear purple) or Gram-negative (cells appear red). Lactic acid bacteria are Gram-positive while acetic acid bacteria are generally Gram-negative cells (older cultures can appear as Gram-variable).

Although the individual reagents can be prepared in-house (Anonymous, 1984), pre-packaged Gram stain kits are available and are sufficient to do hundreds of stains at a reasonable cost. Staining is messy and discolors hands and clothing rather easily. To limit this problem, attach a wooden clothespin to microscope slides to act as a handle. Alternatively, staining racks can be easily prepared by linking two pieces of glass tubing with two short pieces of flexible rubber tubing of sufficient length to reach across the laboratory sink.

An alternative protocol for the Gram stain reaction has been applied to non-wine bacteria (Suslow et al., 1982) and involves placement of two drops of 3% w/v KOH onto a microscopic slide. Bacterial cells are then aseptically transferred from culture media with a flat wooden toothpick and placed into the drop of KOH with rapid, circular agitation. After 5 to 8 sec, the toothpick should be alternately raised/lowered to detect a "stringing" (or slime) effect. Where an increase in viscosity and stringing occurred within 15 sec, the culture is considered to be Gram-negative. Gram-positive microorganisms do not have this reaction.

1. Prepare a smear of the bacterium, making sure to heat-fix the slide prior to staining (Section 12.4). Be sure to use bacterial cultures of the same age (incubation time) and avoid staining very dense preparations.
2. Flood the smear with crystal violet and set aside for 1 min.
3. Lightly wash off excess crystal violet using cold tap water.
4. Flood smear with iodine and allow the iodine to remain on slide for 1 min.
5. Lightly wash off excess iodine using cold tap water.
6. Carefully flood smear with the decolorizing solution until the solvent runs colorlessly from the slide (30 to 60 s).
7. Gently wash the slide with cold tap water.

8. Counter stain with safranin for 30 to 60 s.
9. Wash off excess stain with cold tap water.
10. Carefully blot off excess water and allow to dry.

15.4.7 Ketogenesis

Ketogenesis refers to the ability of some acetic acid bacteria to oxidize glycerol to dihydroxyacetone. This property can also be used to separate two important species of *Acetobacter* found in wines, *A. aceti* and *A. pasteurianus* (Swing, 1992), because *A. aceti* can oxidize glycerol but *A. pasteurianus* cannot.

Because dihydroxyacetone is a reducing sugar alcohol, its presence can be detected using any of several methods for reducing sugars including the widely used Clinitest® (Ough and Amerine, 1988; Zoecklein et al., 1995). Using Clinitest®, formation of a "sharp blue" color suggests *A. pasteurianus* (glycerol was not oxidized) whereas a "light olive-green" indicates *A. aceti* (glycerol was oxidized).

1. Mix and dissolve the following ingredients.
Yeast extract	0.5 g
Glycerol	2 mL
Distilled water	100 mL
2. Transfer 10 mL into 18 × 150 mm test tubes, cap, and sterilize by autoclaving at 121°C/250°F for 15 min.
3. Inoculate each tube with the bacterial isolate (0.1 mL of an actively growing culture) and incubate.
4. Once growth is observed, evaluate glycerol oxidation by detecting the presence of dihydroxyacetone.

15.4.8 Lactate from Glucose

Although the empirical formulae are the same, there are two forms of lactic acid, which differ in their ability to rotate plane-polarized light. The isomers that rotate light to the right are D-lactic acid, whereas L-lactic acid rotates light to the left. Although lactic acid bacteria produce L-lactic acid from L-malic acid, the forms of lactic acid synthesized from carbohydrates are species specific. Here, some species produce only D-lactic acid from glucose, some only the L-form, and others an equal molar mixture of D- and L-lactic acid (indicated as DL). For examples, *Oenococcus* produce only D-lactic acid from glucose (Garvie, 1986a), *Pediococcus* synthesizes either L- or DL-lactate (Garvie, 1986b) and *Lactobacillus* produces D-, L-, or DL-lactic acid (Kandler and Weiss, 1986) depending on species.

The assay described is that of Pilone and Kunkee (1972) and Edwards et al. (1991). The form of lactic acid (D or L) can be quantified by using any commercial assay kits including those available from Boehringer Mannheim GMBH.

1. Mix 0.5 g liver extract with 50 mL distilled water for at least 30 min. Filter the suspension through Whatman no. 1 filter paper.
2. Mix and dissolve the following ingredients.

Tryptone	10 g
Peptone	2.5 g
Yeast extract	2.5 g
Tween 80 (5% w/w solution)	0.5 mL
Distilled water	450 mL

3. Add the liver extract to the medium and adjust pH to 5.5 with 50% v/v H_3PO_4 or 6 M KOH.
4. Divide the medium into two 250 mL portions and autoclave at 121°C/250°F for 15 min.
5. To one portion (250 mL) of the autoclaved medium, add 1 mL of a filter-sterilized (0.45 μm membrane) 5% w/v glucose solution.
6. Distribute the media into sterile 18 × 150 mm tubes at 10 mL per tube.
7. The controls and treatments are as follows.

   Control 1: Medium (tube A)
   Control 2: Medium + glucose (tube B)
   Control 3: Medium + bacteria (tube C)
   Treatment: Medium + bacteria + glucose (tube D)

8. Prepare a heavy inoculum of lactic acid bacteria as follows.
   a. Inoculate 10 mL apple juice Rogosa broth (Section 13.6.1) and incubate for 7 days at 25°C/77°F.
   b. Centrifuge the cultures at 3000 × g for 30 min. Aseptically decant the broth and suspend the pellet in 10 mL phosphate buffer (2.71 g NaH_2PO_4 and 8.12 g $Na_2HPO_4 \bullet 7H_2O$ dissolved in 1000 mL).
   c. Re-centrifuge, decant the phosphate buffer, and resuspend the pellet in 0.1% w/v peptone.
   d. Inoculate 1 mL of culture into each test tube prepared in step 6. Be sure to add 1 mL of 0.1% w/v peptone to all controls without bacteria.
9. Incubate for >2 weeks prior to determining the forms of lactic acid produced from glucose.

10. Calculation: The isomers of lactic acid are quantified using commercially available assay kits that rely on enzymes specific for either D- or L-lactic acid and a spectrophotometer.

> Concentration of D- or L-lactic acid =
> [tube D − (tube A + tube B)] − [tube C − (tube A + tube B)]

Tubes A and B represent concentrations of D- and/or L-lactic acid present in the uninoculated media without and with glucose. Any lactic acid in these tubes must be subtracted from Tubes C and D to calculate the amount of lactic acid produced (refer to step 7). High concentrations of either lactic acid in Tubes A or B indicates contamination or another problem with the assay.

15.4.9 Malate Utilization (Monitoring MLF)

As previously discussed (Section 2.4.3), lactic acid bacteria will metabolize L-malic acid (not D-malic acid) forming L-lactic acid, a conversion that can be detected by paper chromatography. Using this system, organic acids in the wine partition themselves in a chromatographic system according to their relative affinities for the mobile (solvent) and stationary (paper) phases. Usually, the system is operated in an ascending manner (solvent wicks upward into the paper); however, descending chromatography can also be performed using a glass tray suspended above the bottom of the chromatography chamber.

The procedure of Kunkee (1968) is simple, relatively inexpensive to set up, and has an acid indicator present in the solvent. The indicator, bromcresol green, undergoes a color change from yellow to blue in the pH range 3.8 to 5.4. Thus, the presence of an acid will be seen as a yellow spot on a blue background.

Acids are tentatively identified by the distance traveled from the origin (Fig. 15.3). From the baseline to the solvent front, the following order of acids is expected: tartaric, citric (if present), malic, lactic, and succinic. Alternatively, relative front (R_f) values for each spot can be calculated by dividing the distance traveled by the acid from the origin by the distance between the origin and solvent front. R_f values of organic acids separated using solvent 1 are presented in Table 15.3. However, lactic and succinic acids can co-chromatograph, leading to misinterpretations regarding the onset of MLF.

Kunkee (1968) noted that old solvent may cause excessive trailing or smearing of the acid spots due to moisture accumulation. Because trailing makes interpretation of the chromatogram difficult, the author recommended preparing fresh solvent weekly. When not in use, the solvent should be stored in sealed amber bottles away from exposure to direct sunlight. Furthermore, gloves should be worn at all times when handling the chromatography paper because lactic acid is present on human hands.

The detection level for malic acid using this method is approximately 100 mg/L. Thus, the absence of a malic acid spot should not be taken to mean that MLF is fully complete (Section 11.3.4). Conversely, presence of a lactic acid spot may not confirm an ongoing MLF because these bacteria produce the acid from sugars (Section 2.4.1). To confirm MLF completion, it may be necessary to perform additional malic acid analyses using

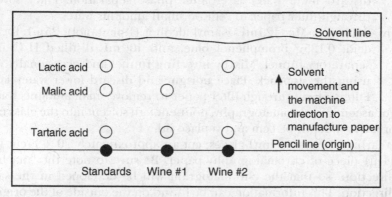

Figure 15.3. Paper chromatogram illustrating the separation of organic acids in the standard acid mixture and in wines no. 1 and no. 2. On this chromatogram, wine no. 2 is undergoing malolactic fermentation as indicated by the disappearance of the malic acid spot.

Table 15.3. R_f values for wine acids separated using paper chromatography and solvent 1.

Organic acid	R_f values
Tartaric	0.28–0.30
Citric	0.42–0.45
Malic	0.51–0.56
Lactic	0.69–0.78
Succinic	0.69–0.78

Adapted from Kunkee (1968) with the kind permission of *Wines and Vines*.

enzymatic or high-performance liquid chromatography techniques (Ough and Amerine, 1988; Zoecklein et al., 1995). Alternatively, Cartesio and Campos (1988) described an improved paper chromatographic method that can detect malic acid at 30 mg/L.

1. Prepare either solvent 1 or 2. Preparation should be done in a fume hood and care taken to avoid breathing the solvents. Chemical-resistant gloves and goggles should also be worn when handling the solvents.
   (a) Solvent 1: Combine 100 mL *n*-butanol, 100 mL distilled water, 10.7 mL stock formic acid, and 15 mL bromcresol green solution (0.1 g water-soluble sodium salt of bromcresol green dissolved in 100 mL distilled water) in a separatory funnel. Mix by inverting funnel and occasionally vent using the stopcock. Place upright and discard lower water layer after phase separation. Filter solvent through filter paper to remove small amounts water.
   (b) Solvent 2: Mix 125 mL *n*-amyl alcohol (1-pentanol), 25 mL formic acid, 0.125 g bromphenol blue, and 100 mL distilled H_2O in a separatory funnel. Mix by inverting funnel and occasionally vent using the stopcock. Place upright and discard lower water layer. Filter solvent through filter paper to remove small amounts water.
2. For ascending chromatography, pour enough solvent into the glass tank or jar to a depth of 1 cm and replace lid.
3. Wearing disposable vinyl gloves, cut an approximately 20 × 35 cm (8 × 14 in) piece of chromatography paper. Be sure to note the "machine direction" so that the chromatogram will be developed in the same direction. This information can be found on the outside of the original box containing the paper. Alternatively, precut paper can be purchased from analytical laboratory and supply houses.
4. Draw a faint pencil line across the long direction, approximately 2 to 2.5 cm above the bottom edge. Organic acid standards and wines will be applied (spotted) along this line (origin).
5. Apply 10 μL of a standard organic acid solution (1% w/v mixture of malic acid, lactic acid, and tartaric acid) on the same spot using a micropipette. Try to avoid spotting all of the 10 μL at once because spots should be as small in diameter as possible (<1 cm). Use of a hair dryer can help quickly dry the paper.
6. Spot 10 μL of each wine in the same way as the standard solution, using a new capillary tube each time. Each spot should be approximately 2 to 2.5 cm from other spots.
7. Allow the spots to fully dry before placing the paper into the tank or jar. For running chromatograms in a jar, staple the ends of the paper together width-wise to form a paper tube and place in the jar. Be sure

that the origin is not immersed in the solvent because this will ruin chromatogram and contaminate the solvent.

8. Develop the chromatogram for 6 to 12 h until the solvent line moves close to the top of the paper.

9. Once developed (e.g., solvent line is near but not at the top of the paper), remove paper from tank and hang to dry under fume hood. Chemical-resistant gloves should be worn to prevent contaminating the paper with hands.

15.4.10 Mannitol from Fructose

Mannitol is produced by some lactic acid bacteria (Section 2.4.4) as per the following reaction:

$$Fructose + NADH + H^+ \rightarrow Mannitol + NAD^+ \text{ (mannitol dehydrogenase)}$$

The formation of mannitol is detected by growing the bacterium in the fructose-enriched medium described by Pilone et al. (1991). Formation of mannitol salt crystals in the dried medium are detected visually as "large rosettes" without further magnification. It is essential to allow a few days for crystal formation at 25°C/77°F after evaporating the water at 37°C/99°F because crystals are often not seen prior to the final drying time.

1. Prepare heterofermentation-arginine broth (Section 13.6.3).
2. Place 9 mL of the HFA broth into 18 × 150 mm test tubes and autoclave at 121°C/250°F for 15 min.
3. Inoculate tubes with 0.1 mL of an actively growing culture from apple juice Rogosa broth.
4. Incubate at 22°C/72°F to 30°C/86°F for up to 3 weeks.
5. After incubation, pour approximately 8 mL of the broth culture into a 90 mm plastic Petri dish.
6. Allow to dry undisturbed at 37°C/99°F for 2 to 3 days and then an additional 2 to 3 days at 25°C/77°F before examination.

15.4.11 Oxidation of Ethanol

Acetobacter and *Gluconobacter* can be separated based on their relative abilities to oxidize ethanol. Over-oxidizers (*Acetobacter*) convert ethanol to acetic acid and then to CO_2 and H_2O. Conversely, acetic acid is an end product for *Gluconobacter*. Two methods commonly used to differentiate these bacteria on this basis involve either the Carr medium or Frateur's medium. Calcium carbonate–ethanol medium (CCE), described below, is closely related in composition to Frateur's medium.

Carr medium (Swings, 1992) includes a pH indicator to facilitate identification of acid-producing bacteria. Bromcresol green is an acid indicator that, when incorporated into the medium, detects presence of acid (yellow) and neutral/alkaline conditions (blue-green). As both *Gluconobacter* and *Acetobacter* will oxidize ethanol to acetic acid, both microorganisms bring about a color change in the medium from blue-green to yellow. However, *Gluconobacter* cannot carry the oxidation from acetic acid to CO_2 and H_2O. In these cases, the color of the medium remains yellow upon additional incubation. Because *Acetobacter* can metabolize acetic acid to CO_2 and H_2O, the color of the medium will slowly return to blue-green from an initial change to yellow with additional incubation.

Using Frateur's or CCE media, acid produced by bacteria neutralizes the insoluble $CaCO_3$ resulting in a "clear" zone or "halo" around the colony. These media differ only in the concentrations of yeast extract (1% w/v for Frateur's and 0.5% w/v for calcium carbonate–ethanol) and ethanol (2% v/v for Frateur's and 3% v/v for CCE).

Both *Gluconobacter* and *Acetobacter* will oxidize ethanol to acetic acid, thereby bringing about a clearing in the media (more transparent) due to neutralization of the $CaCO_3$ by acid production. Redevelopment of an opaque background in previously "cleared" halos is indicative of over-oxidation due to acid oxidation (*Acetobacter*), whereas failure of the medium to re-cloud suggests *Gluconobacter*.

Detection of over-oxidation of ethanol may require an incubation of 3 or more weeks. Under extended incubations, desiccation (loss of water) from media must be limited by maintaining a high humidity around the plates.

15.4.11.1 Carr Medium

1. Mix and dissolve the following ingredients.

Yeast extract	30 g
Bromcresol green	0.022 g
Agar	20 g
Distilled water	800 mL

2. Once dissolved, dilute to volume (1000 mL) with distilled water. Adjust pH to 5.5 with 50% v/v H_3PO_4 or 6M KOH prior to autoclaving at 121°C/250°F for 15 min.
3. Place the container with the medium into a 50°C/122°F water bath. Once the temperature equilibrates, aseptically add "high proof" ethanol or neutral spirits fruit grape (NSFG) to the medium to yield a final concentration of 2% v/v ethanol.

4. Mix the medium thoroughly and pour 25 mL medium per Petri plate.
5. When agar is solidified, streak bacterial culture and incubate until colony development is observed.

15.4.11.2 Calcium Carbonate–Ethanol Medium

1. Mix and dissolve the following ingredients.
Yeast extract	5 g
Calcium carbonate ($CaCO_3$)	20 g
Distilled water	800 mL
Agar	20 g
2. Once dissolved, dilute to volume (1000 mL) with distilled water.
3. After autoclaving at 121°C/250°F for 15 min, place the medium into a 45°C/113°F to 50°C/122°F water bath. Once temperature equilibrates, aseptically add "high proof" ethanol or neutral spirits fruit grape (NSFG) to the medium to yield a final concentration of 3% v/v.
4. Mix the medium thoroughly and pour 25 mL medium per Petri plate. Because $CaCO_3$ tends to settle out of suspension, frequent agitation is required when pouring a large amount of medium.
5. When agar is solidified, streak bacterial culture and incubate until colony development is observed.

15.4.12 Oxidation of Lactate

Another distinctive property of *Acetobacter* is the ability to oxidize D- and L-lactate to CO_2 and H_2O. This test utilizes a medium containing calcium lactate as both a carbon source and indicator. Growth of *Acetobacter* on lactate and subsequent over-oxidation yielding CO_2 and H_2O result in formation of a calcium carbonate precipitate. *Gluconobacter* grows poorly on this medium and does not oxidize enough of the lactate to raise the pH to the point at which $CaCO_3$ forms.

1. Mix and dissolve the following ingredients.
Yeast extract	20 g
Calcium lactate	20 g
Distilled water	800 mL
Agar	15 g
2. Once dissolved, dilute to volume (1000 mL) with distilled water. Adjust to pH 4.5 using 50% v/v H_3PO_4 or 6 M KOH.
3. After autoclaving at 121°C/250°F for 15 min, place the container with the medium into a 45°C/113°F to 50°C/122°F water bath. Once the temperature equilibrates, aseptically add "high proof" ethanol or

neutral spirits fruit grape (NSFG) to the medium to yield a final concentration of 3% v/v.
4. Mix thoroughly and prepare pour plates (25 mL medium per plate).
5. When agar is solidified, streak bacterial culture and incubate until colony development is observed.

CHAPTER 16

OTHER TECHNOLOGIES FOR IDENTIFICATION AND ENUMERATION

16.1 INTRODUCTION

The phenotypic techniques previously described have traditionally served as the basis for identification and enumeration of bacteria and yeasts. Such methods have included determining the fundamental biochemical properties of the viable microorganism including utilization of different forms of carbon and nitrogen, synthesis of specific metabolites, or activity of certain biochemical pathways. However, most of these methods require significant amounts of time for culturing microorganisms, and therefore results can be delayed for periods of days to several weeks. Because of these delays, the usefulness of the information to the winemaker can pass or be of limited value. Clearly, early detection of spoilage microorganisms can minimize the potential for proliferation of undesirable strains leading to improved wine quality.

During the past 20 years, research in molecular biology and genetics has led to the development of techniques capable of detecting and characterizing microorganisms at low population density and in a fraction of the time required for classic isolation and identification (Loureiro and

Querol, 1999; Sancho et al., 2000). Whereas identification based on phenotypic expression requires that the organism be viable, identification based on genetic (molecular) methods do not have this requirement.

Several rapid procedures are described in the following section. Each have their relative merits and deficiencies in terms of interpreting results and all require specialized equipment and skilled personnel. However, developing methods have the potential for being modified and packaged into field kits that can be used outside the laboratory.

16.2 PHENOTYPICAL IDENTIFICATION

Phenotype identifies those physical and physiological characteristics that define the species as being distinct from related microorganisms. Besides the classical methods described in Chapter 15, several additional methods are available to identify microorganisms isolated from grapes or wines.

16.2.1 Biolog System

A product of Biolog Inc. (Hayward, CA), this system characterizes phenotypes based on the oxidation or assimilation of carbohydrates, nitrogen sources, vitamins, and other substrates (Praphailong et al., 1997). After suitable inoculation, growth is evaluated for each well containing a specific substrate, either manually or by a plate-reader, and compared with database information. Although Praphailong et al. (1997) used the system to identify *Saccharomyces cerevisiae*, *Debaryomyces hansenii*, *Kloeckera apiculata*, *Dekkera bruxellensis*, and *Schizosaccharomyces pombe*, these authors reported correct identification of *Zygosaccharomyces bailii*, *Z. rouxii*, and *Pichia membranifaciens* less than half of the time.

16.2.2 Fatty Acid Methyl Ester Analysis

Characterization of cell wall and membrane fatty acid profiles (FAME) has also been used as an identification tool (Tredoux et al., 1987; Malfeito-Ferreira et al., 1989; 1997; Decallonne et al., 1991; Augustyn et al., 1992; Botha and Kock, 1993). Here, fatty acids are extracted and derivatized as methyl esters, compounds that are then separated and quantified by gas-liquid chromatography and results compared with databases (Suzuki, 1993). FAME methods have been successfully applied for fatty acids present in yeasts (Bendova et al., 1991; Rozes et al., 1992; Sancho et al., 2000).

As conditions for growth may affect fatty acid synthesis, the composition of the growth medium as well as incubation time and temperature are normally controlled (Loureiro and Querol, 1999). Identification of unknown isolates requires comparing results (i.e., types and quantities of fatty acids) to those reported in databases. Thus, formation of atypical fatty acids due to changes in growth conditions limits applicability of these methods. In contrast, Loureiro (2000) reported that culture medium was not a significant concern in order to yield reproducible information.

16.2.3 Protein Characterization

Proteins are the unique products of gene expression, either produced on a continuing basis (constitutive proteins) reflecting the ongoing needs or synthesized in response to specific environmental stimuli (inducible proteins). Hence, proteins can serve as "fingerprints" of the nucleic acid segment (gene) that encoded them and, by extrapolation, the sequence of nucleic acids.

Once extracted from the cell, protein separation is performed using one- or two-dimensional polyacrylamide gel electrophoresis (PAGE). Here, the protein extracts are placed onto a semisolid support (gel) and separated based on their relative migration within a "lane" induced by an electrical current. Once separated, the various bands representing different proteins can be stained for visualization. The presence or absence of different proteins can then be used to identify similarities and differences between isolates. These methods have been used in the identification of lactic acid bacteria isolated from wines (Couto and Hogg, 1994; Patarata et al., 1994).

In addition to gel electrophoresis, proteins can also be characterized using matrix-assisted laser desorption ionization time-of-flight mass spectrometry (MALDI-TOF-MS). Developed by Karas et al. (1987), this system uses a laser to irradiate intact cells, which ionizes surface cell proteins that are then characterized. The patterns of proteins yield a profile unique to the microorganism and are therefore used for identification purposes. MALDI-TOF-MS offers the advantage of rapid results with minimal sample preparation and reagents. However, the cost of equipment makes these methods impractical for wineries.

16.2.4 Electrophoretic Characterization of Isozymes (Zymograms)

Isozymes are enzymes that catalyze the same biochemical reactions but differ based on molecular weight. For example, the electrophoretic mobility of lactate dehydrogenases (LDH), enzymes responsible for the transformation of pyruvic acid to lactic acid (Section 2.4.1), has been used to differentiate specific species of *Lactobacillus* (Hensel et al., 1977). For example, two species of *Lactobacillus* found in wines, *L. buchneri* and *L. brevis*, possess very similar physiological characteristics, the major difference being the former can ferment melezitose whereas the latter cannot (Kandler and Weiss, 1986). However, Kandler and Weiss (1986) reported that some strains identified as *L. brevis* by electrophoretic mobility of LDH could ferment melizitose, a finding in agreement with Dicks and Van Vuuren (1988) who studied another strain of *L. brevis*. As a consequence, Kandler and Weiss (1986) recommended that determination of the relative electrophoretic mobilities of LDH enzymes was a more reliable means of differentiation between these two species. The method has also been used for rapid characterization of species and strains of *Saccharomyces* (Duarte et al., 1999).

Compared with other molecular methods, isozyme profiling can be relatively inexpensive to perform (Loureiro and Querol, 1999). Isozyme detection should be performed at various intervals during growth because the proteins being detected may or may not be expressed by the microorganism at similar times.

16.3 IMMUNOCHEMICAL TECHNIQUES

In response to exposure to an antigen, microorganisms and other living entities produce antibodies. Immunological techniques rely on the interaction between an antigen and an antibody specific for that antigen. The antigen may be a microbial cell wall or capsule, a characteristic cellular component, or even a specific metabolite.

16.3.1 Enzyme-Linked Immunosorbent Assay

Among these methodologies, enzyme-linked immunosorbent assay (ELISA) has been used for a number of years. ELISA methods were first developed for use in the medical and veterinary fields in the early 1970s to overcome the often lengthy incubation periods associated with culturing and identifying pathogens (Crowther, 1995). Modern monoclonal antibody systems offer higher degrees of specificity (Dewey et al., 2000), and thus ELISA methods have been extended to other areas where it is necessary to detect, identify, and quantify microorganisms present at low numbers. Because these methods detect only the presence of an antigen, results do not distinguish between live and dead cells.

ELISA utilizes complex ("sandwich-type") and highly specific interactions between immobilized antibody and antigen (microorganism) present in the sample. Addition of a secondary antibody–chromophore complex produces color when the initial coupling is complete. Alternatively, a specific substrate is added such that the attached enzyme produces color (Fig. 16.1). Color development from the modified chromophore is proportional to the concentration (titer) of the primary antigen. Although color formation can be measured visually or by with a colorimeter, automated readers are also available.

Various derivations of these methods that have been evaluated for use in the wine industry include identification/enumeration of the spoilage yeast *Brettanomyces* (Kuniyuki et al., 1984) as well as *Botrytis*-detection on incoming grapes (Ravji et al., 1988; Marois et al., 1993; Dewey et al., 2005). Recent research has reduced the time for immunochromatographic assay to merely minutes (Dewey et al., 2005). The short time required for results as well as reasonable sensitivity may allow application of these methods for analysis in the field and at winery testing facilities. Working with *Brettanomyces*, Kuniyuki et al. (1984) reported a detection limit of 34 cells/mL with virtually no cross-reactivity among the other wine microorganisms studied. Furthermore, the time required for results was only 24h, compared with a week or more for cultures to grow on solidified agar (Section 13.5).

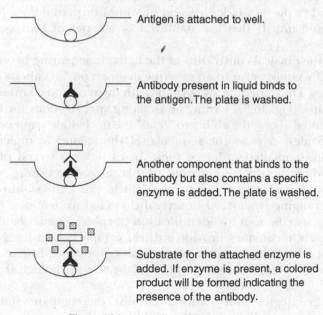

Figure 16.1. A typical ELISA protocol.

16.3.2 Immunochemical Fluorescence Microscopy

As with ELISA, antibodies react with specific antigens within the cell wall or membrane, a reaction linked to a secondary antigen/specific antibody fluorochrome that will then fluoresce. Application of these methods requires the ability to develop specific antibodies to a unique antigenic property of the organism of interest. Additionally, the highly specific nature of the antigen–antibody reaction may preclude fluorescence of closely related strains of the same species against which the antibody was developed (Atlas and Bartha, 1981). Potential difficulties are similar to those noted for general fluorescence microscopy (Section 12.2.4).

16.4 PHYLOGENETIC ANALYSES

Numerous approaches have evolved to characterize microorganisms based upon fundamental similarities or differences of genetic material, either deoxyribonucleic acid (DNA) or ribonucleic acid (RNA). Briefly, DNA consists of two complementary strands (or sequences) that associate with each other to resemble a "twisted ladder." The sides consist of alternating sugars (ribose) and phosphate groups (Fig. 16.2). The bonds between these are covalent, thus conferring a high degree of stability to the molecule. The "rungs" of the ladder consist of nitrogen-containing compounds that are either purines (adenine and guanine) or pyrimidines (cytosine, thymine, and uracil) that are stabilized by a series of hydrogen bonds (Fig. 16.3).

Due to the physical constraints of the helix, base-pairing between each strand of DNA must occur in an exacting manner; purines always pair with pyrimidines. Thus, adenine always pairs with thymine and guanine couples with cytosine (Fig. 16.3). Given the exacting specifications for base-pair ordering noted above, the ability to "read" the nucleotide sequence on one strand provides, *de facto*, the sequence of the second (complementary) strand. The sequential order of these bases in each strand of DNA is unique and creates the "fingerprint" of that microorganism.

Identification of a unique sequence of nucleotide bases within the DNA molecule, ranging from <100 to several thousand nucleotides, can serve as a highly specific tool for identification purposes (Nadal et al., 1996). Further, these techniques provide a direct comparison at the gene level and do not rely on the detection of secondary products (proteins, metabolites) whose expressions may be affected by the physiological status of the cells.

Several strategies have evolved to directly compare similarities/ differences between isolates by examination of their respective genomes

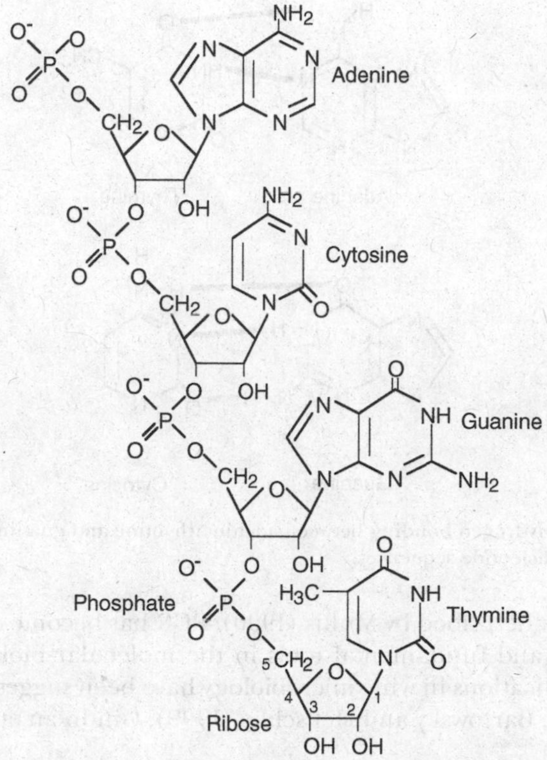

Figure 16.2. Representative polynucleotide illustrating covalent bonds between phosphate and ribose units of individual nucleotides (adenine, cytosine, guanine, and thymine). Removal of the hydroxyl group from carbon 2 on ribose will yield deoxyribose.

(Loureiro, 2000; Deak, 2002; Capece et al., 2003). Each requires initial digestion of the harvested DNA using restriction endonucleases, enzymes that cleave DNA at specific nucleotide sequences unique for that enzyme. The DNA digest is the amplified at specific or randomly selected regions by polymerase chain reaction (PCR). Fragments are subsequently separated electrophoretically and their patterns compared against those of other isolates or databases.

16.4.1 Polymerase Chain Reaction

As the name suggests, polymerase chain reaction (PCR) is a method of amplifying selected segments of target nucleic acid sequences (typically 150–3000 base pairs in length) by several orders of magnitude prior to separation and characterization. Since the pioneering research and

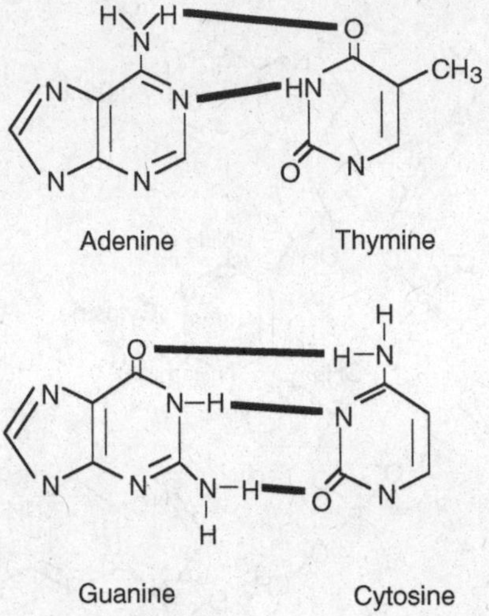

Adenine   Thymine

Guanine   Cytosine

Figure 16.3.   Hydrogen bonding between adenine:thymine and guanine:cytosine on adjacent polynucleotide sequences.

development described by Mullis (1990), PCR has become one of the most widely used and fundamental tools in the molecular biology laboratory. Various applications in wine microbiology have been suggested by Lavallée et al. (1994), Bartowsky and Henschke (1999), Gindreau et al. (2001), and Deak (2002).

PCR involves three steps: (a) denaturation of the helix, (b) binding of primers, and (c) replication of the complementary strains using polymerase (Fig. 16.4). Thermal denaturation of native target DNA is performed by heating the solution to 90°C to 96°C. Primers must be specific for the region of interest on the DNA or else other sections of the molecule will be replicated. The polymerases then start replicating the complementary strands from the primers. However, because the process requires alternating hot and cool reaction cycles that would rapidly denature most enzymes, a polymerase that is stable at elevated temperatures is required. In this regard, *Taq*-DNA polymerase is the enzyme of choice. Assigned an acronym from a theromophilic bacterium (*Thermus aquaticu*), activity of this enzyme is not inhibited at the elevated temperatures of the denaturation phase. The product of this reaction is two new helices, each composed of one of the original strands plus the newly prepared complementary strand.

Because each newly synthesized portion can itself be replicated, target DNA doubles with each replication cycle. Upon completion, the amplified nucleic acid is separated using agarose gel or polyacrylamide gel electro-

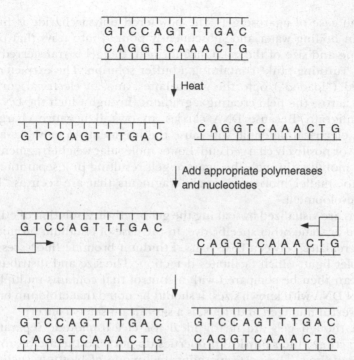

Figure 16.4.   Amplification cycle of a target nucleotides sequence using polymerase chain reaction (PCR).

phoresis (PAGE) and patterns of separation are then compared with known strains or species.

16.4.2 Gel Electrophoresis

As with proteins, gel electrophoresis is used to separate amplified DNA or RNA based on size of the fragments. After partial digestion with restriction endonuclease, amplified fragments are applied onto an agarose gel for electrophoretic separation. In the presence of an electric field, larger fragments move through a gel slower than smaller ones. If a sample contains fragments of several different sizes, these will partition into groups that are visualized as "bands" after staining. Separation then depends on molecular weight, relative affinity for the support matrix, and charge interaction between the fractions and the cathode and anode.

Separation gels are usually made from agar, a polysaccharide extract of seaweed, or other synthetic polymers such as polyacrylamide. The latter find application for high-resolution separations of DNA in the range of tens to hundreds of nucleotides in length.

In the case of agarose gels, the powdered polysaccharide is first dissolved in boiling water and subsequently poured into trays that provide the shape and size of the gel. Once formed, the gel is transferred into a second "running tank" containing a buffer solution. The extracted DNA is placed ("loaded") onto the gel in lanes, and an electrical current is applied across the field creating a gradient through which the DNA fragments migrate. Because DNA carries an overall negative charge, the applied electrical potential will cause the fragments to move toward the cathode or positively charged end. Lower molecular weight fragments have greater mobility through the agarose gel, resulting in a separation from larger to smaller molecular weight fragments that are seen as "bands" upon development.

Bands are visualized by soaking the gel briefly in a solution of ethidium bromide or some other specific dye. In the case of ethidium bromide, the dye intercalates between base pairs. Ethidium bromide fluoresces under ultraviolet light, which facilitates detection. The size and distribution of bands can then be compared with a control that contains multiple fragments of DNA with known sizes. It should be noted that ethidium bromide is a powerful mutagen and, thus, is a significant safety concern.

Even though gels are nearly ideal for electrophoretic separation of DNA fractions, their physical properties do not lend themselves to subsequent handling. The post-separation technique of blotting, such as the Southern blot named after E. M. Southern who developed the method in the 1970s, can then be applied. Here, the partitioned DNA is transferred to a more resilient and easier manipulated nitrocellulose blotting membrane without disturbing the relative position of the banded fragments. Application of the appropriate probes then marks the target sequence. Cocolin et al. (2001) utilized denaturing gradient gel electrophoresis (DGGE) of PCR-amplified ribosomal RNA to characterize native yeast strains during alcoholic fermentation. The authors reported that unlike other similar methods, this modification provided a direct qualitative assessment of the population while eliminating reliance on culture media.

16.4.3 Quantitative Polymerase Chain Reaction

Quantitative PCR (Q-PCR or "real-time" PCR) detects specific DNA amplification products being formed by the polymerases. As with classic spectrophotometric methods, the signal arising from the bound fluorescently-labeled nucleotide probe increases in direct proportion to the amount of PCR product formed during reaction. Because the method is real-time, Q-PCR eliminates the need for post-PCR processing thus shortening the time frame and increasing throughput. In addition to its sensitivity, Q-PCR represents a rapid technique for yeast strain characterization (Lavallée et al., 1994) where 30 or more strains can routinely be identified in a single day.

16.4.4 Hybridization Probes

Knowing the complementary nature of the DNA base pairing, it is possible to develop a detection system based on the use of exogenously applied oligonucleotide sequences that bind with complementary sequence on the DNA sequence of interest. These sequences, known as "probes," are prepared as fragments of labeled DNA or RNA complementary to the area of interest on the chromosome. Thus, hybridization may be DNA–DNA, DNA–RNA or RNA–RNA. Once a compatible probe has been developed, the fidelity of probe–target interaction under controlled test conditions creates a highly specific diagnostic test system for comparison/identification of related strains or specific gene sequences characteristic of a particular microorganism. Several commercial diagnostic kits utilizing DNA hybridization probe technology are available for important food pathogens.

16.4.5 Fluorescence *In Situ* Hybridization

This technique, otherwise known as FISH, uses short sequences of fluorescently labeled nucleic acid–targeted probes to hybridize with the complementary sequence of interest on the chromosome. The reaction depends on localized unwinding of nucleotide chains from the helix, at elevated incubation temperatures, followed by hybridization with complementary probe and, subsequently, rewinding at lower temperatures. The method has the advantage of being carried out on a microscope slide with whole-cell preparations. When hybridization is complete, the fluorescent molecules on the probe can be visualized, and the observer can identify the location of the target DNA on the chromosome. Stender et al. (2001b) report on the use of FISH in identification of *Brettanomyces* in wine, and others used the method for identification of other spoilage yeasts (Kosse et al., 1997).

16.4.6 Ribonucleic Acid Analysis

The process of translating DNA sequences into protein is mediated by RNA. Here, dRNA genes are separated by non-transcribed spacers or intergenic regions. For example, two variable internal transcribed spacers known as ITS1 and ITS2 separate the conserved 18S and 26S genes from the 5.8S gene (Musters et al., 1990). Some regions of the genome are suitable for distinguishing between strains and species (Boekhout et al., 1994), and others are useful in distinguishing between genera (Montrocher et al., 1998).

The sequence of internal spacers can be used to design multiple oligonucleotide probes in order to detect specific microorganisms. Dlauchy et al. (1999) used restriction analysis and amplification of the 18S and ITS1 region of rRNA to identify 128 species of yeast in wine and beer. Egli and Henick-Kling (2001) found the ITS1 and ITS2 regions to be suitable for separation of *Brettanomyces/Dekkera* species and strains.

16.5 PROBE DETECTION SYSTEMS

Several proprietary fluorescence dye-based detection systems that characterize and quantify probe-bound nucleotide sequences have been developed and commercialized in recent years. For example, general and nonspecific DNA dyes can be used that bind with any double-stranded DNA (i.e., the probe–strand complex or otherwise) and are useful in gel electrophoresis. Much more sophisticated systems rely on oligonucleotide probes that incorporate fluorescent dyes that "illuminate" when a match between complementary strands are made. Included among the latter are TaqMan®, molecular beacons, and Scorpion® probes.

16.5.1 TaqMan® Probes

TaqMan® probes are fluorochrome-containing oligonucleotides that hybridize to an internal region of a PCR product (Fig. 16.5). These primers are synthesized with a short-wavelength fluorophore on one end and a long-wavelength fluorophore on the other that quenches the former. After hybridization, the probe is susceptible to degradation by endonucleases of a processing Taq polymerase. Upon removal from the primer, fluorescence of the short-wavelength fluorophore increases whereas the other decreases.

16.5.2 Molecular Beacons

Another type of probe that utilizes a signal system of fluorescent and quenching dyes is the "molecular beacon" (Fig. 16.6). Molecular beacons form a stem-loop structure when free in solution. The close proximity of the fluorochrome and quenching components prevents the probe

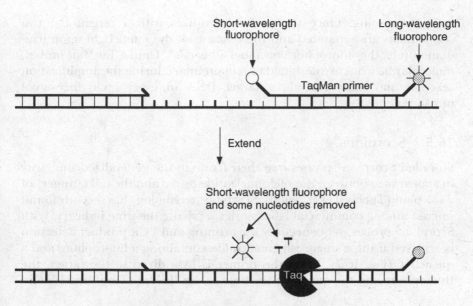

Figure 16.5. Schematic representation of TaqMan® probe reaction with target DNA sequence. Copyright © 2005 and adapted with the kind permission of Invitrogen Corporation.

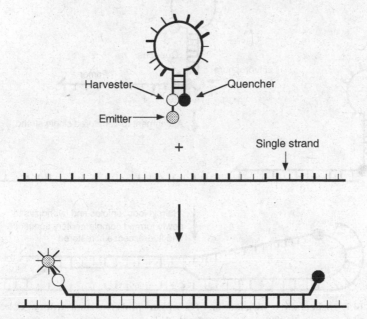

Figure 16.6. Schematic representation of molecular beacon probes reaction with target DNA sequence. Copyright © 2005 and adapted with the kind permission of Invitrogen Corporation.

from fluorescing. Once the probe hybridizes with a target, the two components are separated and the fluorescent dye emits light upon irradiation (i.e., the fluorophore becomes a beacon). Unlike TaqMan probes, molecular beacons are designed to remain intact during the amplification reaction and must rebind to target DNA in every cycle for signal measurement.

16.5.3 Scorpions®

So-called Scorpion® probes owe their name to the generalized similarity in appearance between the oligonucleotide probe and the tail (stinger) of a scorpion (Thelwell et al., 2000). Scorpion® technology has recently found interest among commercial laboratories servicing the wine industry. With Scorpion® probes, sequence-specific priming and PCR product detection is achieved using a single oligonucleotide containing a fluorophore and a quencher (Fig. 16.7). Once the primer is hybridized to the target, the fluorophore and the quencher separate as the "hairpin loop" unfolds and

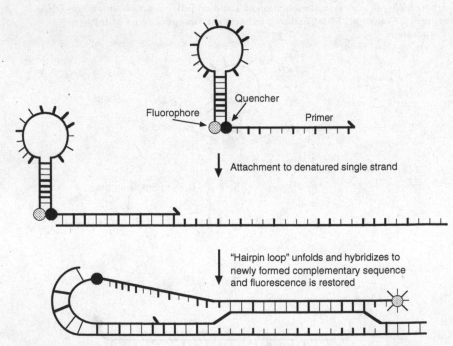

Figure 16.7.   Schematic representation of Scorpion® probe reaction with target DNA sequence. Copyright © 2005 and adapted with the kind permission of Invitrogen Corporation.

hybridizes to the newly formed complementary sequence. This separation leads to an increase in the fluorescence emitted by the fluorophore.

16.5.4 Peptide Nucleic Acid Chemiluminescent *In Situ* Hybridization Probes

Using peptide nucleic acid probes coupled with chemiluminescent hybridization (PNA-CISH), Stender et al. (2001b) and Connell et al. (2002) successfully identified specific target sequences on the 26S ribosomal RNA in microcolonies of *Dekkera bruxellensis.* Stender et al. (2001a; 2001c) reported that PNA probes mimic DNA probes but have greater affinity and specificity compared with the latter.

16.6 OTHER MOLECULAR METHODS

A variety of other PCR-based identification techniques have been developed during the past decade as reviewed by Olive and Bean (1999).

16.6.1 Restriction Fragment Length Polymorphisms

Within a population, slight differences in individual DNA makes one organism uniquely different from others. Referred to as sequence polymorphisms, these are often the result of single base-pair changes that occur in regions of DNA that do not encode a gene but that are recognized and bound by restriction enzymes.

Restriction fragment length polymorphism (RFLP) is a technique in which differences in the DNA between individuals in a population are identified by analysis of patterns derived from cleavage of their respective DNA. Thus, when DNA from two different individuals is cut with a single restriction enzyme, fragments of different lengths are produced, and the pattern of those fragments is unique for different members of a population. The similarities and differences in the patterns generated can then be used to differentiate species and even strains (Johansson et al., 1995; Cocolin et al., 2002).

Isolation of sufficient DNA for RFLP analysis is time-consuming and labor-intensive. Because PCR can be used to amplify very small amounts of DNA in usually in 2 to 3 h, this technique can be used with RFLP analysis (PCR-RFLP). Granchi et al. (1999) demonstrated the usefulness of PCR-RFLP in the detection, quantification, and identification of yeast species. The authors reported that the technique required 30 hours and produced results identical to those obtained by classic isolation, enumeration, and identification methods taking a week or more. More recently, Cocolin et al. (2002) and Baleiras-Couto et al. (2005) also used the technique to monitor yeast ecology during red wine fermentations.

16.6.2 Pulsed Field Gel Electrophoresis

Pulsed field gel electrophoresis (PFGE) utilizes restriction enzymes to digest microbial DNA, which is then subjected to electrophoretic separation (Arbeit et al., 1990; Finney, 1993; Kelly et al., 1993; Maslow et al., 1993). Fragments of the DNA after separation are then compared with known patterns of microorganisms. Although useful, the method separates 10–15 higher molecular weight (10–800 kb) fragments, which may not be sufficient to resolve differences between strains. In addition, the technique is not without health and safety concerns (Richmond and McKinney, 1993). Aside from the use of ethidium bromide to visualize and "develop" banding patterns, PFGE separation of large DNA segments is carried out at higher voltage (5–10 V/cm) than is used in conventional agarose separations of smaller fragments (1–5 V/cm) (Tenover et al., 1995).

16.6.3 Additional Methods

As described by Williams et al. (1990), random amplified polymorphic DNA–PCR (RAPD-PCR) is a variant of PCR that utilizes oligonucleotide probes (9 to 12 base pairs or bp) to amplify several regions of the genome. The amplification products are then separated electrophoretically. Resolution depends upon the primer sequence and reaction conditions. RAPD-PCR can be made more specific by use of highly specific oligonucleotide probes. Holt and Cote (1998) applied this technique toward the identification of dextran-producing *Oenococcus* strains, and Esteve-Zarzoso et al. (1998) were able to identify *Saccharomyces* and *Zygosaccharomyces* species. Quesada and Cenis (1995) used the method to characterize wine yeasts.

The microbial genome contains randomly interspersed repetitive DNA sequences that are strain specific and thus serve as a genotypic fingerprint (Versalovic et al., 1991; 1994; 1998). The value of the repetitive sequence-based polymerase chain reaction (REP-PCR) method is limited only to regions of compatibility (average 10 to 15 bp), and other polymorphic sites are missed.

In multilocus sequencing typing (MLST) methods, several bacterial "housekeeping" genes are compared on the basis of 450 bp internal fragments resulting in each isolate being assigned a unique seven-digit allele combination. Interpretation is based on the probability that identical allelic profiles will be detected by chance alone.

16.7 EXTRACHROMOSOMAL ELEMENTS (SATELLITES)

A wide range of living cells, including wine bacteria and yeast, are infected with extrachromosomal nucleic acid elements known as "variable number of tandem repeats" (VNTR) or "satellites." Their importance to the host cell varies (Cocaign-Bousquet et al., 1996). As these VNTR encode for antigenic variation in some pathogenic microorganisms, they are partially responsible for adaptive evolution (Moxon et al., 1994). In the case of yeast and other bacteria found in wines, their role is less clear. One proposed role is in formation of killer protein among strains of wine yeasts (Schmitt and Neuhausen, 1994). The presence of satellites has also been used to assess genetic diversity in populations (Tautz, 1989).

These intracellular sub-viral elements are present as nucleic acid fragments, either as single-stranded DNA or as single- or double-stranded RNA. As satellites do not contain sufficient nucleic acid to code for the proteins necessary to replicate, they rely upon another "helper virus" to carry out this function. If the second virus does not exist, the satellite is trapped within the host cell.

In yeast, satellite nucleic acid is relatively miniscule (Pupko and Graur, 1999) so these fragments are referred to as microsatellites. These exist as simple sequence tandem repeats of chromosome-associated DNA, ranging in size from 2 to 5 nucleotides (Young et al., 2000). Because these are chromosome-associated, replication is tied to the host.

Since the nucleic acid sequence is distinct from that of the host chromosome, microsatellites may serve as genetic markers and, thus, identification tools (Tautz, 1989). Baleiras-Couto et al. (1996) used PCR-enhanced microsatellite nucleic acid to identify *Zygosaccharomyces bailii* and *Z. bisporous* involved in a case of food spoilage. Gonzalez-Techera et al. (2001) and Perez et al. (2001) also reported differentiation and identification of wine yeasts polymorphic loci containing microsatellite markers.

Microsatellite loci exhibit significant variability and stability, resulting from a high rate of mutation, which may have important consequences in evolution (Sia et al., 1997; Young et al., 2000). Further, not all tandem repeats are polymorphic. As such, care must be taken in using these markers as the basis of relatedness.

As with other molecular methods, separation and identification of satellite protein requires a suitably equipped and staffed molecular biology laboratory. However, methods may eventually be reduced to kit form, where most labs will have the capability for rapid identification of wine microorganisms using their satellite information.

CHAPTER 17

CHEMICAL AND PHYSICAL INSTABILITIES

17.1 INTRODUCTION

On occasion, nonmicrobial sediment and/or hazes develop in bottled wines and juices that can be confused with spoilage microorganisms during microscopic evaluation (Chapter 12). The following protocols can be used to distinguish nonmicrobial and microbial sediments/hazes as well as identify possible cause.

Nonmicrobial instabilities can be classified as crystalline, "crystal-like," fibrous, or amorphous (Table 17.1). To determine the cause of the precipitation, preliminary sample evaluation and review of processing records may be helpful. If small amounts of sediment or haze are present, it may be necessary to first concentrate the material by either centrifugation or benchtop filtration. Alternatively, the bottle may be stood upright for several hours and the precipitate carefully collected for analysis.

17.2 CRYSTALLINE INSTABILITIES

17.2.1 Tartrates

The insoluble salts, potassium bitartrate (KHT) and calcium tartrate (CaT), may be found in bottled wine as crystalline precipitates of varying

Table 17.1. Typical nonmicrobial instabilities in juices and wines.

Visual appearance	Potential cause
Crystalline	Potassium bitartrate, calcium tartrate, or calcium oxalate
"Crystal-like"	Cork fragments or diatomaceous earth
Fibrous	Cellulose, case lint, or asbestos
Amorphous	Protein, phenolic, polysaccharide (glucan, pectin, or starch) or metallic casse

size (fine to course) and color (red to white depending on source). Although cosmetic in nature, many consumers regard their presence as a defect because tartrates can appear "glass-like." Details regarding formation of these precipitates as well as processing protocol for preventing or delaying their formation may be found in Zoecklein et al. (1995).

Crystals can be collected from a wine and then directly examined under either brightfield or phase-contrast microscopy using an oil immersion objective. KHT crystals shown were obtained from red wine (Fig. 17.1A), white wine (Fig. 17.1B), and model wine solution prepared by adding potassium chloride to an ethanolic tartaric acid solution (Fig. 17.1C).

Aside from microscopic examination, chemical diagnostic tests are available. The method described in Section 17.2.1.1 is a general method for detection of tartrates where crystals appear yellow-orange in color after addition of diluted H_2SO_4 and metavanadate (Quinsland, 1978). The method described in Section 17.2.1.2 differentiates between KHT and CaT based on relative distribution of its anions (bitartrate or HT^- and tartrate or T^{2-}) at different pH values. Here, crystal formation at pH 6.0 indicates the presence of CaT, whereas crystals formed at pH 3.6 suggest formation of KHT. The method described in Section 17.2.1.3 confirms the presence of calcium in a wine through observation of elongated, "fiber-like" crystals of calcium sulfate ($CaSO_4$).

17.2.1.1 Tartrates

1. Prepare the following reagents.
   a. Carefully add 1 volume of concentrated sulfuric acid to 3 volumes of distilled water to yield $1 + 3$ H_2SO_4.
   b. In a 100 mL volumetric flask, dissolve 3 g sodium metavanadate ($NaVO_3$) in approximately 70 mL warm, distilled water ($<70°C/$

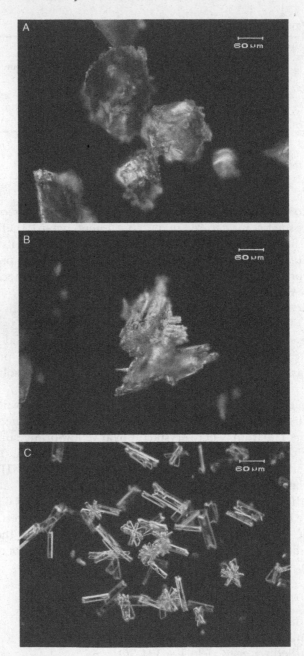

Figure 17.1.   Potassium bitartrate from a red wine (A), white wine (B), and laboratory-prepared (C) as viewed with phase-contrast microscopy at a magnification of 100×. Photograph provided through the courtesy of R. Thornton and E. Akaboshi.

158°F). Cool the solution and bring to volume. Because $NaVO_3$ does not fully dissolve, solutions must be filtered through Whatman no. 2 (or equivalent) filter paper prior to use.
2. Collect a portion of wine sample containing the suspect sediment either by filtration or centrifugation.
3. Rinse the sediment or crystals with a small volume of distilled water, transfer to a membrane filtration apparatus, and apply vacuum.
4. Transfer the membrane to a watch glass. Alternatively, crystals can be collected and placed into a well of a spot plate.
5. Place a drop of $1 + 3$ H_2SO_4 on the precipitate and then a drop of metavanadate solution before examining the color of the sediment.

17.2.1.2 Differentiation Between Potassium and Calcium Tartrates

1. Prepare solutions of 50% w/v sodium hydroxide and 12 M HCl.
2. Collect the precipitate by centrifugation and place into a small test tube.
3. Add enough acid or NaOH to adjust pH to 3.6 and 6.0 and chill.

17.2.1.3 Confirmation of Calcium Tartrate

1. To 10 mL of wine, add a small amount of solid oxalic acid (a "spatula tip" is plenty). Crystal formation may indicate the presence of calcium.
2. The presence of calcium is confirmed by adding several drops of 12 M H_2SO_4 to dissolve the precipitate. Add 3 to 5 mL methanol and heat gently. If calcium is present, a precipitate of calcium sulfate ($CaSO_4$) should appear.

17.3 "CRYSTAL-LIKE" INSTABILITIES

17.3.1 Cork Dust

The recommended procedure for examining samples for cork dust is to use a stain that reacts with lignin, a structural macromolecule in cork. When it is necessary to evaluate color development, the microscope should be operated in brightfield rather than phase-contrast mode.

Microscopically, stained cork debris appears as red, crystal-like aggregations of cells (Fig. 17.2). Although case lint also stains red using this technique, the latter is fibrous in appearance microscopically (Quinsland, 1978).

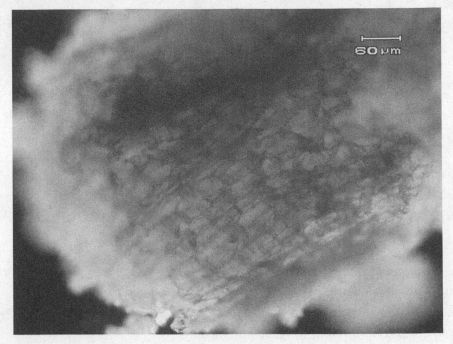

Figure 17.2.   Cork debris as viewed as viewed with phase-contrast microscopy at a magnification of 100×. Photograph provided through the courtesy of R. Thornton and E. Akaboshi.

1. Prepare the stain by mixing 2 g of phloroglucinol (synonym: phloroglucan stain) in 100 mL 12 M HCl. This is a near-saturated mixture, and the supernatant must be decanted from any crystals that do not dissolve. This solution should be made fresh before use.
2. Collect a portion of wine containing the debris by membrane filtration.
3. Wet the filter and sediment with phloroglucinol stain, holding stain in contact with sediment for at least 5 min.
4. Apply vacuum to remove stain and rinse with distilled water. Examine the sediment under a microscope.

17.3.2 Diatomaceous Earth

Diatomaceous earth (DE) is the processed fossilized remains of marine and freshwater algae called diatoms. Commonly used in the winery as a

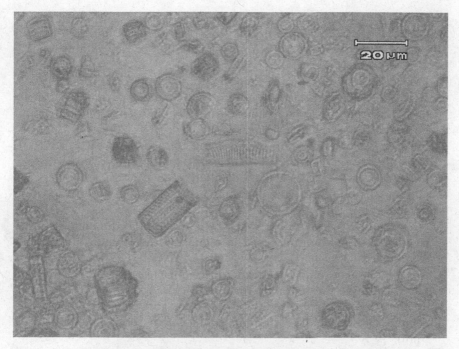

Figure 17.3. Diatomaceous earth as viewed as viewed with phase-contrast microscopy at a magnification of 400×. Photograph provided through the courtesy of R. Thornton and E. Akaboshi.

filtration aid, DE can occasionally "bleed-through" the filtration matrix and be found in the wine. DE can be readily identified by microscopic examination.

1. Collect particulate material by either filtration or centrifugation.
2. Transfer to microscope slide and visually compare with Fig. 17.3.

17.4 FIBROUS INSTABILITIES

Fibrous materials found in bottled wines are usually cellulose, originating from the filter pad matrix or from case lint that may find itself in bottles prior to filling. Case lint containing lignin will appear microscopically as red fibrous material. However, all cellulose material appears light-blue under the zinc/iodine stain. For the zinc/iodine stain (similar to the

Herzberg stain), best results are obtained when the preparation is examined fresh because the color intensity diminishes after 30 min (Quinsland, 1978). Asbestos fibers are not stained using either of these stains.

1. Prepare the stain by mixing 2 g of phloroglucinol in 100 mL of 10% v/v HCl. This is a near-saturated mixture, and the supernatant must be decanted from any crystals that do not dissolve. This solution should be made fresh daily.
2. Prepare the cellulose stain by dissolving 200 g zinc chloride in 100 mL of distilled water. Add 20 mL of an iodine solution prepared by dissolving 10 g KI and 4 g I_2 in 100 mL distilled water.
3. Filter a portion of the wine containing the sediment onto two separate membranes.
4. Stain the first membrane by wetting the filter and sediment with phloroglucinol stain, holding stain in contact with sediment for at least 5 min.
5. Apply vacuum to remove stain and rinse with a few milliliters of distilled water. Examine the sediment under a microscope.
6. To the second membrane, flood the filter and particulates with the cellulose stain, holding stain in contact with sediment for at least 5 min.
7. Apply vacuum to remove stain and rinse with a few milliliters of distilled water. Examine the sediment under a microscope.

17.5 AMORPHOUS INSTABILITIES

When microscopically examined, this group of precipitates lack defined shape and generally assumes a color reflective of the wine. Precipitates in this category include protein and phenolics (and complexes of the two), polysaccharides (glucans, pectin, and starch), and metal casses (copper and iron).

17.5.1 Protein

Proteinaceous materials in wines will stain blue-black using Amido Black 10B and pink to red using Eosine Y as described in method A (Section 17.5.1.1). Method B (Section 17.5.1.2) relies on the interaction of tannin and protein, which forms a visually apparent haze or precipitate. In method B, formation of pronounced haze in the treated sample as compared with a control is indicative of unstable protein (Fig. 17.4).

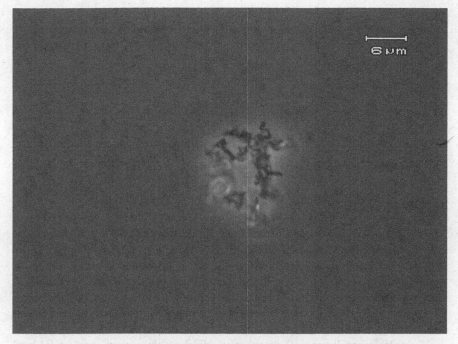

Figure 17.4. Wine protein as viewed as viewed with phase-contrast microscopy at a magnification of 1000×. Photograph provided through the courtesy of R. Thornton and E. Akaboshi.

17.5.1.1 Protein Method A

1. Prepare the following stains.
   a. Dissolve 2 g Amido Black 10B (synonym: naphthol blue black) into 100 mL methanol-acetic acid solution (90 mL methanol + 10 mL concentrated acetic acid). The molarity of the concentrated acetic acid used should be 17.4 M.
   b. Dissolve 2 g Eosin Y (synonym: acid red 87) into 100 mL methanol-acetic acid solution (90 mL methanol + 10 mL concentrated acetic acid).
2. Filter sample containing sediment through a polycarbonate membrane. Cellulose acetate membranes are unstable in presence of the protein stains.
3. Leaving the membrane on the filter housing, wet with stain, and hold for 10 min.

4. Apply vacuum to the membrane. Carefully rinse with extra methanol-acetic acid solvent (90 mL methanol + 10 mL concentrated acetic acid) until the filter is white to remove excess stain prior to examination under a microscope.

17.5.1.2 Protein Method B

1. Add 1 g tannin to a 100 mL volumetric flask and dissolve in 70% v/v ethanol.
2. Add 5 mL of the tannin solution to 100 mL of suspect wine and examine for haze formation.

17.5.2 Phenolics

Many chemical assays for phenolics use the Folin-Ciocalteu reagent. Given the chemicals and time involved in preparation, purchase of commercially prepared reagent is recommended. Whether purchased or made from individual ingredients, the reagent should have a definite yellow coloration without a hint of blue-green. To reoxidize older solutions of Folin-Ciocalteu reagent, add several drops of bromine and reboil under a fume hood using extreme caution.

Phenolic complexes will dissolve to yield a "slate-gray" to "dark-blue" turbid solution when tested using phenolic method A (Section 17.5.2.2). For phenolic method B (Section 17.5.2.3), phenolics (pigments and tannins) present in sample turn dark red. Carbonization, formation of dark, "charcoal-like" material, is suggestive of protein.

17.5.2.1 Folin–Ciocalteu Reagent

1. Transfer 700 mL of distilled water to a 2 L round-bottom boiling flask and add 100 g sodium tungstate and 25 g sodium molybdate.
2. When the sodium tungstate and sodium molybdate are fully dissolved, *carefully* add 50 mL 15 M phosphoric acid and then 100 mL 12 M HCl.
3. Add several glass beads, place boiling flask in a heating mantel, connect reflux condenser, and begin water flow. Reflux for 10 h in an exhaust hood.
4. After refluxing, cool the contents and rinse condenser column residue back into flask with distilled water.
5. Add 150 g of lithium sulfate monohydrate and several drops of bromine. Under an exhaust hood, boil the contents for 15 min. Cool and bring the contents to 1 L final volume with distilled water.
6. Filter the solution through Whatman no. 1 and store the reagent in an amber bottle until used.

17.5.2.2 Phenolic Method A

1. Collect the unidentified sediment by membrane filtration.
2. Prepare or purchase the Folin-Ciocalteu reagent (Section 17.5.2.1).
3. Rinse the sediment with several milliliters distilled water and transfer onto a watch glass using a spatula.
4. Add several drops of diluted (1:10) Folin-Ciocalteu reagent to sediment and note any color formation.

17.5.2.3 Phenolic Method B

1. Collect a portion of sediment in a small test tube.
2. Add 1 mL of concentrated H_2SO_4. *Gently* and *carefully* heat the test tube in a water bath, making certain that the mouth of the test tube is directed toward a safe direction and away from people.
3. Observe test tube for formation of any colored material.

17.5.3 Glucans, Pectins, and Starch

Polysaccharides (glucans, pectins, or starch) may be present in juice and wines where they can contribute to difficulties in clarification, fining, and filtration. Concentrations of glucans greater than 3 mg/L may create clarification and filtration problems.

Glucans are produced by *Botrytis* growth as well as some spoilage lactic acid bacteria, the latter causing a wine defect referred to as "ropiness" (Section 11.3.9). Dubourdieu et al. (1981) proposed two tests for glucans in wine depending on the suspected concentration. Method A (Section 17.5.3.1) is applied when concentrations are assumed to exceed 15 mg/L, whereas method B (Section 17.5.3.2) can be used when the concentrations of glucans are suspected to be lower. In both methods, appearance of filamentous strands suggests the presence of glucans.

The presence of pectins (Section 17.5.3.3) is illustrated by the formation of a gel in alcoholic solution. Finally, a blue-violet color in the test tube indicates presence of starch (Section 17.5.3.4), a color that can dissipate after a few minutes.

17.5.3.1 Glucans Method A

1. In a 100 mL volumetric flask, mix 1 mL 12 M HCL with 90 mL 95% v/v ethanol. Bring to 100 mL volume with 95% v/v ethanol.
2. Transfer 10 mL of the suspect wine into a 18 × 150 mm test tube and add 5 mL of acidulated ethanol.
3. Visually observe formation of filamentous strands.

17.5.3.2 Glucans Method B

1. In a 100 mL volumetric flask, mix 1 mL 12 M HCL with 90 mL 95% v/v ethanol. Bring to 100 mL volume with 95% v/v ethanol.
2. Mix 5 mL of suspect wine with 5 mL acidulated ethanol in a 18 × 150 mm test tube.
3. Incubate at room temperature for 30 min. Centrifuge the mixture at 3000 × g for 20 min.
4. If a precipitate is present, discard the supernatant and redissolve precipitate in 1 mL distilled water.
5. Add 0.5 mL acidulated ethanol and visually examine for filamentous strands.

17.5.3.3 Pectins

1. In a 100 mL graduated cylinder, mix 1 mL 12 M HCL with 90 mL 95% v/v ethanol (or isopropanol). Bring to 100 mL volume with 95% v/v ethanol (or isopropanol).
2. In another 100 mL graduated cylinder, add 25 mL of the wine containing the unidentified haze and 50 mL acidulated ethanol (or isopropanol).
3. Allow to sit for 60 min and examine for gel formation.

17.5.3.4 Starch

1. Dissolve 2 g KI and 0.1 g I_2 in approximately 80 mL distilled water in a 100 mL volumetric flask. Dilute to volume with distilled water.
2. Add 1 mL of the iodine reagent (step 1) into a 16 × 125 mm test tube and add 10 mL suspect juice or wine.
3. Examine for the formation of a blue-violet color. This can be difficult to see in red juices or wines.

17.5.4 Metal Casse

Metal instabilities ("casse") are relatively rare today but when encountered, the metals involved are generally copper or iron. Copper casse is present as an initially white and later a reddish-brown precipitate in bottled or other wines stored under low-oxygen conditions. An iron instability may be present as either ferric phosphate ("white" casse) or ferric tannate ("blue" casse). Even though a ferric phosphate instability is described as "white" casse, it may assume various shades of a blue amorphous precipitate even in white wines. Blue or ferric tannate casse may become a problem in white wines with high levels of iron and after additions of tannin. In red wines, blue casse begins as a blue cloud that eventually yields a similarly colored sediment.

Preliminary acidification of a suspect wine sample using 10% v/v HCl is useful in separation of metal-containing complexes from complexes of protein (Section 17.5.1) and phenolics (Section 17.5.2). If the haze solubilizes (step 5), the problem is probably a metal casse, and the wine can be further characterized for being either copper or iron (Section 17.5.4.2). If the haze remains (step 5), the instability is probably due to protein or complexes of protein, protein–phenolics, or phenolics–phenolics. Finally, the test described in Section 17.5.4.3 can be used to determine if a suspect wine may have the potential develop a casse after bottling.

17.5.4.1 Preliminary Test

1. Place $0.5\,g$ $K_4Fe(CN)_6\bullet3H_2O$ into a $100\,mL$ volumetric flask and add $95\,mL$ distilled water. When completely dissolved, bring to $100\,mL$ final volume with distilled water.
2. Prepare 10% v/v HCl using $12\,M$ HCl.
3. Transfer $15\,mL$ of the suspect wine to a $18 \times 150\,mm$ test tube.
4. Add 3 to $5\,mL$ of 10% v/v HCl (10% v/v) and note whether the haze dissipates (solubilizes) or remains.
5. If the haze dissipates, the problem is probably a metal casse.

17.5.4.2 Confirmation of Casse

1. Place 15 to $20\,mL$ of the suspect wine in a $18 \times 150\,mm$ test tube.
2. Add 5 drops of 30% v/v H_2O_2.
3. If the haze in step 2 dissipates, copper is suspected. Centrifuge the suspect wine and collect sediment on a stainless steel laboratory spatula. Slowly dry the sediment over a Bunsen burner, and when completely dry, attempt to ignite by more intensive exposure to the flame. If the haze consists primarily of complexes of copper and organics, the sediment will partially burn. However, inorganic precipitates (copper sulfide and ferric phosphate) will not burn.
4. If the haze in step 2 remains, transfer $20\,mL$ of turbid wine into two test tubes.
5. To one tube, add $5\,mL$ of potassium ferrocyanide (0.5% w/v). Formation of red coloration is a positive presumptive test for copper and its complexes.

6. To the other tube, add 5 mL of potassium ferrocyanide (0.5% w/v) and 5 mL HCl (10% v/v). Formation of blue coloration is a positive presumptive test for iron.

17.5.4.3 Prebottling

1. Pipette 10 mL of filtered wine into three test tubes and make the following additions:

> Tube A: Add citric acid equivalent to 0.7 g/L and several drops of 3% v/v H_2O_2.
> Tube B: Add several drops of 3% v/v H_2O_2.
> Tube C: Add several drops of 100 mg/L sodium sulfide.

2. Thoroughly aerate tubes A and B by vigorous mixing and examine the next day. Haze and/or sediment suggests the likelihood of future instability due to iron. If the test tube receiving citric acid and H_2O_2 (tube A) shows no sign of haze, addition of citric acid to the wine should be considered. Formation of a haze in tube C indicates a copper concentration in excess of 0.5 mg/L and, thus, a potential problem.

CHAPTER 18

LABORATORY SETUP

18.1 INTRODUCTION

Laboratory space varies in square footage, from small "closets" to well-equipped work areas of $92\,m^2$ ($1000\,ft^2$) or more. General needs for a laboratory include sinks, benchtop areas, utilities (gas, tap and distilled water, electrical power, and vacuum), storage (dry, refrigerated, chemical, media, and glassware), and space for equipment like an autoclave, hoods (fume and/or laminar flow), and incubators (Fig. 18.1).

The biggest challenge facing the microbiologist is microbial contamination, especially airborne microorganisms. To minimize this problem, commercial laboratories will use laminar flow hoods, which maintain a localized sterile environment for manipulation of cultures and microbiological analyses (Section 18.10). Although expensive, their use reduces losses of contaminated media and employee time and effort.

A specifically designated media preparation room within a laboratory (Fig. 18.1) can also reduce airborne microbiological contamination. A key feature of this area is that the air pressure within the room is higher than in other parts of the building. Because of this pressure difference, air will

303

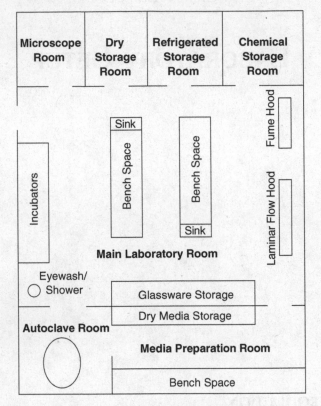

Figure 18.1.   Possible design of a wine microbiology laboratory.

flow from the room, minimizing the influx of airborne microorganisms from other areas within the building. Incoming air to the media preparation room must be filtered before being released.

18.2 MICROSCOPE AND pH METER

Two very important instruments in a wine microbiology laboratory are the microscope and pH meter. As already discussed (Chapter 12), microscopes equipped with phase-contrast are most useful to monitor fermentations and to identify potential spoilage issues. Equally important, a pH meter can be used for routine monitoring and adjustment of must or wine pH as well as providing basic information for calculating the concentration of molecular SO_2 (Section 5.2.1). Furthermore, the pH meter is used in the laboratory to adjust the pH of either solutions or media. Calibration of pH meters is performed using commercial, premade buffers, most commonly pH 4 and 7. A high-quality unit can be purchased for under $2000.

18.3 AUTOCLAVE

The ability to steam-sterilize laboratory media, glassware, and various other supplies is essential to success in the laboratory. Unfortunately, the cost of this piece of equipment parallels that of the microscope, ranging from nearly $3000 for small-capacity, largely manually operated models to more than $10,000 for larger and more automated units. In addition to capacity, other cost features include the ability to vary heat and pressure cycles as well as different exhaust capabilities. Besides the cost for an autoclave, additional supplies necessary to operate the unit include trays, indicator tape that verifies sterilization parameters, autoclavable disposal bags, and gloves for handling hot containers.

Steam is highly corrosive and wears out valves, seals, and other autoclave parts. Because of this problem, it is important to conduct a thorough inspection by a qualified technician at least once a year.

18.4 CENTRIFUGE AND FILTERS

Many laboratories routinely concentrate samples prior to microscopic examination. This can be done by either centrifugation, filtration, or, depending on the density of the suspension and the urgency for response, by simply allowing the bottle to stand upright overnight. In the case of samples suspected of being microbiologically unstable, direct plating of a small sample volume (0.1 or 1.0 mL) without concentration may be sufficient (Section 14.5).

Where the population density is expected to be low (e.g., bottling line), samples can also be concentrated using membrane filtration (Section 14.5.3). Depending on the volume of work to be done, a 300 mL glass funnel with support and a vacuum flask may be adequate. Where greater throughput is needed, manifolded systems of four to six vacuum cups may be purchased at greater cost. Alternatively, several companies market disposable filtration units that are easy to use, relatively inexpensive, and can readily be taken into the winery.

Centrifuges vary in the number and volume of samples that can be run at one time as well as speed control options, timers, and temperature control. Benchtop clinical centrifuges capable of generating a relative centrifugal force of 3000 × *g* or greater range in cost from near $1500 to well over $10,000. To minimize breakage, plastic centrifuge tubes are preferred over their more expensive glass counterparts. As centrifuges hold an even-number of test tubes or vials (≥4), opposite loads across the rotor must be equal in weight to achieve balance. Maximum differences in weights between tubes or vials can be found in the manufacturer's instructions.

18.5 INCUBATORS

General purpose incubators are available in a range of storage capacities, temperatures, and special monitoring equipment (e.g., relative humidity). Fortunately for the wine microbiologist, most microorganisms encountered grow well at room temperature (approximately 25°C/77°F). However, in cases where there are significant temperature shifts during the day, an incubator capable of maintaining a constant temperature is useful.

Most general purpose incubators are constructed to operate at temperatures above ambient. Thus, an incubator set at 27°C/80°F cannot hold that temperature if room temperature warms to 30°C/86°F or more. If storage below ambient temperature is required, low-temperature incubators or environmental chambers are available but at a much higher cost. Although tempting, the use of laboratory drying ovens should not be extended to serve as microbiological incubators as most operate efficiently at temperatures well above growth temperatures for wine microorganisms.

Because some wine microorganisms grow better in the presence of small amounts of oxygen (e.g., lactic acid bacteria), specific incubators can also be purchased that allow for changes in the atmospheric gases by removing O_2. However, these are normally far too expensive to justify their use. Fortunately, suitable alternatives used by wineries include "anaerobe jars" and "candle jars." Anaerobe jars using the GasPak® system (Baltimore Biological Laboratories) are convenient, require little storage space, are portable, and can hold up to 36 Petri plates. These systems rely on chemical reactions to consume O_2 and generate either H_2 or CO_2 gas.

Far less expensive than the GasPak® system are so-called candle jars. These are 1 gallon, wide-mouth glass jars equipped with tight-fitting lids. Once the Petri plates have been placed in the jar, a small candle is ignited. Upon sealing, the candle will burn until most of the O_2 is consumed.

18.6 WATER BATHS

Prior to pouring hot media to prepare either pour (Section 14.5.1) or spread plates (Section 14.5.2), it is necessary to equilibrate or temper the temperature of molten agar media to 45°C/113°F to 50°C/122°F. Tempering is most easily performed using a water bath. Depending on capacity and ability to maintain temperature (±0.5°C/0.9°F), water baths cost from about $1000 to $5000. A large beaker (1 L) containing water and heated on a hot plate can be a substitute for a small water bath.

18.7 GLASS AND PLASTICWARE

In many labs, plasticware (polypropylene, polyethylene, or fluoropolmer construction) has or is rapidly replacing traditional glassware. Aside from being less expensive, plasticware has the advantage of being lightweight, durable, generally chemically inert, and autoclavable. Depending on need, glass-like transparency is available. One disadvantage of plasticware is that the material cannot be exposed to direct heat in the form of hot plates or flame. Maximum exposure tolerances are usually printed on the container.

A frequent debate in microbiology laboratories is whether disposable plastic pipettes are a justifiable expense or an extravagance. Initially, disposable pipettes are less expensive than reusable glass pipettes. However, costs associated with cleaning and sterilizing reusable glass pipettes may or may not be less than that associated with continued purchase of disposables. Disposable pipettes are available in individual packages sterilized prior to shipment.

Another disposable item commonly used in the microbiology laboratory is Petri plates. Glass Petri plates are still available but are quite expensive when costs associated with the intensive cleaning and re-sterilizing are included. Disposable plates may be purchased in individual, pre-sterilized, polyethylene sleeves of 20 and also by the case (usually 500). For general work, standard 100 × 15 mm plates may be most appropriate, although larger (150 mm) and smaller (50 mm) diameters are available.

Other glassware commonly found in wine microbiology laboratories includes test tubes and caps, fixed and variable volume pipettes, and glassware for specific analytical methods. Because wine microbiology laboratories generally are not involved with pathogens, the risks associated with cleaning and reusing test tubes are largely those associated with potential breakage during the operation.

Fixed and variable volume pipettes find application in microbiology labs for preparation of dilutions as well as accurate dispensing of small (μL) volumes of reagents. Generally, fixed volume pipettes are used for routine laboratory purposes such as dilution, whereas variable volume pipettes are used in enzymatic assays requiring small-volume transfers.

Finally, specific application glassware, such as a Cash still for measurement of volatile acidity, the aeration/oxidation apparatus for SO_2, and distillation glassware are normally needed.

18.8 MEDIA

Several commercially available preformulated agar media serve as the basis for much of the laboratory work discussed in Chapter 13. Where large amounts of media are prepared and utilized on a regular basis, it can be cost effective to prepare media using individual ingredients. However, most small wineries may wish to purchase preformulated media that require relatively minimal preparation prior to autoclaving and pouring.

Depending on the microorganism(s) of interest, pre-packaged "field kits" are also available. These consist of sterile Petri plates, 0.45 μm grided membranes, and absorbent media-impregnated membrane support pads (or the pad with growth media in ampoules). Such kits were originally developed to be taken to the site(s) where testing was to be performed rather than having to collect samples and return to the laboratory. Although more expensive, these methods can be useful because the annual cost would be low for wineries that infrequently perform these analyses.

18.9 PHOTOMETERS

Aside from measuring absorbance/transmittance of soluble compounds, spectrophotometric and nephelometric methods have also been used to estimate particle (microbial) density in a suspension. In the microbiology laboratory, the spectrophotometer find most frequent application in enzymatic assays (e.g., malic or lactic acid) whereas nephelometers quantify haze (turbidity) in a sample. Both spectrophotometers and nephelometers measure the interaction of light with soluble compounds or particles (Section 14.7).

18.10 LAMINAR FLOW HOODS

Typically, winery laboratories harbor dense populations of microorganisms ranging from those associated with the winemaking process to airborne species carried into the area through open windows and ventilation systems. An unusual but interesting and troublesome example of the latter

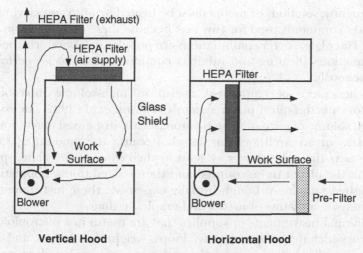

Figure 18.2. Air flows for vertical and horizontal laminar flow hoods.

is airborne mites that can end up contaminating Petri plates as they wander across the agar foraging on microbiol colonies.

The concerns for sterility can be addressed by installation of laminar flow sterile air filters that provide a constant source of "clean" air to the work area. Mechanically, room air is channeled through a prefilter and then a sterile high-efficiency particulate air (HEPA) filter prior to flowing across the work surface at uniform velocity (Fig. 18.2). Fully equipped laminar flow safety cabinets are very expensive (>$8000) and may not be justified in the winery microbiology laboratory. However, tabletop workstations that incorporate HEPA-filtered air are available at considerably less cost ($3000) and serve as good alternatives.

18.11 MISCELLANEOUS

An important element in wine laboratories is the source of water to be used to prepare reagents. Due to varying amounts of chlorine or minerals, tap water should not be used to prepare reagents or for rinsing glassware. Rather, purified water is best for these purposes. Small distillation stills are available that provide distilled water necessary for most laboratory needs. Other systems produce deionized water, and there are some units that rely on reverse osmosis. As expected, costs vary widely but depend on the needed output of the system as well as the quality of the incoming water.

Frequently, solutions or media must be heated prior to autoclaving. Hot plates are commonly used for this task because these do not rely on open flames. Hot plate/stirrer combinations are popular in laboratories because both functions (heating and mixing) commonly need to be performed simultaneously.

Another piece of equipment useful in high-volume microbiology laboratories is the spiral plater system (Swanson et al., 1992). In essence, a known volume of juice or wine is automatically dispensed onto a rotating agar plate in an Archimedean spiral. Because the amount of sample decreases as the stylus moves away from the center of the plate, specific zones on the plate can be counted and then related to the population in the original sample. Although initially expensive, these instruments can save extensive amounts of media and employee time.

Additional instruments or supplies that are useful in a microbiological laboratory include pumps, transfer loops, weighing balances, and flame sources. Small pumps, commonly peristaltic, can be used to prepare multiple 9 mL dilution blanks (Section 14.3) rather than having to hand-pipette. Many different types with varying costs are available. Loops are required for the transfer of microorganisms. Although disposable plastic loops are available presterilized, metal loops made of Nichrome or platinum are commonly used because these can be re-flamed and reused. Balances, normally "top loaders" with an accuracy of ±0.1 g, work well for media preparation. Finally, Bunsen burners are important for aseptic technique and can be designed to burn either natural gas or alcohol, the latter of which can cause serious injuries or fire if tipped over. Some newer natural gas models have pilot lights available to reduce gas usage.

CHAPTER 19

LABORATORY SAFETY

19.1 INTRODUCTION

Today, worksite safety has become far more than a cliché; it is the law. Significant penalties, both criminal and civil, may be imposed upon supervisors or managers who fail to maintain a safe working environment or repeatedly ignore employee concerns. Although a legal concern, practicing sound and proper safety and health procedures makes good business sense as well.

Personnel in the microbiology laboratory normally have the same safety concerns as those in the chemistry laboratory but, additionally, may face challenges unique to their workplace. These include the potential for unexpected exposure to billions of potentially pathogenic microorganisms, the use of high-pressure sterilization equipment, open flames for aseptic techniques, and extremely toxic media ingredients (e.g., cycloheximide, DMDC, etc.).

19.2 INJURY AND ILLNESS PREVENTION PROGRAM

As with any workplace, proactive management is the key and most fundamental component in establishing and maintaining a safe environment.

The first step in establishing a viable safety program is development of an Injury and Illness Prevention Program (IIPP). This document identifies the winery's program to develop, maintain, and update policy and procedures that addresses employee safety and health while on the job. The IIPP document must be on file at the winery and available for examination upon request. Normally, this document is prepared with input from both management and employees, possibly through a formation of a safety committee. The IIPP should be thought of as a working document and be regularly revised as potential hazards are identified.

Key elements of any IIPP are the availability of training programs and information, communication, and installation of safety equipment and systems such as eyewash stations and safety showers, fire alarms and prevention, first aid kits, and personal protective equipment.

19.2.1 Training

Each employee, regardless of position, has the legal right to know the potential health and safety risks associated with their worksite. To this end, supervisors and safety coordinators are required to conduct regular training sessions. All employees should receive training in prevention of back injury, fire extinguisher operation, prevention of slip and falls, and the use and care of personal protective equipment. Furthermore, employees may need to receive additional training depending on their job description. Such training should include flagging and traffic control, forklift operation, working with hazardous chemicals, and motor vehicle operation. Further, maintenance of crushers, washers, stemmers, presses, pumps, other equipment, respiratory protective equipment, ladders, welding equipment safety, and working in confined spaces or high locations are also training priorities. A safety orientation checklist should be developed and placed in each employee's employment file in order to document specific training (Fig. 19.1).

19.2.2 Information

All employees are entitled to have readily available documentation describing the physical and chemical characteristics as well as health implications of each chemical or solution that they may be exposed to at their worksite. All the information that fulfills the employee's right-to-know is found in the Materials Safety Data Sheets (MSDS). All chemicals shipped in the United States are required to have these documents. In the event of accidental exposure, MSDS documents provide emergency response

Employee Name	Hire Date	Orientation Date

Position/Job Assignment

Check One: ( ) New Employee ( ) Transfer ( ) Rehire ( ) Part-Time

Check items discussed:

( ) 1. Purpose of orientation.
( ) 2. Reporting accidents to supervisor immediately.
( ) 3. First aid.
    a. Obtaining treatment
    b. Location and use of emergency equipment (first aid kits, eyewashes)
    c. Location and names of first aid trained employees.
( ) 4. Potential hazards on the job.
    a. Identification of hazards andprocedures to maintain safety.
    b. Use and care of required personal protective equipment.
( ) 5. Emergencies.
    a. Exit locations and evacuation routes.
    b. Locations and operation of fire alarms and extinguishers.
    c. Specific procedures for medical, chemical, firee mergencies.
( ) 6. Injury and Illness Prevention Program.
    a. Function of safety committee and names of representatives.
    b. Importance of safety policy and rules.
( ) 7. Personal work habits.
    a. Proper lifting techniques and avoiding slips and falls.
    b. Good housekeeping and smoking policies.
    c. Safe work procedures.
( ) 8. Specific additional training required for position.

    a. _____
    b. _____

I have instructed this employee on the items checked.

Date _____ Supervisor_____

I have received orientation on the items checked.

Date _____ Employee _____

Figure 19.1. An example of a safety orientation checklist.

personnel and poison control centers with information needed to treat the patient. The compound does not need to be overtly hazardous to require a MSDS. For example, seemingly benign compounds such as glucose, tartaric acid, and even distilled water require their own MSDS.

As safety regulations require that employees must have easy access to MSDS at any time of day, most laboratories place these in clearly identifiable three-ring binders and in alphabetical order. Bright yellow binders are recommended due to their ease of identification. Enough copies of these binders should be prepared to be located in every laboratory or work area. MSDS sheets can also be found on the Internet at various sites.

19.2.3 Communication

Many accidents can be either prevented or minimized through communication. For instance, formation of a safety committee allows employees and management to discuss and formulate safety policies and procedures that would be incorporated into the IIPP. The safety committee also creates a mechanism where employees can identify potential hazards, thereby reducing the risk of accidents.

Because accidents must be reported to supervisors immediately, all laboratories must post a list of emergency telephone numbers. Although "generic" accident report forms are available (Fig. 19.2), it is recommended that the winery develop its own forms to document accidents as well as those used by supervisors conducting accident investigations. Required information includes names and telephone numbers of the injured and witnesses, the date, time, and location of the incident, a description of the incident, the involved department(s), and the contract person and telephone number.

Another form of communication is the use of hazard warning placards. The posting of placards or signs within a laboratory alerts the employees to safety and health concerns as well as rapid response in the event of accidents.

19.2.4 Eyewash Stations and Safety Showers

Even though personnel should always wear protective eyewear, splatter on the face may flow into eyes. Eyewash stations are designed to wash chemicals from the eyes in the event of an accident in the laboratory. Showers are designed to wash irritating chemicals from exposed body areas in the event of accident. Most are designed to deliver a specific amount of water in a specific amount of time. Because skin and eye irritation are both associated with lab accidents, safety showers and eyewash stations are sometimes combined into a single unit. If they are purchased separately, both stations should be in close proximity to each other.

Eyewash stations must be within 15 m (50 ft) from any workstation in the laboratory. Because the individual's vision may be impaired during

Employee Name		Accident Date	
Supervisor Name		Investigation Date	
Check all factors contributing to the accident:			
☐ HUMAN		☐ EQUIPMENT/TOOLS	
☐ TIME FACTORS		☐ POLICIES/PROCEDURES	
☐ SITE CONDITIONS		☐ OCCUPATIONAL EXPOSURES	
Explain all checked factors in the space below. Use additional space and/or diagrams if necessary.			
List recommended corrective action. Use additional space and/or diagrams if necessary.			
Actual corrective action performed and date completed			
Supervisor signature		Date	
Employee signature		Date	

Figure 19.2.   An example of a supervisor's accident investigation report.

these types of accidents, there should never be any obstacles in front of the station. Like all safety equipment, shower and eyewash stations must be tested regularly. Each station should have an attached certification card that requires the name of the inspector and the date of testing.

19.2.5 Fire Alarms and Prevention

Fire alarms and overhead sprinklers should be checked regularly to ensure proper operation. Sprinklers should be at least 18 inches away from stored materials. Decorative items (e.g., mobiles or other artwork) should never be hung from sprinkler heads.

All labs should be equipped with fire extinguishers. These should be clearly identified and wall-mounted near exits. Multipurpose extinguishers (ABC type) are generally utilized because these are effective against the most common types of fires. Minimally, extinguishers should be inspected annually. In addition to fire extinguishers, sand and other adsorbents are also a fast and easy way of stopping small fires and should be easily accessible.

19.2.6 First Aid

First aid training is critical when dealing with accidents and must be required of all employees. Refresher training exercises should be available to employees on an annual basis. Attendance at these exercises as well as any safety meetings should be documented within the employee's personal file.

Each laboratory should have a first aid kit that contains a variety of items to deal with injuries. The size and extent of the kits should depend on the number of personnel normally assigned to the worksite. In general, kits should contain adhesive bandages, bandage compresses, scissors and tweezers, triangular bandage, antiseptic soap or pads, stretch roller gauze, latex gloves, and cardiopulmonary resuscitation (CPR) face shields. Other elements could include eye dressings, adhesive tape, chemical cold pack, bee sting swabs, knuckle and finger bandages, disposable thermometer, and saline eyewash solution. Kits should be regularly inventoried and restocked as needed. Oral medications should not be available in first aid kits due to the possibility of abuse or unexpected medical reaction.

19.2.7 Personal Protective Equipment

Personal protective equipment (PPE) must be consistent with workplace hazard assessment. PPE can protect the employee against burns, absorption, abrasions, airborne hazards, punctures, or other hazards. In most cases, such equipment involves the use of gloves, goggles, laboratory coats, and protective shoes (Table 19.1). In each case, workers need to be trained regarding both the need for PPE as well as proper use. Damaged or defective PPE should never be used.

Table 19.1. Minimum personal protection equipment (ppe) required for different hazards.

Hazard	Body location		
	Eyes	Face	Body/hands
Any laboratory or general use of chemicals	Safety glasses		Normal work attire (no open-toed shoes or short pants). Over-garments (laboratory coats) and/or gloves may be needed depending on situation.
Corrosive chemicals, strong oxidizing agents, carcinogens, or mutagens.	Safety glasses or chemical splash goggles	Full-face shield over goggles	Appropriately resistant gloves, apron, and over-garment. A full protective suit may be required depending on chemical volume.
Sharp objects, glass, insertion of glass into stoppers	Safety glasses		Heavy cloth barrier or leather gloves.
Temperature extremes (extremely hot or cold liquids or material)	Safety glasses or chemical splash goggles	Full-face shield over goggles	Insulated gloves for materials over 100°C or below –1°C.

Perhaps one of the most important PPE is protective eyewear. Prescription or loose-fitting safety glasses offer only limited protection against chemical splash. Here, goggles that are snug-fitting are preferred because these offer a measure of protection against impact and splash incidents in the laboratory.

Proper dress in the laboratory also minimizes the risk of skin contact. Among those regularly working in the laboratory, appropriate coats should be required. As the long sleeves of a lab coat are meant to protect the forearms from splashes, these should not be rolled up. Likewise, open-toed shoes or sandals should be avoided. Workers who regularly work with sanitizing chemicals in the winery should utilize water-repellant aprons and boots, in addition to goggles and appropriate gloves.

Hand protection is achieved by wearing disposable latex or nitrile gloves. Protective gloves are lightweight, inexpensive, and offer a high degree of dexterity and tactile sensitivity. Because these gloves have limited chemical resistance, they should be intended to afford momentary splash protection only. For those sensitive to talc, latex gloves can be purchased without the slipping agent.

Probably the least understood piece of safety equipment is the respirator. Although dust masks may protect workers against many airborne particulates, these offer little or no protection against volatile chemicals. Respirators should be considered for use only by trained and certified employees, only under special circumstances, and only after all other controls of airborne contaminants have been implemented. In the laboratory, exhaust fans and fume hoods represent the first line of defense to contain and/or exhaust volatile hazardous materials. Without appropriate training, certification, medical approval, and careful fitting to the user's face, the use of respirators may, by itself, pose a significant health hazard.

19.3 EXAMPLES OF SAFETY ISSUES

One foundation of any safety program is "good housekeeping." Here, it is generally accepted that fewer accidents occur in a well-organized worksite. Equipment, instruments, glassware, extra reagents, chemicals, and media should be stored properly when no longer in use. Clutter should not be allowed to accumulate on benchtops, and chemicals should be stored only in approved cabinets. Chipped or broken glassware should be disposed of in clearly designated containers. Most broken glass can be recycled, the exception being borosilicate (Pyrex) which should be treated as laboratory waste.

19.3.1 Biohazard and Chemical Waste

Compliance with biohazard and chemical waste management regulations continues to be one of the most significant issues in any laboratory. Chemical waste, either stock or reagent, should never be poured down laboratory drains. Instead, it should be collected into appropriate and clearly identified ("HAZARDOUS WASTE") containers prior to removal by waste management services. Chemicals and reaction products should never be mixed. Not only is the practice dangerous but also creates additional costs in terms of dealing with the waste.

Used Petri plates, plastic containers and pipette tips should be stored in approved autoclave bags ("BIOHAZARD") for autoclaving along with cultures prior to disposal. Any item capable of causing puncture wounds, such as hypodermic syringes, glass or plastic pipettes, or razor blades, should be placed in covered contained and labeled "SHARPS."

Broken containers with small volumes of active microbial cultures should be covered with paper towels soaked in lab disinfectant. If broken glass is present, collect using a lab brush and dust pan (not fingers) and transfer to disposal bags. Larger volumes of liquid culture can be mopped up after disinfecting. It is recommended that the mop and bucket be soaked in disinfectant for several hours.

19.3.2 Electricity

Electrical issues represent one of the most frequently encountered hazards in the laboratory. Although multioutlet plugs are common to most worksites, it is recommended that these not be used unless a built-in circuit breaker is installed. The total amperage on any given circuit should not exceed the rating printed on the outlet box (commonly, 15 amps). The area immediately surrounding circuit breaker boxes must be kept clear as a work area for maintenance personnel. In most cases, display placards on the doors can be used to indicate the need for cleared space around circuit boxes.

Extension cords may be used but only on a temporary basis. These cords should be at least 16 gauge, 3-wire/3-prong types. If the piece of equipment is in regular use (e.g., water bath or incubator), extension cords should be replaced by direct wiring. In such cases, installations should always be performed by qualified electricians. Worn or frayed electrical cords on equipment should be replaced.

19.3.3 Heat and Steam

Most laboratories have an assortment of equipment designed to sterilize growth media and supplies by application of heat or heat and steam.

Examples include hot water baths, drying ovens, and autoclaves (Chapter 18). Repeated use of equipment where heat, hot water, and/or steam are involved may lead to accidents where burns are the major safety concern. Specific safety issues associated with normal autoclave use are reviewed in Chapter 13. Above all, consult manufacturer-instruction manuals for more information regarding proper use.

19.3.4 Machinery Safeguards

Although machine hazards are minimized through engineering design, areas that require guards include rotating components (connecting rods, etc.), cutting and shearing movements, belts, rollers, gears, or moving cylinders. To minimize accidents, controls for the equipment must be within easy reach. Furthermore, the operator must be able to turn off the power to the machine without leaving the place from where the machine is to be operated.

19.3.5 Storage

Improper storage of chemicals or equipment is a serious safety hazard. For instance, high-pressure gas cylinders used in the laboratory can literally become "missiles" if stored without appropriate caps or not clamped to a wall or laboratory bench. Gas cylinders should be stored with their cap in place and properly chained or clamped at all times. In addition, chemicals stored together must be compatible in order to minimize cross reactions (e.g., acids should not be stored with bases).

Household refrigerators are not designed to handle flammable substances. Rather, flammable chemicals should be stored in specially designed cabinets, which properly vent vapors and are explosion proof. If a flammable chemical requires refrigeration, only explosion-proof refrigerators or storage cabinets should be used. Such units have enclosed electrical parts to eliminate sparking. To limit mistakes, each refrigerator should be clearly labeled as to storage compatibility.

Food or beverage items including coffeemakers and soft drinks should never be consumed in the laboratory or stored in laboratory refrigerators. Prominent warning signs should be posted on the refrigerator. A separate employee break or dining area complete with a refrigerator and microwave should be provided for this purpose.

19.3.6 Ultraviolet Light

Ultraviolet light (UVL) has a long history of use in the microbiology laboratory where it finds primarily application as the energy source for germicidal lamps. Here, cell death results from high-energy radiation (180 to 250 nm) that brings about irreversible alteration to cellular DNA. These lamps are used in biological safety cabinets and in laminar air flow hoods to kill microorganisms and sterilize the environmental air.

In that such lamps emit harmful radiation, personnel must take care to avoid direct exposure to unprotected areas of the body. The severity of reaction depends on exposure time and the body surface affected. Hands, arms, and face may sunburn, whereas irreversible damage may result from exposure to eyes, and longer term exposure may result in skin cancer. Overexposure to UVL is not immediately noticeable and, hence, the injury may have already occurred before the individual develops a response. Thus, laboratory training and education is crucial to minimizing employee health and safety risks.

Lab personnel working with UVL should always wear protective garments such as lab coats and gloves to minimize direct skin contact. In that normal eyeglasses or contact lenses offer only limited protection, the best protection is a full-face shield. Signs warning of the effects of UVL should be posted at workstations where it is used.

GLOSSARY

Absorption: The movement of a chemical from the site of contact across a biological barrier.

Active transport: A biochemical process requiring energy where compounds are moved across the cell membrane, commonly against a concentration gradient.

Adenosine triphosphate (ATP): A high-energy containing molecule that allows organisms to transfer energy.

Aerobe: A microorganism whose growth requires the presence of air (oxygen). *Acetobacter* is an example of an obligate ("must have") aerobe.

Aerobic respiration: A sequential series of biochemical processes (glycolysis, Krebs cycle, and oxidative phosphorylation) where glucose is oxidized to CO_2, H_2O and energy in the forth of ATP.

Agar: A complex carbohydrate material refined from marine algae and used to produce semisolid media. Solidified agar has the texture of a gelatin dessert. It solidifies at temperatures between 40°C/104°F and 45°C/113°F and will not remelt until it is boiled. Most liquid media can be solidified by the addition of 1.5% to 2% w/v agar.

Amino acid: Organic compound that contains both acid (COOH) and amine (NH_2) groups. Amino acids serve as the basic unit within proteins and enzymes.

Anaerobe: A microorganism that grows only or best in the absence of air (oxygen). Many lactic acid bacteria are considered to be facultative ("not required") anaerobes; these microorganisms grow well under anaerobic conditions but can also grow in the presence of some oxygen.

Anaerobic jars: As most laboratories do not have anaerobic incubators due to cost, a good substitute is plastic jars that contain racks that hold disposable CO_2 generators. Alternatively, some winemakers rely on "candle jars" in which a candle is lit inside of the jar prior to being sealed (the flame will remove the oxygen that is present before being extinguished).

Anamorph: Asexual or "imperfect" form of a yeast. Although anamorphs and teleomorphs are the same microorganism, these forms differ in their inability (anamorphs) or ability (teleomorphs) to form spores. Microbiologists assign separate names (genus and sometimes species) to differentiate the ability of a yeast to produce (or not) spores.

Ascospores: The sexual spore that many yeasts can produce as a mean of reproduction.

Ascus: A structure formed by many yeasts that contains ascospores.

Aseptic technique: Any technique or procedure in which precautions against microbial contamination is taken. Once media or instruments are sterile, they are kept free of microorganisms using this technique.

Autoclave: A pressure vessel capable of reaching temperatures in excess of 100°C/212°F by using steam. Autoclaves are used to sterilize media and instruments.

Autotroph: A microorganism that can produce through biochemical processes all required organic components from inorganic sources.

Bactericidal: Chemicals that are lethal to bacteria.

Bacteriophage: A virus that infects bacteria. Phage are strain specific and are lethal to bacteria.

Bacteriostatic: Chemicals that inhibit bacteria without necessarily being lethal.

Biohazard: Biological or infectious material such as a microbiological pathogen (e.g., *Salmonella*).

°Brix (°Balling): Measurement used to express the concentration of soluble solids, primarily sugars, and expressed on a weight/weight basis (g sucrose per 100 g liquid).

Brownian motion: Random movement and motion of microorganisms when viewed in a wet mount under a microscope.

Budding: Asexual reproduction of yeasts involving the formation of daughter cells (buds) off of the mother cell. Yeasts can reproduce by bipolar (buds only appear at end of cell) or multipolar/mulilaterial (buds can appear on all surfaces of cell).

Carcinogen: A chemical or physical agent capable of causing cancer in animals or humans.

Chemical compatibility: The ability of different reagents to potentially react with one another resulting in a safety concern. Some chemicals should not be stored in close proximity due to their reactivity (acids and bases).

Cocci: Cells that microscopically appear as being "round" or slightly "oval."

Coenzyme: A compound that "assists" enzymes catalyzing reactions, commonly by donating or receiving electrons. For instance, nicotinamide adenine dinucleotide can exist in the oxidized form (NAD^+) or the reduced form with an added electron ($NADH + H^+$).

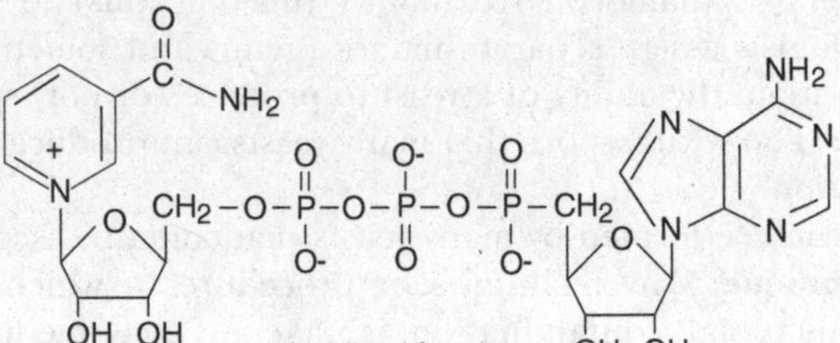

(oxidized form)

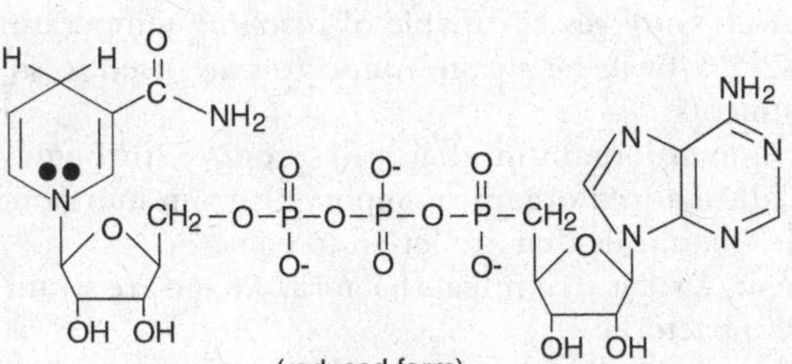

(reduced form)

Colonies: When a viable cell is deposited on the surface of a solidified agar medium, it reproduces and forms a cluster (or colony) of cells, which are counted for enumeration. However, these colonies may also initially arise from pairs, chains, or clumps of cells depending on the morphology of the microorganism. In these cases, the population should be recorded as "colony-forming units" per mL (CFU/mL) rather than cells per mL.

Contamination: The result of materials, chemical substances, or microbes entering systems causing a reduction in quality or a safety hazard.

Corrosive: A substance that causes visible damage to humans at the site of contact.

Dilution blank: A sterilized solution consisting of 0.1% w/v peptone water used to dilute samples that contain large numbers of viable microorganisms. Samples are normally serially diluted (1:10, 1:100, 1:1000, 1:10,000, etc.) prior to plating using solidified agar.

Durham tube: A small test tube (9.5×50 mm) placed inverted in a liquid broth and used to detect CO_2 formation by physically trapping the gas.

Ecology: The study of the interrelationships between an organism and the environment.

Enzyme: A protein catalyst that causes changes in other molecules without undergoing any alterations itself. For example, the enzyme catalase acts on hydrogen peroxide (H_2O_2) as a substrate to form water (H_2O) and oxygen (O_2). Enzyme names commonly have the suffix "ase" (e.g., catalase, pectinase, etc.).

Eukaryote: An organism (e.g., yeast) that has cellular organization including a membrane-bound nucleaus and internal organelles.

Fastidious: Microorganisms that require many different nutrients due to their inability to synthesize these compounds.

Fermentation: The breakdown of organic molecules such as sugars into other products, commonly under anaerobic conditions.

Genus: A taxonomic category of related organisms, usually containing several species. The genus is the first name of an organism within the binomial system of taxonomy.

Glycolysis: The anaerobic process of breaking down glucose to form pyruvic acid or lactic acid and energy in the form of ATP.

Gram stain: A differential staining procedure that classifies microorganisms as either Gram-positive or Gram-negative based on retention of a specific dye (crystal violet). Using brightfield microscopy, Gram-positive cells appear "purple" whereas Gram-negative ones appear "red."

Haploid: A single cell that has only one set of chromosomes, as opposed to diploids which have two sets.

Hazard: Any physical, chemical, or biological material or situation that can potentially cause injury or death. Examples are flammable or corrosive chemicals or some biological agents.

Heterothallic: Organisms in which the two sexes reside in different individuals. By contrast, homothallic organisms possess both sexes in the same individual.

Hexose: A six-carbon sugar. In grape musts, glucose and fructose as the most common examples.

Hockey stick: A glass rod bent into the shape of an "L" that is used to evenly spread samples onto solidified media.

Hyphae: The filaments or threads formed by a number of yeasts and molds that form a mat (mycelium).

Inoculum: A small amount of viable microorganisms introduced into a medium or juice with the goal of growing the cells.

Krebs cycle: Also known as the citric acid (or tricarboxylic acid) cycle, this is a series of aerobic reactions by which the pyruvic acid produced in glycolysis is converted into energy (i.e., ATP and reduced molecules NADH and $FADH_2$), CO_2, and H_2O.

Long-Term Exposure Limits (LTEL): The maximum concentration to which a worker can be exposed to continuously for a defined long period (8 hour day or 40 hour week) without experiencing ill effects.

Loop: A device with a handle and a metal loop on the end used to transfer microorganisms. It is also known as a "laboratory loop" or a "transfer loop" and is sterilized by holding the metal loop in an open flame.

Material Safety Data Sheets (MSDS): Mandatory printed in formation provided by the manufacturer/distributor that identifies hazard/physical properties of chemicals as well as first aid treatment for exposure.

Medium: A formulation composed of various ingredients that will support the growth of different microorganisms. Such ingredients include glucose, peptone, yeast extract, liver extract, and others. A medium can be prepared as a liquid or solid, the latter with the addition of agar.

Methylene blue: A differential stain that can be added to a liquid suspension to evaluate yeast viability. Live yeasts will reduce the dye to a colorless form, whereas dead cells appear blue/black as viewed using a microscope.

Microorganism: A very small organism that can only be seen using a microscope. Although most microbes appear microscopically as a single cell, some form pairs or chains of many cells.

Milk dilution bottle: A square-sided bottle that is commonly used for larger dilution blanks. Though total volume can be up to 160 mL, these bottles most commonly contain 99 mL. Many brands will have an etched line on the side of the bottle that represents a mark for 99 ± 1 mL volume.

Mold: A fungus that has a filamentous physical structure.

Mycelium: An interwoven mass of individual fungal filaments or hyphae.

NAD$^+$/NADH: See "coenzyme."

Needle: Needles attached to wooden handles are used to prepare stabs and can be resterilized using an open flame.

Pentose: A five-carbon sugar such as arabinose or ribose. Normally, these sugars are found in grape musts at low concentrations, <0.05%.

Personal protective equipment (PPE): Equipment used by employees to reduce or eliminate the potential for exposure to hazardous materials or situations. Such examples would include eye goggles, protective gloves, and the like.

Petri plates or dishes: Sterilized glass or plastic dishes with covers that are used to hold solidified media.

Phase (growth): Microbial growth in juice, wine, or a medium has four distinctive phases: (a) lag, (b) logarithmic, (c) stationary, and (d) death. During lag phase (A), cells are adjusting to the new environment and increase in size (no increase in cell numbers). Logarithmic phase (B) is the period when cell numbers rapidly increase. At some point, the growth rate decreases and the cells enter stationary phase (C) where growth and death rates are approximately equal. The accumulation of toxic wastes and the decreasing availability of nutrients eventually result in death of the cells (D).

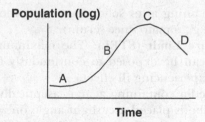

Phase-contrast microscope: Rather than staining a culture in order to visualize the microorganisms using a brightfield microscope, phase-contrast microscopes allow direct viewing of microbes in liquid cultures. Light passing through a denser medium (a microbial cell) than another medium (the liquid) will be retarded. Phase-contrast enhances differences in refractive index between these two media (a cell and the liquid).

Pour plates: A pour plate is one in which the liquid sample to be microbiologically enumerated is aseptically transferred into a Petri dish, and cooled (yet liquid) agar medium is added to the culture. The medium and the liquid sample must be well mixed prior to solidification of the agar gel.

Prokaryote: An organism whose cellular organization lacks a true nucleus and other internal organelles (e.g., bacteria).

Pseudohyphae: Elongated forms or filaments between yeast buds that resemble hyphae. Constrictions between buds in the pseudohyphae differentiate these from true hyphae.

Pseudomycelium: Collective mass of pseudohyphae.

Redox: An abbreviation for reduction/oxidation, redox refers to simultaneous reactions in which one agent is oxidized (loss of an electron) and one is reduced (gain of an electron).

Residual: Chemicals remaining on equipment or floor surfaces after cleaning or sanitizing.

Rod: Cells that microscopically appear as being "rectangular" with two parallel sides of the microbe being longer that the other two sides.

Sanitation: Sanitation refers to reducing microbial populations in the winery and keeping the populations as low as possible. In contrast, sterilization implies destruction of all microorganisms.

Secondary containment: A method by which additional leakage or spills are minimized if the primary container breaks. An example would be a tray in which a bottle of waste chemicals is placed.

Selective agent: This is a chemical that is added to a medium such that undesirable microorganisms will not grow while desirable microbes will grow.

Serum: Liquid remaining after solids have been removed, normally by centrifugation (e.g., tomato juice serum).

Short-Term Exposure Limit (STEL): The maximum concentration to which a worker can be exposed to continuously for a defined short period without experiencing ill effect.

Slant: Liquefied media containing agar is aseptically placed into a test tube. The test tube is placed at a slight angle on a tabletop while the agar media cools and sets. Slants are used to aseptically store yeast and bacteria cultures for longer periods of time.

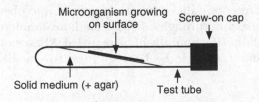

Microorganism growing on surface   Screw-on cap

Solid medium (+ agar)   Test tube

sp. or spp.: Normally placed after a genus name, these designations refer to one (sp.) or more (spp.) unidentified species.

Spore: A nonvegetative structure formed by some microorganisms that is resistant to stresses such as heat. Spores are capable of development into an individual viable microorganism when conditions are favorable for growth.

Spread plates: Spread plates are ones in which the sample is aseptically added onto the surface of media already solidified (gel formed) in Petri dishes. The sample, normally 0.1 mL, is evenly spread on the surface of the media by using a "hockey stick."

Stab: Similar to slant, stabs are used to store microorganisms for longer periods of time, especially those microbes that do not require oxygen.

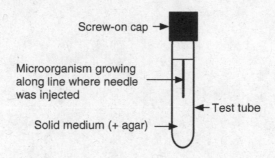

Screw-on cap →

Microorganism growing along line where needle was injected →

← Test tube

Solid medium (+ agar) →

Strain: Microorganisms that share sufficient biochemical, physiologial, and genetic characteristics to be assigned the same species name but that possess minor but consistent variation(s) in certain properties. Such properties may include (but not necessarily) fermentation of a specific sugar, pH tolerance, and so forth.

Teleomorph: Refers to the sexual or spore-forming form of a given yeast (see anamorph).

Temper: Media containing agar needs to temperature equilibrate prior to pouring into Petri plates. Most microbiologists will place media into a water bath for an hour at 45°C/113°F to 50°C/122°F after autoclaving.

Too-Numerous-To-Count (TNTC): When counting colonies on pour or spread plates, the total count must be between 25 and 250 colonies. If greater than 250 colonies, the plate is deemed to be TNTC and a higher dilution plate should be examined.

Ullage: The empty space within a barrel above the wine surface. Ullage normally increases as wine is aged in barrels due to evaporative losses of ethanol and water.

Viable–But–Not–Culturable (VBNC): A physiological state of micro-organisms where growth on conventional media is not observed but the microorganisms remain intact and remain viable.

Vortex (vortexer): A special type of device that rapidly mixes samples in centrifuge or test tubes by creating a vortex.

w/w or v/v: Symbols that refer to concentrations either on a weight/weight or a volume/volume basis.

REFERENCES

ABRUNHOSA, L., R.R.M. PATERSON, Z. KOZAKIEWICZ, N. LIMA, and A. VENÂNCIO. 2001. Mycotoxin production from fungi isolated from grapes. *Lett. Appl. Microbiol.* 32: 240–242.

ACREE, T.E., E.P. SONOFF, and D.F. SPLITTSTOESSER. 1972. Effect of yeast strain and type of sulfur compound on hydrogen sulfide production. *Am. J. Enol. Vitic.* 23: 6–9.

ADAMS, M.R. 1998. Vinegar. In: *Microbiology of Fermented Foods.* B.J.B. Wood (Ed.), 2nd edition, Volume 1, Chapter 1, pp. 1–44. Blackie Academic and Professional, London.

AGENBACH, W.A. 1977. A study of must nitrogen content in relation to incomplete fermentations, yeast production and fermentation activity. In: *Proceedings of the South African Society for Enology and Viticulture.* E.F. Beukman (Ed.), pp. 66–88. Stellenbosch, South Africa.

AGUILAR USCANGA, M.G., M.-L. DÉLIA, and P. STREHAIANO. 2003. *Brettanomyces bruxellensis*: effect of oxygen on growth and acetic acid production. *Appl. Microbiol. Biotechnol.* 61: 157–162.

ALDERCREUTZ, P. 1986. Oxygen supply to immobilized cells. Theoretical calculations and experimental data for oxidation of glycerol by immobilized *Gluconobacter oxydans* with oxygen or *p*-benzoquinone as electron acceptors. *Biotechnol. Bioeng.* 28: 223–232.

331

ALEXANDRE, H. and C. CHARPENTIER. 1998. Biochemical aspects of stuck and sluggish fermentations in grape must. *J. Indust. Microbiol. Biotechnol.* 20: 20–27.

ALEXANDRE, H., D. HEINTZ, D. CHASSAGNE, M. GUILLOUX-BENATIER, C. CHARPENTIER, and M. FEUILLAT. 2001. Protease A activity and nitrogen fractions released during alcoholic fermentation and autolysis in enological conditions. *J. Indust. Microbiol. Biotechnol.* 26: 235–240.

ALEXANDRE, H., P.J. COSTELLO, F. REMIZE, J. GUZZO, and M. GUILLOUX-BENATIER. 2004. *Saccharomyces cervisiae—Oenococcus oeni* interactions in wine: current knowledge and perspectives. *Int. J. Food Microbiol.* 93: 141–154.

ALUR, M.D. 2000. *Botrytis.* In: *Encyclopedia of Food Microbiology.* R.K. Robinson, C.A. Batt, and P.D. Patel (Eds.), Volume 1, pp. 279–283. Academic Press, New York, NY.

ÁLVAREZ-RODRÍGUEZ, M.L., C. BELLOCH, M. VILLA, F. URUBURU, G. LARRIBA, and J.-J.R. COQUE. 2003. Degradation of vanillic acid and production of guaiacol by microorganisms isolated from cork samples. *FEMS Microbiol. Lett.* 220: 49–55.

AMACHI, T. 1975. Chemical structure of a growth factor (TJF) and its physical significance for malo-lactic bacteria. In: *Lactic Acid Bacteria in Beverages and Foods.* J.G. Carr, C.V. Cutting, and G.C. Whiting (Eds.), pp. 103–118. Academic Press, London.

AMERINE, M.A. and R.E. KUNKEE. 1968. Microbiology of winemaking. *Ann. Rev. Microbiol.* 22: 323–358.

AMERINE, M.A., H.W. BERG, and W.V. CRUESS. 1972. *The Technology of Winemaking,* 3rd edition. AVI Publishing, Westport, CT.

AN, D. and C.S. OUGH. 1993. Urea excretion and uptake by wine yeasts as affected by various factors. *Am. J. Enol. Vitic.* 44: 35–40.

ANONYMOUS. 1983. *The Merck Index.* M. Windholz (Ed.), 10th edition. Merck and Co., Inc., Rahway, NJ.

ANONYMOUS. 1984. *Difco Manual,* 10th edition. Difco Laboratories, Detroit, MI.

ANONYMOUS. 1987. Velcorin Cold Sterilant Process Information Guide. Mobsy Corporation, Pittsburgh, PA.

ANONYMOUS. 2002. Juice HAACP Training Curriculum. The Juice HAACP Alliance (P. Slade, chair), National Center for Food Safety and Technology, Illinois Institute of Technology, Chicago, IL.

ARAGON, P., J. ATIENZA, and M.D. CLIMENT. 1998. Influence of clarification, yeast type, and fermentation temperature on the organic acid and higher alcohols of Malvasia and Muscatel wines. *Am. J. Enol. Vitic.* 49: 211–219.

ARBEIT, R.D., M. ARTHUR, R. DUNN, C. KIM, R.K. SELANDER, and R. GOLDSTEIN. 1990. Resolution of recent evolutionary divergence among *Escherichia coli* from related lineages: the application of pulsed field gel electrophoresis to molecular epidemiology. *J. Infect. Diseases* 161: 230–235.

ARENA, M.E. and M.C. MANCA DE NADRA. 2001. Biogenic amine production by *Lactobacillus. J. Appl. Microbiol.* 90: 158–162.

ARENDT, E.K. and W.P. HAMMES. 1992. Isolation and characterization of *Leuconostoc oenos* phages from German wines. *Appl. Microbiol. Biotechnol.* 37: 643–646.

ARRIAGADA-CARRAZANA, J.P., C. SÁEZ-NAVARRETE, and E. BORDEU. 2005. Membrane filtration effects on aromatic and phenolic quality of Cabernet Sauvignon wines. *J. Food Engin.* 68: 363–368.

ATLAS, R.M. and R. BARTHA. 1981. *Microbial Ecology: Fundamentals and Applications.* Addison-Wesley Publishing, Reading, MA.

AUGUSTYN, O.P.H., J.L.F. KOCK, and D. FERREIRA. 1992. Differentiation between yeast species and strains within species by cellular fatty acid analysis 5. A feasible technique? *Sys. Appl. Microbiol.* 15: 105–115.

AXELSSON, L. 1998. Lactic acid bacteria: classification and physiology. In: *Lactic Acid Bacteria. Microbiology and Functional Aspects.* S. Salminen and A. von Wright (Eds.), pp. 1–72. Marcel Dekker Inc., New York, NY.

AYESTARÁN, B.M., M.C. ANCÍN, A.M. GARCÍA, A. GONZÁLEZ, and J.J. GARRIDO. 1995. Influence of prefermentation clarification on nitrogenous contents of musts and wines. *J. Agric. Food Chem.* 43: 476–482.

BAE, S., G.H. FLEET, and G.M. HEARD. 2004. Occurrence and significance of *Bacillus thuringiensis* on wine grapes. *Int. J. Food Microbiol.* 94: 301–312.

BALDWIN, G. 1993. Treatment and prevention of spoilage films on wines. *Aust. Grape. Wine.* No. 352, pp. 55–56.

BALEIRAS COUTO, M.M., R.G. REIZINHO, and F.L. DUARTE. 2005. Partial 26S rDNA restriction analysis as a tool to characterize non-*Saccharomyces* yeasts present during red wine fermentations. *Int. J. Food Microbiol.* 102: 49–56.

BALEIRAS-COUTO, M.M., B.J. HARTOG, J.H.H. HUIS IN'T VELD, H. HOFSTRA, and J.M.B.M. VAN DER VOSSEN. 1996. Identification of spoilage yeasts in a food-production chain by microsatellite polymerase chain reaction fingerprinting. *Food Microbiol.* 13: 59–67.

BARBOUR, E.A. and F.G. PRIEST. 1988. Some effects of *Lactobacillus* contamination in Scotch whisky fermentations. *J. Inst. Brew.* 94: 89–92.

BARTOWSKY, E.J. and P.A. HENSCHKE. 1995. Malolactic fermentation and wine flavour. *Aust. Grape. Wine.* 378: 83–94.

BARTOWSKY, E.J. and P.A. HENSCHKE. 1999. Use of polymerase chain reaction for specific detection of the malolactic fermentation bacterium *Oenococcus oeni* (formerly *Leuconostoc oenos*) in grape juice and wine samples. *Aust. J. Grape Wine Res.* 5: 39–44.

BARTOWSKY, E.J., I.L. FRANCIS, J.R. BELLON, and P.A. HENSCHKE. 2002. Is buttery aroma perception in wines predictable from the diacetyl concentration? *Aust. J. Grape Wine Res.* 8: 180–185.

BARTOWSKY, E.J., D. XIA, R.L. GIBSON, G.H. FLEET, and P.A. HENSCHKE. 2003. Spoilage of bottled red wine by acetic acid bacteria. *Lett. Appl. Microbiol.* 36: 307–314.

BARTOWSKY, E.J. and P.A. HENSCHKE. 2004a. The "buttery" attribute of wine-diacetyl-desirability, spoilage and beyond. *Int. J. Food Microbiol.* 96: 235–252.

BARTOWSKY, E.J. and P.A. HENSCHKE. 2004b. The "buttery" attribute of wine-diacetyl. Desirability, spoilage and beyond. Butter or no butter. In: *Proceedings of the XVI's Entretiens Scientifiques Lallemand.* pp. 11–17. Oporto, Portugal.

BATTILANI, P. and A. PIETRI. 2002. Ochratoxin A in grapes and wine. *Eur. J. Plant Path.* 108: 639–643.

BATTILANI, P., A. LOGRIECO, P. GIORNI, G. COZZI, T. BERTUZZI, and A. PIETRI. 2004. Ochratoxin A production by *Aspergillus carbonarius* on some grape varieties grown in Italy. *J. Sci. Food Agric.* 84: 1736–1740.

BAUCOM, T.L., M.H. TABACCHI, T.H.E. COTTRELL, and B.S. RICHMOND. 1986. Biogenic amine content of New York State wines. *J. Food Sci.* 51: 1376–1377.

BAUER, R., H.A. NEL, and L.M.T. DICKS. 2003. Pediocin PD-1 as a method to control growth of *Oenococcus oeni* in wine. *Am. J. Enol. Vitic.* 54: 86–91.

BAUZA, T., A. BLAISE, P.L. TEISSEDRE, J.P. MESTRES, F. DAUMAS, and J.C. CABANIS. 1995. Changes in biogenic amines content in musts and wines during the winemaking process. *Sci. Aliments* 15: 559–570.

BEECH, F.W. 1993. Yeasts in cider-making. In: *The Yeasts*. A.H. Rose and J.S. Harrison (Eds.), 2nd edition. 5: 169–213. Academic Press, London.

BEECH, F.W., L.F. BURROUGHS, C.F. TIMBERLAKE, and G.C. WHITING. 1979. Progrès récents sur l'aspect chimique et l'action anti-microbienne de l'anhydride sulfureux (SO_2). *Bull. O.I.V.* 52: 1001–1022.

BEELMAN, R.B. 1982. Development and utilization of starter cultures to induce malo-lactic fermentation in red table wines. In: *Proceedings of the University of California, Davis, Grape and Wine Centennial Symposium.* pp. 109–117. University of California, Davis, CA.

BEELMAN, R.B. and R.E. KUNKEE. 1985. Inducing simultaneous malolactic and alcoholic fermentation in red table wines. In: *Proceedings of the Australian Society for Viticulture and Oenology Seminar on Malolactic Fermentation*, T.H. Lee (Ed.), pp. 97–112. Australian Wine Research Institute, Adelaide, South Australia.

BEELMAN, R.B., A. GAVIN III, and R.M. KEEN. 1977. A new strain of *Leuconostoc oenos* for induced malo-lactic fermentation in Eastern wines. *Am. J. Enol. Vitic.* 28: 159–165.

BEELMAN, R.B., F.J. MCARDLE, and G.R. DUKE. 1980. Comparison of *Leuconostoc oenos* strains ML-34 and PSU-1 to induce malolactic fermentation in Pennsylvania red table wines. *Am. J. Enol. Vitic.* 31: 269–276.

BEELMAN, R.B., R.M. KEEN, M.J. BANNER, and S.W. KING. 1982. Interactions between wine yeast and malolactic bacteria under wine conditions. *Dev. Indust. Microbiol.* 23: 107–121.

BELL, A.A., C.S. OUGH, and W.M. KLIEWER. 1979. Effects on must and wine composition, rates of fermentation, and wine quality of nitrogen fertilization of *Vitis vinifera* var. Thompson Seedless grapevines. *Am. J. Enol. Vitic.* 30: 124–129.

BELLÍ, N., S. MARÍN, A. DUAIGÜES, A.J. RAMOS, and V. SANCHIS. 2004. Ochratoxin A in wines, musts and grape juices from Spain. *J. Sci. Food Agric.* 84: 591–594.

BELTRAN, G., B. ESTEVE-ZARZOSO, N. ROZÈS, A. MAS, and J.M. GUILLAMÓN. 2005. Influence of the timing of nitrogen additions during synthetic grape must fermentations on fermentation kinetics and nitrogen consumption. *J. Agric. Food Chem.* 53: 996–1002.

BENDOVA, O., V. RICHTER, B. JANDEROVA, and J. HAUSLER. 1991. Identification of industrial yeast strains of *Saccharomyces cerevisiae* by fatty acid profiles. *Appl. Microbiol. Biotechnol.* 35: 810–812.

BENEDUCE, L., G. SPANO, A. VERNILE, D. TARANTINO, and S. MASSA. 2004. Molecular characterization of lactic acid populations associated with wine spoilage. *J. Basic Microbiol.* 44: 10–16.

BERG, H.W., F. FILIPELLO, E. HINREINER, and A.D. WEBB. 1955. Evaluation of thresholds and minimum difference concentrations for various constituents of wines. I. Water solutions of pure substances. *Food Technol.* 9: 23–26.

BERTHELS, N.J., R.R. CORDERO OTERO, F.F. BAUER, J.M. THEVELEIN, and I.S. PRETORIUS. 2004. Discrepancy in glucose and fructose utilisation during fermentation by *Saccharomyces cerevisiae* wine yeast strains. *FEMS Yeast Res.* 4: 683–689.

BEYER, W.F. and I. FRIDOVICH. 1985. Pseudocatalase from *Lactobacillus plantarum*: evidence for a homopentameric structure containing two atoms of manganese per subunit. *Biochem.* 24: 6460–6467.

BISSON, L.F. 1991. Influence of nitrogen on yeast and fermentation of grapes. In: *Proceedings of the International Symposium on Nitrogen in Grapes and Wine.* J.M. Rantz (Ed.), pp. 78–89. American Society for Enology and Viticulture, Davis, CA.

BISSON, L.F. 1999. Stuck and sluggish fermentations. *Am. J. Enol. Vitic.* 50: 107–119.

BISSON, L.F. and R.E. KUNKEE. 1991. Microbial interactions during wine production. In: *Mixed Cultures in Biotechnology.* J.G. Zeikus and E.A. Johnson (Eds.), pp. 37–68. McGraw-Hill, New York, NY.

BISSON, L.F. and C.E. BUTZKE. 2000. Diagnosis and rectifications of stuck and sluggish fermentations. *Am. J. Enol. Vitic.* 51: 168–177.

BLESA, J., J.M. SORIANO, J.C. MOLTÓ, and J. MAÑES. 2004. Concentration of ochratoxin A in wines from supermarkets and stores of Valencian Community (Spain). *J. Chromato. A.* 1054: 397–401.

BLONDIN, B., R. RATOMAHENINA, A. ARNAUD, and P. GALZY. 1982. A study of cellobiose fermentation by a *Dekkera* strain. *Biotechnol. Bioeng.* 24: 2031–2037.

BOEKHOUT, T., C.P. KURTZMAN, K. O'DONNELL, and M.T. SMITH. 1994. Phylogeny of the yeast genera *Hanseniaspora* (anamorph *Kloeckera*), *Dekkera* (anamorph *Brettanomyces*), and *Eeniella* as inferred from partial 26S ribosomal DNA nucleotide sequences. *Int. J. Syst. Bacteriol.* 44: 781–786.

BOIDO, E., A. LLORET, K. MEDINA, F. CARRAU, and E. DELLACASSA. 2002. Effect of β-glycosidase activity of *Oenococcus oeni* on the glycosylated flavor precursors of Tannat wine during malolactic fermentation. *J. Agric. Food Chem.* 50: 2344–2349.

BOTHA, A. and J.L.F. KOCK. 1993. Application of fatty acid profiles in the identification of yeasts. *Int. J. Food Microbiol.* 19: 39–51.

BOULTON, R.B., V.L. SINGLETON, L.F. BISSON, and R.E. KUNKEE. 1996. *Principles and Practices of Winemaking.* Chapman and Hall Publishers, New York, NY.

BOUSBOURAS, G.E. and R.E. KUNKEE. 1971. Effect of pH on malo-lactic fermentation in wine. *Am. J. Enol. Vitic.* 22: 121–126.

BRECHOT, P., J. CHAUVET, P. DUPUY, M. CROSON, and A. RABATU. 1971. Acide oleanoique, facteur de croissance anaerobie de la levure du vin. *C.R. Acad. Sci.* 272: 890–893.

BRETON, A. and Y. SURDIN-KERJAN. 1977. Sulfate uptake in *Saccharomyces cerevisiae*: biochemical and genetic study. *J. Bacteriol.* 132: 224–232.

BRITZ, T.J. and R.P. TRACEY. 1990. The combination effect of pH, SO_2, ethanol and temperature on the growth of *Leuconostoc oenos*. *J. Appl. Bacteriol.* 68: 23–31.

BULIT, J. and R. LAFON. 1970. Quelques aspects de la Biologie du *Botrytis cinerea* Pers. agent de la pourriture grise des raisins. *Conn. Vigne Vin* 4: 159–174.

BUSER, H.R., C. ZANIER, and H. TANNER. 1982. Identification of 2,4,6-trichloro-anisole as a potent compound causing cork taint in wine. *J. Agric. Food Chem.* 30: 359–362.

BUTZKE, C.E. 1998. Survey of yeast assimilable nitrogen status in musts from California, Oregon, and Washington. *Am. J. Enol. Vitic.* 49: 220–224.

CABAÑES, F.J., F. ACCENSI, M.R. BRAGULAT, M.L. ABARCA, G. CASTELLÁ, S. MINGUEZ, and A. PONS. 2002. What is the source of ochratoxin A in wine? *Int. J. Food Microbiol.* 79: 213–215.

CABAROGLU, T., S. SELLI, A. CANBAS, J.-P. LEPOUTRE, and Z. GÜNATA. 2003. Wine flavor enhancement through the use of exogenous fungal glycosidases. *Enz. Microbial Technol.* 33: 581–587.

CAILLET, M.M. and Y. VAYSSIER. 1984. Utilisation des biomass de bactéries lactiques pour le déclenchement de la fermentation malo-lactique. *Rev. Fran. D'Oenol.* 24(95): 63–69.

CANAS, B.J., D.C. HAVERY, L.R. ROBINSON, M.P. SULLIVAN, F.L. JOE JR., and G.W. DIACHENKO. 1989. Ethyl carbamate levels in selected fermented foods and beverages. *J. Assoc. Off. Anal. Chem. Int.* 72: 873–876.

CANNON, M.C. and G.J. PILONE. 1993. Interactions between commercial wine yeast and malolactic bacteria. In: *The New Zealand Grape and Wine Symposium*. D.T. Jordan (Ed.), pp. 85–95. The New Zealand Society for Viticulture and Oenology, Auckland, NZ.

CAPECE, A., G. SALZANO, and P. ROMANO. 2003. Molecular typing techniques as a tool to differentiate non-*Saccharomyces* wine species. *Int. J. Food Microbiol.* 84: 33–39.

CAPUCHO, I. and M.V. SAN RAMAO. 1994. Effect of ethanol and fatty acids on malolactic activity of *Leuconostoc oenos*. *Appl. Microbiol. Biotechnol.* 42: 391–395.

CARIDI, A. and V. CORTE. 1997. Inhibition of malolactic fermentation by cryotolerant yeasts. *Biotech. Lett.* 19: 723–726.

CARLILE, M.J., S.C. WATKINSON, and G.W. GOODAY. 2001. *The Fungi*, 2nd edition. Academic Press, New York, NY.

CARR, F.J., D. CHILL, and N. MAIDA. 2002. The lactic acid bacteria: a literature survey. *Crit. Rev. Microbiol.* 28: 281–370.

CARR, J.G., P.A. DAVIES, and A.H. SPARKS. 1976. The toxicity of sulphur dioxide towards certain lactic acid bacteria from fermented apple juice. *J. Appl. Bacteriol.* 40: 201–212.

CARR, J.G., P.A. DAVIES, F. DELLAGLIO, M. VESCOVO, and R.A.D. WILLIAMS. 1977. The relationship between *Lactobacillus mali* from cider and *Lactobacillus yamanashiensis* from wine. *J. Appl. Bacteriol.* 42: 219–228.

CARRETE, R., M.T. VIDAL, A. BORDONS, and M. CONSTANTI. 2002. Inhibitory effect of sulfur dioxide and other stress compounds in wine on the ATPase activity of *Oenococcus oeni*. *FEMS Microbiol. Lett.* 211: 155–159.

CARTESIO, M.S. and T.V. CAMPOS. 1988. Malolactic fermentation in wine: improvement in paper chromatographic techniques. *Am. J. Enol. Vitic.* 39: 188–189.

CASEY, G.P., C.A. MAGNUS, and W.M. INGLEDEW. 1984. High-gravity brewing: effects of nutrition on yeast composition, fermentative ability, and alcohol production. *Appl. Environ. Microbiol.* 48: 639–646.

CASTOR, J.G.B. and J.F. GUYMON. 1952. On the mechanism of formation of higher alcohols during alcoholic fermentation. *Science* 115: 147–149.

CAVAZZA, A., E. POZNANSKI, and G. TRIOLI. 2004. Restart of fermentation of simulated stuck wines by direct inoculation of active dry yeasts. *Am. J. Enol. Vitic.* 55: 160–167.

CAVIN, J.F., V. ANDIOC, P.X. ETIEVANT, and C. DIVIES. 1993. Ability of wine lactic acid bacteria to metabolize phenol carboxylic acids. *Am. J. Enol. Vitic.* 44: 76–80.

CHALFAN, Y., I. GOLDBERG, and R.I. MATELES. 1977. Isolation and characterization of malo-lactic bacteria from Israeli red wines. *J. Food Sci.* 42: 939–943, 968.

CHAMPAGNE, C.P., N. GARDNER, and A. LAFOND. 1989. Production of *Leuconostoc oenos* in apple juice media. *Lebensm.-Wiss. Technol.* 22: 376–381.

CHANG, P.-K., D. BHATNAGAR, and T.E. CLEVELAND. 2000. *Aspergillus.* In: *Encyclopedia of Food Microbiology.* R.K. Robinson, C.A. Batt, and P.D. Patel (Eds.), Volume 1, pp. 62–66. Academic Press, New York, NY.

CHARPENTIER, C. and M. FEUILLAT. 1993. Yeast autolysis. In: *Wine Microbiology and Biotechnology.* G.H. Fleet (Ed.), Chapter 7, pp. 225–242. Harwood Academic Publishers, Chur, Switzerland.

CHARPENTIER, C., A.M. DOS SANTOS, and M. FEUILLAT. 2004. Release of macromolecules by *Saccharomyces cerevisiae* during ageing of French flor sherry wine "Vin jaune." *Int. J. Food Microbiol.* 96: 253–262.

CHASSAGNE, D., M. GUILLOUX-BENATIER, H. ALEXANDRE, and A. VOILLEY. 2005. Sorption of wine volatile phenols by yeast lees. *Food Chem.* 91: 39–44.

CHATONNET, P., D. DUBOURDIEU, J.-N. BOIDRON, and M. PONS. 1992. The origin of ethylphenols in wines. *J. Sci. Food Agric.* 60: 165–178.

CHATONNET, P., D. DUBOURDIEU, and J.N. BOIDRON. 1995. The influence of *Brettanomyces/Dekkera* sp. yeasts and lactic acid bacteria on the ethylphenol content of red wines. *Am. J. Enol. Vitic.* 46: 463–468.

CHEN, E.C.H. 1978. The relative contribution of Ehrlich and biosynthetic pathways to the formation of fusel alcohols. *Am. Soc. Brew. Chem.* 36: 39–43.

CHISHOLM, M.G. and J.M. SAMUELS. 1992. Determination of the impact of the metabolites of sorbic acid on the odor of a spoiled red wine. *J. Agric. Food Chem.* 40: 830–833.

CIANI, M. and L. FERRARO. 1997. Role of oxygen on acetic acid production by *Brettanomyces/Dekkera* in winemaking. *J. Sci. Food Agric.* 75: 489–495.

CIANI, M. and L. FERRARO. 1998. Combined use of imobilized *Candida stellata* cells and *Saccharomyces cerevisiae* to improve the quality of wines. *J. Appl. Microbiol.* 85: 247–254.

CIANI, M. and F. MACCARELLI. 1998. Oenological properties of non-*Saccharomyces* yeasts associated with wine-making. *World J. Microbiol. Biotechnol.* 14: 199–203.

CIANI, M. and F. FATICHENTI. 1999. Selective sugar consumption by apiculate yeasts. *Lett. Appl. Microbiol.* 28: 203–206.

CIANI, M. and V. PEPE. 2002. The influence of pre-fermentative practices on the dominance of inoculated yeast starter under industrial conditions. *J. Sci. Food Agric.* 82: 573–578.

CIANI, M., L. FERRARO, and F. FATICHENTI. 2000. Influence of glycerol production on the aerobic and anaerobic growth of the wine yeast *Candida stellata*. *Enz. Microbiol. Technol.* 27: 698–703.

CLAISSE, O. and A. LONVAUD-FUNEL. 2000. Assimilation of glycerol by a strain of *Lactobacillus collinoides* isolated from cider. *Food Microbiol.* 17: 513–519.

CLEENWERCK, I., K. VANDEMEULEBROECKE, D. JANSSENS, and J. SWINGS. 2002. Re-examination of the genus *Acetobacter*, with descriptions of *Acetobacter cerevisiae* sp. nov. and *Acetobacter malorum* sp. nov. *Int. J. Sys. Evol. Microbiol.* 52: 1551–1558.

CLEMENTE-JIMENEZ, J.M., L. MINGORANCE-CAZORLA, S. MARTÍNEZ-RODRÍGUEZ, F.J. LAS HERAS-VÁZQUEZ, and F. RODRÍGUEZ-VICO. 2004. Molecular character-ization and oenological properties of wine yeasts isolated during spontaneous fermentation of six varieties of grape must. *Food Microbiol.* 21: 149–155.

CLESCERI, L.S., A.E. GREENBERG, and R.R. TRUSSELL (Eds.). 1989. Intralaboratory quality control guidelines. In: *Standard Methods for the Examination of Water and Wastewater.* 17th edition, Part 9020B, pp. 9–5 to 9–23. American Public Health Association. Washington, DC.

COCAIGN-BOUSQUET, M., C. GARRIGUES, P. LOUBIERE, and N.D. LINDLEY. 1996. Physiology of pyruvate metabolism in *Lactococcus lactis*. *Antonie van Leeuwenhoek* 70: 253–267.

COCOLIN, L. and D.A. MILLS. 2003. Wine yeast inhibition by sulfur dioxide: a comparison of culture-dependent and independent methods. *Am. J. Enol. Vitic.* 54: 125–130.

COCOLIN, L., A. HEISEY, and D.A. MILLS. 2001. Direct identification of the indigenous yeasts in commercial wine fermentations. *Am. J. Enol. Vitic.* 52: 49–53.

COCOLIN, L., M. MANZANO, S. REBECCA, and G. COMI. 2002. Monitoring of yeast population changes during a continuous wine fermentation by molecular methods. *Am. J. Enol. Vitic.* 53: 24–27.

COFRAN, D.R. and Bro. J. MEYER. 1970. The effect of fumaric acid on malolactic fermentation. *Am. J. Enol. Vitic.* 21: 189–192.

COGAN, T.M. and K.N. JORDAN. 1994. Metabolism of *Leuconostoc* bacteria. *J. Dairy Sci.* 77: 2704–2717.

COLLINS, E.B. 1972. Biosynthesis of flavor compounds by microorganisms. *J. Dairy Sci.* 55: 1022–1028.

COLVIN, R.J., L. CHENE, L.C. SOWDEN, and M. TAKAI. 1977. Purification and properties of a soluble polymer of glucose from cultures of *Acetobacter xylinum*. *Can. J. Biochem.* 55: 1057–1063.

COMITINI, F., J.I. DE, L. PEPE, I. MANNAZZU, and M. CIANI. 2004. *Pichia anomala* and *Kluyveromyces wickerhamii* killer toxins as new tools against *Dekkera/ Brettanomyces* spoilage yeasts. *FEMS Microbiol. Lett.* 238: 235–240.

COMITINI, F., R. FERRETTI, F. CLEMENTI, I. MANNAZZU, and M. CIANI. 2005. Interactions between *Saccharomyces cerevisiae* and malolactic bacteria: preliminary characterization of a yeast proteinaceous compound(s) active against *Oenococcus oeni. J. Appl. Microbiol.* 99: 105–111.

CONNELL, L., H. STENDER, and C.G. EDWARDS. 2002. Rapid detection of *Brettanomyces* from winery air samples based on peptide nucleic acid analysis. *Am. J. Enol. Vitic.* 53: 322–324.

CONNER, A.J. 1983. The comparative toxicity of vineyard pesticides to wine yeasts. *Am. J. Enol. Vitic.* 34: 278–279.

CONNER, D.E. 1993. Naturally occurring compounds. In: *Antimicrobials in Foods.* P.M. Davidson and A.L. Branen (Eds.), 2nd edition, Chapter 13, pp. 441–468. Marcel Dekker, Inc., New York, NY.

CONSTANTÍ, M., C. REGUANT, M. POBLET, F. ZAMORA, A. MAS, and J.M. GUILLAMÓN. 1998. Molecular analysis of yeast populations dynamics: effect of sulphur dioxide and inoculum on must fermentation. *Int. J. Food Microbiol.* 41: 169–175.

COSTELLO, P. 1988. The conduct of malolactic fermentation under commercial conditions. II. Use of commercial freeze-dried and frozen starter cultures of *Leuconostoc oenos. Tech. Rev. Aust. Wine Res. Inst.* 51: 9–12.

COSTELLO, P.J. and P.A. HENSCHKE. 2002. Mousy off-flavor of wine: precursors and biosynthesis of the causative N-heterocycles 2-ethyltetrahydropyridine, 2-acetyltetrahydropyridine, and 2-acetyl-1-pyrroline by *Lactobacillus hilgardii* DSM 20176. *J. Agric. Food Chem.* 50: 7079–7087.

COSTELLO, P.J., G.J. MORRISON, T.H. LEE, and G.H. FLEET. 1983. Numbers and species of lactic acid bacteria in wines during vinification. *Food Technol. Aust.* 35: 14–18.

COSTELLO, P.J., T.H. LEE, and P.A. HENSCHKE. 2001. Ability of lactic acid bacteria to produce N-heterocycles causing mousy off-flavour in wine. *Aust. J. Grape Wine Res.* 7: 160–167.

COTON, E., G. ROLLAN, A. BERTRAND, and A. LONVAUD-FUNEL. 1998. Histamine-producing lactic acid bacteria in wines: early detection, frequency, and distribution. *Am. J. Enol. Vitic.* 49: 199–204.

COUTO, J.A. and T.A. HOGG. 1994. Diversity of ethanol-tolerant lactobacilli isolated from Douro fortified wine: clustering and identification by numberical analysis of electrophoretic protein profiles. *J. Appl. Bacteriol.* 76: 487–491.

COUTO, J.A., F. NEVES, F. CAMPOS, and T. HOGG. 2005. Thermal inactivation of the wine spoilage yeasts *Dekkera/Brettanomyces. Int. J. Food Microbiol.* 104: 337–344.

COX, D.J. and T. HENICK-KLING. 1989. Chemiosmotic energy from malolactic fermentation. *J. Bacteriol.* 171: 5750–5752.

COX, D.J. and T. HENICK-KLING. 1995. Proton motive force and ATP generation during malolactic fermentation. *Am. J. Enol. Vitic.* 46: 319–323.

CRAIG, J.T. and T. HERESZTYN. 1984. 2-Ethyl-3,4,5,6-tetrahydropyridine-an assessment of its possible contribution to the mousy off-flavour of wines. *Am. J. Enol. Vitic.* 35: 46–48.

CROWELL, E.A. and J.F. GUYMON. 1963. Influence of aeration and suspended material on higher alcohols, acetoin, and diacetyl during fermentation. *Am. J. Enol. Vitic.* 14: 214–222.

CROWELL, E.A. and J.F. GUYMON. 1975. Wine constituents arising from sorbic acid addition and identification of 2-ethoxyhexa-3,5-diene as a source of geranium-like off-odor. *Am. J. Enol. Vitic.* 26: 97–102.

CROWTHER, J.R. 1995. ELISA. Theory and Practice. In: *Methods in Molecular Biology.* J.M. Walker (Ed.), Volume 42. Humana Press, Totowa, NJ.

D'INCECCO, N., E. BARTOWSKY, S. KASSARA, A. LANTE, P. SPETTOLI, and P. HENSCHKE. 2004. Release of glycosidically bound flavor compounds of Chardonnay by *Oenococcus oeni* during malolactic fermentation. *Food Microbiol.* 21: 257–264.

DAESCHEL, M.A., D.-S. JUNG, and B.T. WATSON. 1991. Controlling wine malolactic fermentation with nisin and nisin-resistant strains of *Leuconostoc oenos. Appl. Environ. Microbiol.* 57: 601–603.

DAESCHEL, M.A., T. MUSAFIJA-JEKNIC, Y. WU, D. BIZZARRI, and A. VILLA. 2002. High-performance liquid chromatography analysis of lysozyme in wine. *Am. J. Enol. Vitic.* 53: 154–157.

DARRIET, P., M. PONS, S. LAMY, and D. DUBOURDIEU. 2000. Identification and quantification of geosmin, an earthy odorant contaminating wines. *J. Agric. Food Chem.* 48: 4835–4838.

DARRIET, P., S. LAMY, S. LA GUERCHE, M. PONS, D. DUBOURDIEU, D. BLANCARD, P. STELIOPOULOS, and A. MOSANDL. 2001. Stereodifferentiation of geosmin in wine. *Eur. Food Res. Technol.* 213: 122–125.

DAUDT, C.E. and C.S. OUGH. 1980. Action of dimethyldicarbonate on various yeasts. *Am. J. Enol. Vitic.* 31: 21–23.

DAVIS, C.R., N.F.A. SILVEIRA, and G.H. FLEET. 1985a. Occurrence and properties of bacteriophage of *Leuconostoc oenos* in Australian wines. *Appl. Environ. Microbiol.* 50: 872–876.

DAVIS, C.R., D. WIBOWO, R. ESCHENBRUCH, T.H. LEE, and G.H. FLEET. 1985b. Practical implications of malolactic fermentation: a review. *Am. J. Enol. Vitic.* 36: 290–301.

DAVIS, C.R., D.J. WIBOWO, T.H. LEE, and G.H. FLEET. 1986a. Growth and metabolism of lactic acid bacteria during and after malolactic fermentations of wines at different pH. *Appl. Environ. Microbiol.* 51: 539–545.

DAVIS, C.R., D.J. WIBOWO, T.H. LEE, and G.H. FLEET. 1986b. Growth and metabolism of lactic acid bacteria during fermentation and conservation of some Australian wines. *Food Technol. Aust.* 38: 35–40.

DAVIS, C.R., D. WIBOWO, G.H. FLEET, and T.H. LEE. 1988. Properties of wine lactic acid bacteria: their potential enological significance. *Am. J. Enol. Vitic.* 39: 137–142.

DE LEY, J. and J. SCHELL. 1959. Oxidation of several substrates by *Acetobacter aceti. J. Bacteriol.* 77: 445–451.

DE LEY, J. and J. SWINGS. 1984. Genus II. *Gluconobacter.* In: *Bergey's Manual of Systematic Bacteriology.* J.G. Holt (Ed.), Volume 1, pp. 275–278. Williams and Wilkins Co., Baltimore, MD.

DE LEY, J., J. SWINGS, and F. GOSSELÉ. 1984. Genus I. *Acetobacter. Bergey's Manual of Systematic Bacteriology.* N.R. Krieg (Ed.), Volume 1, pp. 268–274. Williams and Wilkins Co., Baltimore, MD.

DE REVEL, G., N. MARTIN, L. PRIPIS-NICOLAU, A. LONVAUD-FUNEL, and A. BERTRAND. 1999. Contribution to the knowledge of malolactic fermentation influence on wine aroma. *J. Agric. Food Chem.* 47: 4003–4008.

DE ROSA, T., G. MARGHERI, I. MORET, G. SCARPONI, and G. VERSINI. 1983. Sorbic acid as a preservative in sparkling wine. Its efficacy and adverse flavor effect associated with ethyl sorbate formation. *Am. J. Enol. Vitic.* 34: 98–102.

DE VUYST, L. and E.J. VANDAMME. 1994. Antimicrobial potential of lactic acid bacteria. In: *Bacteriocins of Lactic Acid Bacteria.* L. De Vuyst and E.J. Vandamme (Eds.), pp. 91–142. Blackie Academic and Professional, Glasgow.

DEAK, T. 2002. Application of molecular techniques in wine microbiology. *Acta Alimentaria* 31: 37–44.

DEAK, T. and L.R. BEUCHAT. 1996. *Handbook of Food Spoilage Yeasts.* CRC Press, Inc., New York, NY.

DECALLONNE, J., M. DELMEE, P. WAUTHOZ, M. EL-LIOUI, and R. LAMBERT. 1991. A rapid procedure for the identification of lactic acid bacteria based on the gas chromatographic analysis of the cellular fatty acids. *J. Food Prot.* 54: 217–224.

DELAGE, N., A. D'HARLINGUE, B.C. CECCALDI, and G. BOMPEIX. 2003. Occurrence of mycotoxins in fruit juices and wine. *Food Cont.* 14: 225–227.

DELAQUIS, P., M. CLIFF, M. KING, B. GIRARD, J. HALL, and A. REYNOLDS. 2000. Effect of two commercial malolactic cultures on the chemical and sensory properties of Chancellor wines vinified with different yeasts and fermentation temperatures. *Am. J. Enol. Vitic.* 51: 42–48.

DELCROIX, A., Z. GUNATA, J.C. SAPIS, J.M. SALMON, and C. BAYONOVE. 1994. Glycosidase activities of three enological yeast strains during winemaking: effect on the terpenol content of Muscat wine. *Am. J. Enol. Vitic.* 45: 291–296.

DELFINI, C. 1989. Ability of wine malolactic bacteria to produce histamine. *Sci. Aliments* 9: 413–416.

DELFINI, C. and A. COSTA. 1993. Effects of the grape must lees and insoluble materials on the alcoholic fermentation rate and the production of acetic acid, pyruvic acid, and acetaldehyde. *Am. J. Enol. Vitic.* 44: 86–92.

DELFINI, C., C. COCITO, S. RAVAGLIA, and L. CONTERNO. 1993. Influence of clarification and suspended grape solid materials on sterol content of free run and pressed grape musts in the presence of growing yeast cells. *Am. J. Enol. Vitic.* 44: 452–458.

DELFINI, C., L. CONTERNO, G. CARPI, P. ROVERE, A. TABUSSO, C. COCITO, and A. AMATI. 1995. Microbiological stabilisation of grape musts and wine by hydrostatic pressures. *J. Wine Res.* 6: 143–151.

DELFINI, C.M. CERSOSIMA, V. DEL PRETE, M. STRANO, G. GAETANO, A. PAGLIARA, and S. AMBRÒ. 2004. Resistance screening essay of wine lactic acid bacteria on lysozyme: efficacy of lysozyme in unclarified grape musts. *J. Agric. Food Chem.* 52: 1861–1866.

DEWEY, F.M., S.E. EBELER, D.O. ADAMS, A.C. NOBLE, and U.M. MEYER. 2000. Quantification of *Botrytis* in grape juice determined by a monoclonal antibody-based immunoassay. *Am. J. Enol. Vitic.* 51: 276–282.

DEWEY, F.M., U. MEYER, and C. DANKS. 2005. Rapid immunoassays for stable *Botrytis* antigens in pre- and postsymptomatic grape berries, grape juice and wines. Abstr. 56th American Society for Enology Viticulture Annual Meeting, Seattle, WA. *Am. J. Enol. Vitic.* 56: 302A–303A.

DHARMADHIKARI, M.R. and K.L. WILKER. 1998. Deacidification of high malate must with *Schizosaccharomyces pombe*. *Am. J. Enol. Vitic.* 49: 408–412.

DIAS, L., S. DIAS, T. SANCHO, H. STENDER, A. QUEROL, M. MALFEITO-FERREIRA, and V. LOUREIRO. 2003a. Identification of yeasts isolated from wine-related environments and capable of producing 4-ethylphenol. *Food Microbiol.* 20: 567–574.

DIAS, L., S. PEREIRA-DA-SILVA, M. TAVAARES, M. MALFEITO-FERREIRA, and V. LOUREIRO. 2003b. Factors affecting the production of 4-ethylphenol by the yeast *Dekkera bruxellensis* in enological conditions. *Food Microbiol.* 20: 377–384.

DICK, K.J., P.C. MOLAN, and R. ESCHENBRUCH. 1992. The isolation from *Saccharomyces cerevisiae* of two antibacterial cationic proteins that inhibit malolactic bacteria. *Vitis* 31: 105–116.

DICKS, L.M.T. and H.J.J. VAN VUUREN. 1988. Identification and physiological characteristics of heterofermentative strains of *Lactobacillus* from South African red wines. *J. Appl. Bacteriol.* 64: 505–513.

DICKS, L.M.T., F. DELLAGLIO, and M.D. COLLINS. 1995. Proposal to reclassify *Leuconostoc oenos* as *Oenococcus oeni* [corrig.] gen. nov., comb. nov. *Int. J. Syst. Bacteriol.* 45: 395–397.

DITTRICH, H.H. 1977. *Mikrobiologie des Weines. Handbuch der Getranketechnologie.* Ulmer, Stuttgart.

DITTRICH, H.H., W.R. SPONHOLZ, and W. KAST. 1974. Vergleichende Untersuchungen von Mosten und Weinen aus gesunden und aus Botrytis-infizierten Traubenbeeren. I. Saeurestoffwechsel, Zuchkerstofwechselprodukte, Leucoanthocyangehalte. *Vitis* 13: 36–49.

DITTRICH, H.H., W.R. SPONHOLZ, and H.G. GOEBEL. 1975. Vergleichende Untersuchungen von Mosten und Weinen aus gesunden und aus Botrytis-infizierten Traubenbeeren II. Modellversuche zur Veranderung des Mostes durch Botrytis-Infektion und ihre Konsequenzen fuer die Nebenproduktbildung bei der Gaerung. *Vitis* 13: 336–347.

DIVOL, B. and A. LONVAUD-FUNEL. 2005. Evidence for viable but nonculturable yeasts in botrytis-affected wine. *J. Appl. Microbiol.* 99: 85–93.

DLAUCHY, D., J. TORNAI-LEHOCZLI, and G. PETER. 1999. Restriction enzyme analysis of PCR amplified rDNA as a taxonomic tool in yeast identification. *Sys. Appl. Microbiol.* 22: 445–453.

DOCO, T., P. VUCHOT, V. CHEYNIER, and M. MOUTOUNET. 2003. Structural modification of wine arabinogalactans during aging on lees. *Am. J. Enol. Vitic.* 54: 150–157.

DOIGNON, F. and N. ROZÈS. 1992. Effect of triazole fungicides on lipid metabolism of *Saccharomyces cerevisiae. Lett. Appl. Microbiol.* 15: 172–174.

DONÈCHE, B.J. 1993. Botrytized wines. In: *Wine Microbiology and Biotechnology.* G.H. Fleet (Ed.), Chapter 11, pp. 327–351. Harwood Academic Publishers, Chur, Switzerland.

DONNELLY, D.M. 1977. Airborne microbial contamination in a winery bottling room. *Am. J. Enol. Vitic.* 28: 176–181.

DOORES, S. 1993. Organic acids. In: *Antimicrobials in Foods.* P.M. Davidson and A.L. Branen (Eds.), 2nd edition, pp. 95–136. Marcel Dekker, Inc., New York, NY.

DOTT, W., M. HEINZEL, and H.G. TRÜPER. 1976. Sulfite formation by wine yeasts. I. Relationships between growth, fermentation and sulfite formation. *Arch. Microbiol.* 107: 289–292.

DOUGLAS, H.C. and W.V. CRUESS. 1936. A *Lactobacillus* from California wine: *Lactobacillus hilgardii. Food Res.* 1: 113–119.

DRYSDALE, G.S. and G.H. FLEET. 1985. Acetic acid bacteria in some Australian wines. *Food Technol. Aust.* 37: 17–20.

DRYSDALE, G.S. and G.H. FLEET. 1988. Acetic acid bacteria in winemaking: a review. *Am. J. Enol. Vitic.* 39: 143–154.

DRYSDALE, G.S. and G.H. FLEET. 1989a. The effect of acetic acid bacteria upon the growth and metabolism of yeasts during the fermentation of grape juice. *J. Appl. Bacteriol.* 67: 471–481.

DRYSDALE, G.S. and G.H. FLEET. 1989b. The growth and survival of acetic acid bacteria in wines at different concentrations of oxygen. *Am. J. Enol. Vitic.* 40: 99–105.

DU PLESSIS, L.D.W. 1963. The microbiology of South African winemaking. Part V. Vitamin and amino acid requirements of lactic acid bacteria from dry wines. *S. Afr. J. Agric. Sci.* 6: 485–494.

DU PLESSIS, L.D.W. and J.A. VAN ZYL. 1963a. The microbiology of South African winemaking. Part IV. The taxonomy and incidence of lactic acid bacteria from dry wines. *S. Afr. J. Agric. Sci.* 6: 261–273.

DU PLESSIS, L.D.W. and J.A. VAN ZYL. 1963b. The microbiology of South African winemaking. Part VI. Fermentation of D-glucose, D-fructose, D-xylose, and L-arabinose by lactic acid bacteria from dry wines. *S. Afr. J. Agric. Sci.* 6: 673–688.

DU PLESSIS, H.W., L.M.T. DICKS, I.S. PRETORIUS, M.G. LAMBRECHTS, and M. DU TOIT. 2004. Identification of lactic acid bacteria isolated from South African brandy base wines. *Int. J. Food Microbiol.* 91: 19–29.

DU TOIT, M. and I.S. PRETORIUS. 2000. Microbial spoilage and preservation of wine: using weapons from nature's own arsenal—a review. *S. Afr. J. Enol. Vitic.* 21: 74–96.

DU TOIT, W.J. and I.S. PRETORIUS. 2002. The occurrence, control and esoteric effect of acetic acid bacteria in winemaking. *Ann. Microbiol.* 52: 155–179.

DU TOIT, W.J. and M.G. LAMBRECHTS. 2002. The enumeration and identification of acetic acid bacteria from South African red wine fermentations. *Int. J. Food Microbiol.* 74: 57–64.

Du Toit, W.J., I.S. Pretorius, and A. Lonvaud-Funel. 2005. The effect of suphur dioxide and oxygen on the viability and culturability of a strain of *Acetobacter pasteurianus* and a strain of *Brettanomyces bruxellensis* isolated from wine. *J. Appl. Microbiol.* 98: 862–871.

Duarte, F.L., C. Pais, I. Spencer-Martins, and C. Leao. 1999. Distinctive electrophoretic isoenzyme profiles in *Saccharomyces sensu stricto*. *Int. J. Sys. Bacteriol.* 49: 1907–1913.

Dubernet, M., P. Ribéreau-Gayon, H.R. Lerner, E. Harel, and A.M. Mayer. 1977. Purification and properties of laccase from *Botrytis cinerea*. *Phytochem.* 16: 191–193.

Dubourdieu, D., P. Ribéreau-Gayon, and B. Fournet. 1981. Structure of the extracellular β-D-glucan from *Botrytis cinerea*. *Carbohydr. Res.* 93: 294–299.

Dubourdieu, D., C. Grassin, C. Deruche, and P. Ribéreau-Gayon. 1984. Mise au point d'une mesure rapide de l'activite laccase dans les mouts et dan les vins par la methode a la syringaldazine. Application a l'appreciation de l'etat sanitaire des vendages. *Conn. Vigne Vin* 18: 237–252.

Duenas, M., A. Irastorza, K. Fernandez, and A. Bilbao. 1995. Heterofermentative lactobacilli causing ropiness in Basque Country ciders. *J. Food Prot.* 58: 76–80.

Duke, G.R. 1979. Factors influencing the survival and utilization of lyophilized cultures of *Leuconostoc oenos* PSU-1 for inoculation of wines. Master of Science Thesis. The Pennsylvania State University, Department of Food Science, University Park, PA.

Dupin, I.V.S., B.M. McKinnon, C. Ryan, M. Boulay, A.J. Markides, G.P. Jones, P.J. Williams, and E.J. Waters. 2000. *Saccharomyces cerevisiae* mannoproteins that protect wine from protein haze: their release during fermentation and lees contact and a proposal for their mechanism of action. *J. Agric. Food Chem.* 48: 3098–3105.

Edinger, W.D. 1986. Reducing the use of sulfur dioxide in winemaking. Part I. *Vineyard Winery Manage.* 12 (Nov/Dec): 24–27.

Edinger, W.D. and D.F. Splitistoesser. 1986. Production by lactic acid bacteria of sorbic alcohol, the precursor of geranium odor compound. *Am. J. Enol. Vitic.* 37: 34–38.

Edwards, C.G. 2005. *Illustrated Guide to Microbes and Sediments in Wine, Beer, and Juice.* WineBugs Publishing LLC, Pullman, WA.

Edwards, C.G. and R.B. Beelman. 1987. Inhibition of the malolactic bacterium, *Leuconostoc oenos* (PSU-1) by decanoic acid and subsequent removal of the inhibition by yeast ghosts. *Am. J. Enol. Vitic.* 38: 239–242.

Edwards, C.G. and R.B. Beelman. 1989. Inducing malolactic fermentation in wines. In: *Biotechnology Advances.* M. Moo-Young (Ed.), 7: 333–360. Pergamon Press, Oxford.

Edwards, C.G. and K.A. Jensen. 1992. Occurrence and characterization of lactic acid bacteria from Washington state wines: *Pediococcus* spp. *Am. J. Enol. Vitic.* 43: 233–238.

EDWARDS, C.G. and J.C. PETERSON. 1994. Sorbent extraction and analysis of volatile metabolites synthesized by lactic acid bacteria isolated from wines. *J. Food Sci.* 59: 192–196.

EDWARDS, C.G., R.B. BEELMAN, C.E. BARTLEY, and A.L. MCCONNELL. 1990. Production of decanoic acid and other volatile compounds and the growth of yeast and malolactic bacteria during vinification. *Am. J. Enol. Vitic.* 41: 48–56.

EDWARDS, C.G., K.A. JENSEN, S.E. SPAYD, and B.J. SEYMOUR. 1991. Isolation and characterization of native strains of *Leuconostoc oenos* from Washington state wines. *Am. J. Enol. Vitic.* 42: 219–226.

EDWARDS, C.G., J.R. POWERS, K.A. JENSEN, K.M. WELLER, and J.C. PETERSON. 1993. *Lactobacillus* spp. from Washington State wines: isolation and characterization. *J. Food Sci.* 58: 453–458.

EDWARDS, C.G., J.C. PETERSON, T.D. BOYLSTON, and T.D. VASILE. 1994. Interactions between *Leuconostoc oenos* and *Pediococcus* spp. during vinification of red wines. *Am. J. Enol. Vitic.* 45: 49–55.

EDWARDS, C.G., K.M. HAAG, M.D. COLLINS, R. HUTSON, and Y.C. HUANG. 1998a. *Lactobacillus kunkeei* sp. nov., a spoilage organism associated with grape juice fermentations. *J. Appl. Microbiol.* 84: 698–702.

EDWARDS, C.G., K.M. HAAG, and M.D. COLLINS. 1998b. Identification of some lactic acid bacteria associated with sluggish/stuck fermentations. *Am. J. Enol. Vitic.* 49: 445–448.

EDWARDS, C.G., A.G. REYNOLDS, A.V. RODRIGUEZ, M.J. SEMON, and J.M. MILLS. 1999a. Implication of acetic acid in the induction of slow/stuck grape juice fermentations and inhibition of yeast by *Lactobacillus* sp. *Am. J. Enol. Vitic.* 50: 204–210.

EDWARDS, C.G., K.M. HAAG, M.J. SEMON, A.V. RODRIGUEZ, and J. MILLS. 1999b. Evaluation of processing methods to control the growth of *Lactobacillus kunkeei*, a microorganism implicated in sluggish alcoholic fermentations of grape musts. *S. Afr. J. Enol. Vitic.* 20: 11–19.

EDWARDS, C.G., M.D. COLLINS, P.A. LAWSON, and A.V. RODRIGUEZ. 2000. *Lactobacillus nagelii* sp. nov., an organism isolated from a partially fermented wine. *Int. J. Syst. Evol. Microbiol.* 50: 699–702.

EGLI, C.M. and T. HENICK-KLING. 2001. Identification of *Brettanomyces/Dekkera* species based on polymorphism in the rRNA internal transcribed spacer region. *Am. J. Enol. Vitic.* 52: 241–247.

EGLI, C.M. W.D. EDINGER, C.M. MITRAKUL, and T. HENICK-KLING. 1998. Dynamics of indigenous and inoculated yeast populations and their effect on the sensory character of Riesling and Chardonnay wines. *J. Appl. Microbiol.* 85: 779–789.

EGLINTON, J.M. and P.A. HENSCHKE. 1996. *Saccharomyces cerevisiae* strains AWRI 838, Lalvin EC1118 and Maurivin PDM do not produce excessive sulfur dioxide in white wine fermentations. *Aust. J. Grape Wine Res.* 2: 77–83.

EL-GENDY, S.M., H. ABDEL-GALIL, Y. SHAHIN, and F.Z. HEGAZI. 1983. Acetoin and diacetyl production by homo- and heterofermentative lactic acid bacteria. *J. Food Prot.* 46: 420–425.

ENGLISH, J.T., A.M. BLEDSOE, J.J. MAROIS, and W.M. KLIEWER. 1990. Influence of grapevine canopy management on the evaporative potential in the fruit zone. *Am. J. Enol. Vitic.* 41: 137–141.

ERASMUS, D.J., M. CLIFF, and H.J.J. VAN VUUREN. 2004. Impact of yeast strain on the production of acetic acid, glycerol, and the sensory attributes of icewine. *Am. J. Enol. Vitic.* 55: 371–378.

ERTEN, H. 2002. Relations between elevated temperatures and fermentation behaviour of *Kloeckera apiculata* and *Saccharomyces cerevisiae* associated with winemaking in mixed cultures. *World J. Microbiol. Biotechnol.* 18: 373–378.

ESAU, P. 1967. Pentoses in wine. I. Survey of possible sources. *Am. J. Enol. Vitic.* 18: 210–216.

ESCHENBRUCH, R. 1974. Sulfite and sulfide formation during wine making. A review. *Am. J. Enol. Vitic.* 25: 157–161.

ESCHENBRUCH, R. and P. BONISH. 1976. Production of sulphite and sulphide by low and high-sulphite forming wine yeasts. *Arch. Microbiol.* 107: 299–302.

ESCHENBRUCH, B. and H.H. DITTRICH. 1986. Metabolism of acetic-acid bacteria in relation to their importance to wine quality. *Zentralbl. Mikrobiol.* 141: 279–289.

ESCHENBRUCH, R., P. BONISH, and B.M. FISHER. 1978. The production of H_2S by pure culture wine yeasts. *Vitis* 17: 67–74.

ESSIA NGANG, J.J., F. LETOURNEAU, E. WOLNIEWICZ, and P. VILLA. 1990. Inhibition of beet molasses alcoholic fermentation by lactobacilli. *Appl. Microbiol. Biotechnol.* 33: 490–493.

ESTEBAN, A., M. LOURDES ABARCA, M. ROSA BRAGULAT, and F. JAVIER CABAÑES. 2004. Effects of temperture and incubation time on production of ochratoxin A by black aspergilli. *Res. Microbiol.* 155: 861–866.

ESTEVE-ZARZOSO, B., P. MANZANARES, D. RAMÓN, and A. QUEROL. 1998. The role of non-*Saccharomyces* yeasts in industrial winemaking. *Int. Microbiol* 1: 143–149.

EWART, A.J.W., N.J. HASELGROVE, J.H. SITTERS, and R. YOUNG. 1989. The effect of *Botrytis cinerea* on the color of *Vitis vinifera* cv. Pinot Noir. *Proceedings of the Seventh Australian Wine Industry Technology Conference.* P.J. Williams D.M. Davidson, and T.H. Lee (eds.). Australian Wine Research Institute, pp. 229–230.

FERRANDO, M., C. GÜELL, and F. LÓPEZ. 1998. Industrial wine making: comparison of must clarification treatments. *J. Agric. Food Chem.* 46: 1523–1528.

FEUILLAT, M. 2003. Yeast macromolecules: origin, composition, and enological interest. *Am. J. Enol. Vitic.* 54: 211–213.

FEULLAT, M., M. GUILLOUX-BENATIER, and V. GERBAUX. 1985. Essais d'activation de la fermentation malolactique dans les vins. *Sci. Aliments* 5: 103–122.

FINNEY, M. 1993. Pulsed-field gel electrophoresis. In: *Current Protocols in Molecular Biology.* F.M. Ausubel, R. Brent, R.E. Kingston, D.D. Moore, J.G. Seidman, J.A. Smith and K. Struhl (Eds.), Volume 1, pp. 2.5.9–2.5.17. Green-Wiley, New York, NY.

FIRME, M.P., M.C. LEITAO, and M.V. SAN RAMAO. 1994. The metabolism of sugar and malic acid by *Leuconostoc oenos*: effect of malic acid, pH, and aeration conditions. *J. Appl. Bacteriol.* 76: 173–181.

FLEET, G.H. 2003. Yeast interactions and wine flavour. *Int. J. Food Microbiol.* 86: 11–22.

FLEET, G.H. and G.M. HEARD. 1993. Yeasts—Growth during fermentation. In: *Wine Microbiology and Biotechnology*. G.H. Fleet (Ed.), Chapter 2, pp. 27–55. Harwood Academic Publishers, Chur, Switzerland.

FLEET, G.H., S. LAFON-LAFOURCADE, and P. RIBEREAU-GAYON. 1984. Evolution of yeasts and lactic acid bacteria during fermentation and storage of Bordeaux wines. *Appl. Environ. Microbiol.* 48: 1034–1038.

FLOWERS, R.S., J.S. GECAN, and D.J. PUSCH. 1992. Laboratory quality assurance. In: *Compendium of Microbiological Methods for the Examination of Foods*. C. Vanderzant and D.F. Splittstoesser (Eds.), 3rd edition, Chapter 1, pp. 1–23. American Public Health Association, Washington, DC.

FONSECA, A., J.W. FELL, C.P. KURTZMAN, and I. SPENCER-MARTINS. 2000. *Candida tartarivorans* sp. nov., an anamorphic ascomycetous yeast with the capacity to degrade L(+)- and meso-tartaric acid. *Int. J. System. Evol. Microbiol.* 50: 389–394.

FORNACHON, J.C.M. 1957. The occurrence of malolactic fermentation in Australian wines. *Aust. J. Appl. Sci.* 8: 120–129.

FORNACHON, J.C.M. 1963. Inhibition of certain lactic acid bacteria by free and bound sulphur dioxide. *J. Sci. Food Agric.* 14: 857–862.

FORNACHON, J.C.M. 1968. Influence of different yeasts on the growth of lactic acid bacteria in wine. *J. Sci. Food Agric.* 19: 374–378.

FORNACHON, J.C.M. and B. LLOYD. 1965. Bacterial production of diacetyl and acetoin in wine. *J. Sci. Food Agric.* 16: 710–716.

FORNACHON, J.C.M., H.C. DOUGLAS, and R.H. VAUGHN. 1949. *Lactobacillus trichodes* nov. spec., a bacterium causing spoilage in appetizer and dessert wines. *Hilgardia* 19: 129–132.

FOY, J.J. 1994a. Use and manufacturing of active dry wine yeast cultures. In: *Proceedings of the New York Wine Industry Workshop*. T. Henick-Kling (Ed.), pp. 21–28. Geneva, NY.

FOY, J.J. 1994b. Evaluation of commercial active dry wine yeast. In: *Proceedings of the New York Wine Industry Workshop*. T. Henick-Kling (Ed.), pp. 29–38. Geneva, NY.

FRANK, J.F. and R. CHMIELEWSKI. 2001. Influence of surface finish on the cleanability of stainless steel. *J. Food Prot.* 64: 1178–1182.

FRANTA, B.D., L.R. MATICK, and J.W. SHERBON. 1986. The analysis of pentoses in dry wine by high performance liquid chromatography with post-column derivatization. *Am. J. Enol. Vitic.* 37: 269–274.

FREER, S.N. 2002. Acetic acid production by *Dekkera/Brettanomyces* yeasts. *World J. Microbiol. Biotechnol.* 18: 271–275.

FUGELSANG, K.C. and B.W. ZOECKLEIN. 1993. MLF Survey. *Pract. Winery Vineyard* 14 (May/June): 12–18.

FUGELSANG, K.C. and B.W. ZOECKLEIN. 2003. Population dynamics and effects of *Brettanomyces bruxellensis* strains on Pinot noir (*Vitis vinifera* L.) wines. *Am. J. Enol. Vitic.* 54: 294–300.

FUGELSANG, K.C., M.M. OSBORN, and C.J. MULLER. 1993. *Brettanomyces* and *Dekkera*. Implications in winemaking. In: *Beer and Wine Production*. B.H. Gump (Ed.), 536: 110–129. American Chemical Society, Washington, DC.

FUMI, M.D., G. TRIOLI, M.G. COLOMBI, and O. COLAGRANDE. 1988. Immobilization of *Saccharomyces cerevisiae* in calcium alginate gel and its application to bottle-fermented sparkling wine production. *Am. J. Enol. Vitic.* 39: 267–272.

G-ALEGRÍA, E., I. LÓPEZ, J.I. RUIZ, J. SÁENZ, E. FERNÁNDEZ, M. ZARAZAGA, M. DIZY, C. TORRES, and F. RUIZ-LARREA. 2004. High tolerance of wild *Lactobacillus plantarum* and *Oenococcus oeni* strains to lyophilisation and stress environmental conditions of acid pH and ethanol. *FEMS Microbiol. Lett.* 230: 53–61.

GALLANDER, J.F. 1977. Deacidification of Eastern table wines with *Schizosaccharomyces pombe*. *Am. J. Enol.Vitic.* 28: 65–68.

GAMBARO, A., E. BOIDO, A. ZLOTEJABLKO, K. MEDINA, A. LLORET, E. DELLACASSA, and F. CARRAU. 2001. Effect of malolactic fermentation on the aroma properties of Tannat wine. *Aust. J. Grape Wine Res.* 7: 27–32.

GAO, C. and G.H. FLEET. 1988. The effects of temperature and pH on the ethanol tolerance of the wine yeasts, *Saccharomyces cerevisiae*, *Candida stellata*, and *Kloeckera apiculata*. *J. Appl. Bacteriol.* 65: 405–409.

GARVIE, E.I. 1967a. *Leuconostoc oenos* sp. nov. *J. Gen. Microbiol.* 48: 431–438.

GARVIE, E.I. 1967b. The growth factor and amino acid requirements of species of the genus *Leuconostoc*, including *Leuconostoc paramesenteroides* (sp. nov.) and *Leuconostoc oenos*. *J. Gen. Microbiol.* 48: 439–447.

GARVIE, E.I. 1974. Nomenclatural problems of the pediococci. Request for an opinion. *Int. J. Sys. Bacteriol.* 24: 301–306.

GARVIE, E.I. 1976. Hybridization between the deoxyribonucleic acids of some strains of heterofermentative lactic acid bacteria. *Int. J. Sys. Bacteriol.* 26: 116–122.

GARVIE, E.I. 1984. Separation of species of the genus *Leuconostoc* and differentiation of the leuconostocs from other lactic acid bacteria. *Meth. Microbiol.* 16: 147–178.

GARVIE, E.I. 1986a. Genus *Leuconostoc*. In: *Bergey's Manual of Systematic Bacteriology*. P.H.A. Sneath, N.S. Mair, M.E. Sharpe, and J.G. Holt (Eds.), pp. 1071–1075. The Williams and Wilkins Co., Baltimore, MD.

GARVIE, E.I. 1986b. Genus *Pediococcus*. In: *Bergey's Manual of Systematic Bacteriology*. P.H.A. Sneath, N.S. Mair, M.E. Sharpe, and J.G. Holt (Eds.), pp. 1075–1079. The Williams and Wilkins Co., Baltimore, MD.

GARVIE, E.I. and L.A. MABBITT. 1967. Stimulation of growth of *Leuconostoc oenos* by tomato juice. *Arch. Microbiol.* 55: 398–407.

GERBAUX, V., A. VILLA, C. MONAMY, and A. BERTRAND. 1997. Use of lysozyme to inhibit malolactic fermentation and to stabilize wine after malolactic fermentation. *Am. J. Enol. Vitic.* 48: 49–54.

GERBAUX, V., B. VINCENT, and A. BERTRAND. 2002. Influence of maceration temperature and enzymes on the content of volatile phenols in Pinot noir wines. *Am. J. Enol. Vitic.* 53: 131–137.

GERGELY, S., E. BEKASSY-MOLNAR, and GY. VATAI. 2003. The use of multiobjective optimization to improve wine filtration. *J. Food Eng.* 58: 311–316.

GIANNAKOPOULOS, P.I., P. MARKAKIS, and G.S. HOWELL. 1984. The influence of malolactic strain on the fermentation and wine quality of three Eastern red wine grape cultivars. *Am. J. Enol. Vitic.* 35: 1–4.

GIBSON, T. and Y. ABDEL-MALEK. 1945. The formation of carbon dioxide by lactic acid bacteria and *Bacillus licheniformis* and a cultural method of detecting the process. *J. Dairy Res.* 14: 35–44.

GIL, J.V., J.J. MATEO, M. JIMENEZ, A. PASTOR, and T. HUERTA. 1996. Aroma compounds in wine as influenced by apiculate yeasts. *J. Food Sci.* 61: 1247–1266.

GILLILAND, R.B. and J.P. LACEY. 1964. Lethal action by an *Acetobacter* on yeasts. *Nature* 202: 727–728.

GINDREAU, E., E. WALLING, and A. LONVAUD-FUNEL. 2001. Direct polymerase chain reaction detection of ropy *Pediococcus damnosus* strains in wine. *J. Appl. Microbiol.* 90: 535–542.

GINI, B. and R.H. VAUGHN. 1962. Characteristics of some bacteria associated with the spoilage of California dessert wines. *Am. J. Enol. Vitic.* 13: 20–31.

GLÒRIA, M.B.A., B.T. WATSON, L. SIMON-SARKADI, and M.A. DAESCHEL. 1998. A survey of biogenic amines in Oregon Pinot noir and Cabernet Sauvignon wines. *Am. J. Enol. Vitic.* 49: 279–282.

GODSHALL, M.A. 1997. How carbohydrates influence food flavor. *Food Tech.* 51: 63–67.

GOLDSTEIN, A. and J.O. LAMPEN. 1975. β-D-Fructofuranoside fructohydrolase from yeast. *Meth. Enzymol.* 42: 504–511.

GOÑI, D.T. and C.A. AZPILICUETA. 2001. Influence of yeast strain on biogenic amines content in wines: relationship with the utilization of amino acids during fermentation. *Am. J. Enol. Vitic.* 52: 185–190.

GONCALVES, F., A. HEYRAUD, M. NORBERTA DE PINHO, and M. RINAUDO. 2002. Characterization of white wine mannoproteins. *J. Agric. Food Chem.* 50: 6097–6101.

GONZÁLEZ, A., N. HIERRO, M. POBLET, A. MAS, and J.M. GUILLAMÓN. 2005. Application of molecular methods to demonstrate species and strain evolution of acetic acid bacteria population during wine production. *Int. J. Food Microbiol.* 102: 295–304.

GONZALEZ-TECHERA, A., S. JUBANY, F.M. CARRAU, and C. GAGGERO. 2001. Differentiation of industrial wine yeast strains using microsatellite markers. *Lett. Appl. Microbiol.* 33: 71–75.

GORGA, A., O. CLAISSE, and A. LONVAUD-FUNEL. 2002. Organisation of the genes encoding glycerol dehydratase of *Lactobacillus collinoides*, *Lactobacillus hilgardii* and *Lactobacillus diolivorans*. *Sci. Aliments* 22: 151–160.

GOTO, S., K. TAKAYAMA, and T. SHINOHARA. 1989. Occurrence of molds in wine storage cellars. *J. Ferm. Bioeng.* 68: 230–232.

GOTTSCHALK, G. 1986. *Bacterial Metabolism*, 2nd edition. Springer-Verlag, New York, NY.

GRADWOHL, R.B.H. 1948. *Clinical Laboratory Methods and Diagnosis*. 4th edition, Volume II, pp. 1400–1401. The C.V. Mosby Company, St. Louis, MO.

GRANCHI, L., M. BOSCO, A. MESSINI, and M. VINCENZINI. 1999. Rapid detection and quantification of yeast species during spontaneous wine fermentation by PCR-RFLP analysis of the rDNA ITS region. *J. Appl. Microbiol.* 87: 949–956.

GRBIN, P.R. and P.A. HENSCHKE. 2000. Mousy off-flavor production in grape juice and wine by *Dekkera* and *Brettanomyces* yeasts. *Aust. J. Grape Wine Res.* 6: 255–262.

GREENE, A.K., P.J. VERGANO, B.K. FEW, and J.C. SERAFINI. 1994. Effect of ozonated water sanitization on gasket materials used in fluid food processing. *J. Food Eng.* 21: 439–446.

GRIMALDI, A., H. MCLEAN, and V. JIRANEK. 2000. Identification and partial characterization of glycosidic activities of commercial strains of the lactic acid bacterium, *Oenococcus oeni*. *Am. J. Enol. Vitic.* 51: 362–369.

GROAT, M. and C.S. OUGH. 1978. Effects of insoluble solids added to clarified musts on fermentation rate, wine composition, and wine quality. *Am. J. Enol. Vitic.* 29: 112–119.

GUERZONI, E. and R. MARCHETTI. 1987. Analysis of yeast flora associated with grape sour rot and of the chemical disease markers. *Appl. Environ. Microbiol.* 53: 571–576.

GUIDICI, P. and R.E. KUNKEE. 1994. The effect of nitrogen deficiency and sulfur containing amino acids on the reduction of sulfate to hydrogen sulfide by wine yeasts. *Am. J. Enol. Vitic.* 44: 107–112.

GUILLOUX-BENATIER, M., M. FEUILLAT, and B. CIOLFI. 1985. Contribution á l'étude de la dégradation de l'acide L-malique par les bactéries lactiques isolées du vin: effet stimulant des autolysate du leuvres. *Vitis* 24: 59–74.

GUITART, A., P.H. ORTE, and J. CACHO. 1998. Effect of different clarification treatments on the amino acid content of Chardonnay musts and wines. *Am. J. Enol. Vitic.* 49: 389–396.

GÜNATA, Y.Z., C.L. BAYONOVE, R.L. BAUMES, and R.E. CORDONNIER. 1986. Stability of free and bound fractions of some aroma components of grapes cv. Muscat during the wine processing: preliminary results. *Am. J. Enol. Vitic.* 37: 112–114.

GÜNATA, Y.Z., C.L. BAYONOVE, C. TAPIERO, and R.E. CORDONNIER. 1990. Hydrolysis of grape monoterpenyl β-D-glucosides by various β-glucosidases. *J. Agric. Food Chem.* 38: 1232–1236.

GUYMON, J.F., J.L. INGRAHAM, and E.A. CROWELL. 1961. Influence of aeration upon the formation of higher alcohols by yeasts. *Am. J. Enol. Vitic.* 12: 60–66.

HALLINAN, C.P., D.J. SAUL, and V. JIRANEK. 1999. Differential utilisation of sulfur compounds for H_2S liberation by nitrogen-starved wine yeasts. *Aust. J. Grape Wine Res.* 5: 82–90.

HAMPSON, B. 2000. Use of ozone for winery and environmental sanitation. *Pract. Winery Vineyard* 20 (Jan/Feb): 27–30.

HANSEN, E.H., P. NISSEN, P. SOMMER, J.C. NIELSEN, and N. ARNEBORG. 2001. The effect of oxygen on the survival of non-*Saccharomyces* yeasts during mixed culture fermentations of grape juice with *Saccharomyces cerevisiae*. *J. Appl. Microbiol.* 91: 541–547.

HARRIGAN, W.F. 1998. *Laboratory Methods in Food Microbiology*, 3rd edition. Academic Press, New York, NY.

HARTMAN, P.A., B. SWAMINATHAN, M.S. CURIALE, R. FIRSTENBERG-EDEN, A.N. SHARPE, N.A. COX, D.Y.C. FUNG, and M.C. GOLDSCHMIDT. 1992. Rapid methods and automation. In: *Compendium of Microbiological Methods for the Examination of*

Foods. C. Vanderzant and D.F. Splittstoesser (Eds.), 3rd edition, Chapter 39, pp. 665–746. American Public Health Association, Washington, DC.

HAWKER, J.S., H.P. RUFFNER, and R.R. WALKER. 1976. The sucrose content of some Australian grapes. *Am. J. Enol. Vitic.* 27: 125–129.

HAYMAN, D.C. and P.R. MONK. 1982. Starter culture preparation for the induction of malolactic fermentation in wine. *Food Tech. Aust.* 34: 14–18.

HEARD, G.M. and G.H. FLEET. 1985. Growth of natural yeast flora during the fermentation of inoculated wines. *Appl. Environ. Microbiol.* 50: 727–728.

HEARD, G.M. and G.H. FLEET. 1986. Occurrence and growth of yeast species during the fermentation of some Australian wines. *Food Technol. Aust.* 38: 22–25.

HEARD, G.M. and G.H. FLEET. 1988. The effects of temperature and pH on the growth of yeast species during the fermentation of grape juice. *J. Appl. Bacteriol.* 65: 23–28.

HEINTZE, K. 1976. Über die gegenseitige Beeinflussung von Sorbinsäure und schwefliger Säure. *Die Industrielle Obst- und Gemüseverwertung* 61: 555–556.

HENICK-KLING, T. 1993. Malolactic fermentation. In: *Wine Microbiology and Biotechnology.* G.H. Fleet (Ed.), Chapter 10, pp. 286–326. Harwood Academic Publishers, Chur, Switzerland.

HENICK-KLING, T. 1994. Nitrogen requirement by yeasts during fermentation. In: *Proceedings of the New York Wine Industry Workshop.* T. Henick-Kling (Ed.), pp. 65–69. Geneva, NY.

HENICK-KLING, T. 1995. Control of malo-lactic fermentation in wine: energetics, flavour modification and methods of starter culture preparation. *J. Appl. Bacteriol. Symp. Supp.* 79: 29S–37S.

HENICK-KLING, T. and Y.H. PARK. 1994. Considerations for the use of yeast and bacterial starter cultures: SO$_2$ and timing of inoculation. *Am. J. Enol. Vitic.* 45: 464–469.

HENICK-KLING, T., T.H. LEE, and D.J.D. NICHOLAS. 1986a. Inhibition of bacterial growth and malolactic fermentation in wine by bacteriophage. *J. Appl. Bacteriol.* 61: 287–293.

HENICK-KLING, T., T.H. LEE, and D.J.D. NICHOLAS. 1986b. Characterization of the lytic activity of bacteriophages of *Leuconostoc oenos* isolated from wine. *J. Appl. Bacteriol.* 61: 525–534.

HENICK-KLING, T., T.E. ACREE, S.A. KRIEGER, M.H. LAURENT, and W.D. EDINGER. 1994. Modification of wine flavor by malolactic fermentation. In: *Proceedings from the New York Wine Industry Workshop.* T. Henick-Kling (Ed.), pp. 120–138. Cornell University, Geneva, NY.

HENICK-KLING, T., W. EDINGER, P. DANIEL, and P. MONK. 1998. Selective effects of sulfur dioxide and yeast starter culture addition on indigenous yeast populations and sensory characteristics of wine. *J. Appl. Microbiol.* 84: 865–876.

HENSCHKE, P.A. and V. JIRANEK. 1991. Hydrogen sulfide formation during fermentation: effect of nitrogen composition in model grape must. In: *Proceedings of the International Symposium on Nitrogen in Grapes and Wine.* J.M. Rantz (Ed.), pp. 172–184. American Society for Enology and Viticulture, Davis, CA.

HENSCHKE, P.A. and V. JIRANEK. 1993. Yeasts—Metabolism of nitrogen compounds. In: *Wine Microbiology and Biotechnology.* G.H. Fleet (Ed.), Chapter 4, pp. 77–165. Harwood Academic Publishers, Chur, Switzerland.

HENSCHKE, P.A., J.M. EGLINTON, P.J. COSTELLO, I.L. FRANCIS, H. GOCKOWIAK, A. SODEN, and P.B. HØJ. 2002. Winemaking with selected strains of non-*Saccharomyces cerevisiae* yeasts. Influence of *Candida stellata* and *Saccharomyces bayanus* on Chardonnay wine composition and flavour. In: *Proceedings of the 13th International Enology Symposium.* H. Trogus, J. Gafner, and A. Sütterlin (Eds.), pp. 459–481, Institut National de la Recherche Agronomique (INRA), Montpellier, France (June 9–12).

HENSEL, R., U. MAYR, K.O. STETTER, and O. KANDLER. 1977. Comparative studies of lactic acid dehydrogenases in lactic acid bacteria. I. Purification and kinetics of the allosteric L-lactic acid dehydrogenase from *Lactobacillus casei* spp. *casei* and *Lactobacillus curvatus. Arch. Microbiol.* 112: 81–93.

HERESZTYN, T. 1986. Formation of substituted tetrahydropyridines by species of *Brettanomyces* and *Lactobacillus* isolated from mousy wines. *Am. J. Enol. Vitic.* 37: 127–132.

HERNANDEZ-ORTE, P., A. GUITART, V. FERREIRA, J. GARCIA, and J. CACHO. 1998. Effect of maceration time and the addition of enzymes on the amino acid composition of musts and wines and its influence on wine aroma. *Food Sci. Tech. Int.* 4: 407–418.

HERRAIZ, T., G. REGLERO, M. HERRAIZ, P.J. MARTIN-ALVAREX, and M.D. CABEZUDO. 1990. The influence of the yeast and type of culture on the volatile composition of wines fermented without sulfur dioxide. *Am. J. Enol. Vitic.* 41: 313–318.

HOLLOWAY, P. and R.E. SUBDEN. 1991. Volatile metabolites produced in a Riesling must by wild yeast isolates. *Can. Inst. Sci. Technol. J.* 24: 57–59.

HOLLOWAY, P., R.E. SUBDEN, and M.A. LACHANCE. 1990. The yeasts in a Riesling must from the Niagara grape-growing region of Ontario Canada. *Can. Inst. Food Sci. Technol. J.* 23: 212–216.

HOLT, S.M. and G.L. COTE. 1998. Differentiation of dextran-producing *Leuconostoc* strains by a modified randomly amplified polymorphic DNA protocol. *Appl. Environ. Microbiol.* 64: 3096–3098.

HOLT, J.G., N.R. KRIEG, P.H.A. SNEATH, J.T. STALEY, and S.T. WILLIAMS. 1994. Genus *Acetobacter* and *Gluconobacter.* In: *Bergey's Manual of Determinative Bacteriology.* J.G. Holt (Ed.), pp. 71–84. The Williams and Wilkins Co., Baltimore, MD.

HOLZAPFEL, W.H. and U. SCHILLINGER. 1992. The genus *Leuconostoc.* In: *The Prokaryotes.* A. Balows, H.G. Trüper, M. Dworkin, W. Harder, and K.-H. Schleifer (Eds.), 2nd edition, Volume II, Chapter 69, pp. 1508–1534. Springer-Verlag, New York, NY.

HOOD, A. 1983. Inhibition of growth of wine lactic-acid bacteria by acetaldehyde-bound sulphur dioxide. *Aust. Grapegrow. Wine.* 232: 34–43.

HOUTMAN, A.C., J. MARAIS, and C.S. DU PLESSIS. 1980a. The possibilities of applying present-day knowledge of wine aroma components: influence of several juice factors on fermentation rate and ester production during fermentation. *S. Afr. J. Enol. Vitic.* 1: 27–33.

HOUTMAN, A.C., J. MARAIS, and C.S. DU PLESSIS. 1980b. Factors affecting the reproducibility of fermentation of grape juice and of the aroma composition of wines. I. Grape maturity, sugar, inoculum concentration, aeration, juice turbidity and ergosterol. *Vitis* 19: 37–54.

HUANG, Z. and C.S. OUGH. 1989. Effect of vineyard locations, varieties, and rootstocks on the juice amino acid composition of several cultivars. *Am. J. Enol. Vitic.* 40: 135–139.

HUANG, Z. and C.S. OUGH. 1991. Amino acid profiles of commercial grape juices and wines. *Am. J. Enol. Vitic.* 42: 261–267.

HUANG, Z. and C.S. OUGH. 1993. Identification of N-carbamyl amino acids in wines and in yeast cells. *Am. J. Enol. Vitic.* 44: 49–55.

HUANG, Y.-C., C.G. EDWARDS, J.C. PETERSON, and K.M. HAAG. 1996. Relationship between sluggish fermentations and the antagonism of yeast by lactic acid bacteria. *Am. J. Enol. Vitic* 47: 1–10.

HURST, A. and D.G. HOOVER. 1993. Nisin. In: *Antimicrobials in Foods.* P.M. Davidson and A.L. Branen (Eds.), 2nd edition, Chapter 10, pp. 369–394. Marcel Dekker, Inc., New York, NY.

IBEAS, J.I., I. LOZANO, F. PERDIGONES, and J.F. JIMENEZ. 1996. Detection of *Dekkera-Brettanomyces* strains in sherry by a nested PCR method. *Appl. Environ. Microbiol.* 62: 998–1003.

ILAGAN, R.D. 1979. *Studies on the Sporulation of* Dekkera. Master of Science Thesis, University of California, Davis, CA.

INGLEDEW, W.M. and R.E. KUNKEE. 1985. Factors influencing sluggish fermentations of grape juice. *Am. J. Enol. Vitic.* 36: 65–76.

IZUAGBE, Y.S., T.P. DOHMAN, W.E. SANDINE, and D.A. HEATHERBELL. 1985. Characterization of *Leuconostoc oenos* isolated from Oregon wines. *Appl. Environ. Microbiol.* 50: 680–684.

JACK, R.W., J.R. TAGG, and B. RAY. 1994. Bacteriocins of Gram-positive bacteria. *Microbiol. Rev.* 59: 171–200.

JACKSON, R.S. 2000. *Wine Science. Principles, Practices, Perception*, 2nd edition. Academic Press, New York, NY.

JACOBS, C.J. and H.J.J. VAN VUUREN. 1991. Effects of different killer yeasts on wine fermentations. *Am. J. Enol. Vitic.* 42: 295–300.

JARISCH, R. and F. WANTKLE. 1996. Wine and headache. *Int. Arch. Allergy Immunol.* 110: 7–12.

JENNINGS, W.G. 1965. Theory and practice of hard surface cleaning. *Adv. Food Res.* 14: 325–359.

JENSEN, K.A. and C.G. EDWARDS. 1991. Modification of the API rapid CH system for characterization of *Leuconostoc oenos*. *Am. J. Enol. Vitic.* 42: 274–277.

JIRANEK, V., P. LANGRIDGE, and P.A. HENSCHKE. 1995a. Regulation of hydrogen sulfide liberation in wine-producing *Saccharomyces cerevisiae* strains by assimilable nitrogen. *Appl. Environ. Microbiol.* 61: 461–467.

JIRANEK, V., P. LANGRIDGE, and P.A. HENSCHKE. 1995b. Amino acid and ammonium utilization by *Saccharomyces cerevisiae* wine yeasts from a chemically defined medium. *Am. J. Enol. Vitic.* 46: 75–83.

JIRANEK, V., P. LANGRIDGE, and P.A. HENSCHKE. 1995c. Validation of bismuth-containing indicator media for predicting H_2S-producing potential of

Saccharomyces cerevisiae wine yeasts under enological conditions. *Am. J. Enol. Vitic.* 46: 269–273.

JIRANEK, V., P. LANGRIDGE, and P.A. HENSCHKE. 1996. Determination of sulphite reductase activity and its response to assimilable nitrogen status in a commercial *Saccharomyces cerevisiae* wine yeast. *J. Appl. Bacteriol.* 81: 329–336.

JOHANSSON, M.L., M. QUEDNAU, G. MOLIN, and S. AHRNE. 1995. Randomly amplified polymorphic DNA (RAPD) for rapid typing of *Lactobacillus plantarum* strains. *Lett. Appl. Microbiol.* 21: 155–159.

JOHNSTON, M.A. and E.A. DELWICHE. 1962. Catalase of the Lactobacillaceae. *J. Bacteriol.* 83: 936–938.

JOLLY, N.P., O.P.H. AUGUSTYN, and I.S. PRETORIUS. 2003. The use of *Candida pulcherrima* in combination with *Saccharomyces cerevisiae* for the production of Chenin blanc wine. *S. Afr. J. Enol. Vitic.* 24: 63–69.

JOYEUX, A., S. LAFON-LAFOURCADE, and P. RIBÉREAU-GAYON. 1984a. Metabolism of acetic acid bacteria in grape must: consequences on alcoholic and malolactic fermentation. *Sci. Aliments* 4: 247–255.

JOYEUX, A., S. LAFON-LAFOURCADE, and P. RIBÉREAU-GAYON. 1984b. Evolution of acetic acid bacteria during fermentation and storage of wine. *Appl. Environ. Microbiol.* 48: 153–156.

JULIEN, A., J.-L. ROUSTAN, L. DULAU, and J.-M. SABLAYROLLES. 2000. Comparison of nitrogen and oxygen demands of enological yeasts: technological consequences. *Am. J. Enol. Vitic.* 51: 215–222.

KALATHENOS, P., J.P. SUTHERLAND, and T.A. ROBERTS. 1995. Resistance of some wine spoilage yeasts to combinations of ethanol and acids present in wine. *J. Appl. Bacteriol.* 78: 245–250.

KAMINSKI, E., S. STAWICKI, and E. WASOWICZ. 1974. Volatile flavor compounds produced by molds of *Aspergillus*, *Penicillium*, and *Fungi imperfecti*. *Appl. Microbiol.* 27: 1001–1004.

KANDER, O. 1983. Carbohydrate metabolism in lactic acid bacteria. *Antonie van Leeuwenhoek* 49: 209–224.

KANDLER, O. and N. WEISS. 1986. Genus *Lactobacillus*. In: *Bergey's Manual of Systematic Bacteriology*. P.H.A. Sneath, N.S. Mair, M.E. Sharpe, and J.G. Holt (Eds.), pp. 1209–1234. The Williams and Wilkins Co., Baltimore, MD.

KARAGIANNIS, S. and P. LANARIDIS. 1999. The effect of various vinification parameters on the development of several volatile sulfur compounds in Greek white wines of the cultivars Batiki and Muscat of Hamburg. *Am. J. Enol. Vitic.* 50: 334–342.

KARAS, M., D. BACHMANN, U. BAHR, and F. HILLENKAMP. 1987. Matrix-assisted ultraviolet laser desorption of non-volatile compounds. *Int. J. Mass Spec. Ion Process.* 78: 53–68.

KELLY, W.J., R.V. ASMUNDSON, and D.H. HOPCROFT. 1989. Growth of *Leuconostoc oenos* under anaerobic conditions. *Am. J. Enol. Vitic.* 40: 277–282.

KELLY, W.J., C.M. HUANG, and R.V. ASMUNDSON. 1993. Comparison of *Leuconostoc oenos* strains by pulsed-field gel electrophoresis. *Appl. Environ. Microbiol.* 59: 3969–3972.

KHADRE, M.A., A.E. YOUSEF, and J.-G. KIM. 2001. Microbiological aspects of ozone applications in food: a review. *J. Food Sci.* 66: 1242–1252.

KING, A.D., J.D. PONTING, D.W. SANSHUCK, R. JACKSON, and K. MIHARA. 1981. Factors affecting death of yeast by sulfur dioxide. *J. Food Prot.* 44: 92–97.

KING, S.W. and R.B. BEELMAN. 1986. Metabolic interactions between *Saccharomyces cerevisiae* and *Leuconostoc oenos* in a model grape juice/wine system. *Am. J. Enol. Vitic.* 37: 53–60.

KITAHARA, K., T. KANEKO, and O. GOTO. 1957. Taxonomic studies on the hiochi-bacteria, specific saprophytes of sake. II. Identification and classification of hiochi-bacteria. *J. Gen. Appl. Microbiol.* 3: 111–120.

KITOS, P.A., C.H. WANG, B.A. MOHLER, T.E. KING, and V.H. CHELDELIN. 1958. Glucose and gluconate dissimilation in *Acetobacter suboxydans*. *J. Biol. Chem.* 233: 1295–1298.

KLINGSHIRN, L.M., J.R. LIU, and J.F. GALLANDER. 1987. Higher alcohol formation in wines as related to the particle size profiles of juice insoluble solids. *Am. J. Enol. Vitic.* 38: 207–210.

KODAMA, S., T. SUZUKI, S. FUJINAWA, P. DE LA TEJA, and F. YOTSUZUKA. 1994. Urea contribution to ethyl carbamate formation in commercial wines during storage. *Am. J. Enol. Vitic.* 45: 17–24.

KONO, Y. and I. FRIDOVICH. 1983. Isolation and characterization of the pseudo-catalase of *Lactobacillus*. *J. Biol. Chem.* 258: 6015–6019.

KOSSE, D., H. SEILER, R. AMANN, W. LUDWIG, and S. SCHERER. 1997. Identification of yoghurt-spoiling yeasts with 18S rRNA-targeted oligonucleotide probes. *Sys. Appl. Microbiol.* 20: 468–480.

KOSSEVA, M., V. BESCHKOV, J.F. KENNEDY, and L.L. LLOYD. 1998. Malolactic fermentation in Chardonnay wine by immobilized *Lactobacillus casei* cells. *Process Biochem.* 33: 793–797.

KOTSERIDIS, Y. and R. BAUMES. 2000. Identification of impact odorants in Bordeaux red grape juices, in the commercial yeast used for its fermentation, and in the produced wine. *J. Agric. Food Chem.* 48: 400–406.

KRIEGER, S.A., W.P. HAMMES, and T. HENICK-KLING. 1990. Management of malolactic fermentation using starter cultures. *Vineyard Winery Manage.* 16 (Nov/Dec): 45–50.

KRIEGER, S.A., W.P. HAMMES, and T. HENICK-KLING. 1993. How to use malolactic starter cultures in the winery. *Wine Ind. J.* (May): 153–160.

KRUMPERMAN, P.H. and R.H. VAUGHN. 1966. Some lactobacilli associated with decomposition of tartaric acid in wine. *Am. J. Enol. Vitic.* 17: 185–190.

KUDO, M., P. VAGNOLI, and L.F. BISSON. 1988. Imbalance of pH and potassium concentration as a cause of stuck fermentation. *Am. J. Enol. Vitic.* 49: 295–301.

KUMAR, C.G. and S.K. ANAND. 1998. Significance of microbial biofilms in food industry: a review. *Int. J. Food Microbiol.* 42: 9–27.

KUNIYUKI, A.H., C. ROUS, and J.L. SANDERSON. 1984. Enzyme-linked immunosorbent assay (ELISA) detection of *Brettanomyces* contaminants in wine production. *Am. J. Enol. Vitic.* 35: 143–145.

KUNKEE, R.E. 1967a. Control of malo-lactic fermentation induced by *Leuconostoc citrovorum*. *Am. J. Enol. Vitic.* 18: 71–77.

KUNKEE, R.E. 1967b. Malo-lactic fermentation. *Adv. Appl. Microbiol.* 9: 235–279.

KUNKEE, R.E. 1968. A simplified chromatographic procedure for detection of malo-lactic fermentation. *Wines Vines* 49 (Mar): 23–24.

KUNKEE, R.E. 1974. Malo-lactic fermentation and winemaking. In: *Chemistry of Winemaking.* A.D. Webb (Ed.), pp. 151–170. American Chemical Society, Washington, DC.

KUNKEE, R.E. 1984. Selection and modification of yeasts and lactic acid bacteria for wine fermentations. *Food Microbiol.* 1: 315–332.

KUNKEE, R.E. 1996. Several decades of wine microbiology: have we changed or have the microbes? In: *Proceedings of the Wine Spoilage Microbiology Conference.* pp. 44–49. California State University Fresno, CA.

KUNKEE, R.E. and F. NERADT. 1974. A rapid method for detection of viable yeasts in wines. *Wine Vines* 55: 36–39.

KUNKEE, R.E., C.S. OUGH, and M.A. AMERINE. 1964. Induction of malo-lactic fermentation by inoculation of must and wine with bacteria. *Am. J. Enol. Vitic.* 15: 178–183.

KUPINA, S.A. 1984. Simultaneous quantitation of glycerol, acetic acid and ethanol in grape juice by high performance liquid chromatography. *Am. J. Enol. Vitic.* 35: 59–62.

KURTZMAN, C.P. 1998a. *Issatchenkia* Kudryavtev emend. Kurtzman, Smiley, and Johnson. In: *The Yeasts.* C.P. Kurtzman and J.W. Fell (Eds.), 4th edition, Chapter 35, pp. 221–226. Elsevier, New York, NY.

KURTZMAN, C.P. 1998b. *Pichia* E.C. Hansen emend. Kurtzman. In: *The Yeasts.* C.P. Kurtzman and J.W. Fell (Eds.), 4th edition, Chapter 42, pp. 273–352. Elsevier, New York, NY.

KURTZMAN, C.P. 1998c. *Zygosaccharomyces* Barker. In: *The Yeasts.* C.P. Kurtzman and J.W. Fell (Eds.), 4th edition, Chapter 57, pp. 424–432. Elsevier, New York, NY.

LAFON-LAFOURCADE, S. 1983. Wine and brandy. In: *Biotechnology. Food and Feed Production with Microorganisms.* H.J. Rehm and G. Reed (Eds.), Volume 5, pp. 81–163. Verlag Chemie, Weinheim, West Germany.

LAFON-LAFOURCADE, S. and P. RIBÉREAU-GAYON. 1984. Les alterations des vins par les bacteries acetiques et les bacteries lactiques. *Conn. Vigne Vin* 18: 67–82.

LAFON-LAFOURCADE, S., F. LARUE, and P. RIBÉREAU-GAYON. 1979. Evidence for the existence of "survival factors" as an explanation for some peculiarities of yeast growth, especially in grape must of high sugar concentration. *Appl. Environ. Microbiol.* 38: 1069–1073.

LAFON-LAFOURCADE, S., E. CARRE, A. LONVAUD-FUNEL, and P. RIBÉREAU-GAYON. 1983a. Induction de la fermentation malolactique des vins par inoculation d'une biomasse industrielle congelée de *L. oenos* après réactivation. *Conn. Vigne Vin* 17: 55–71.

LAFON-LAFOURCADE, S., E. CARRE, and P. RIBÉREAU-GAYON. 1983b. Occurrence of lactic acid bacteria during different stages of the vinification and conservation of wines. *Appl. Environ. Microbiol.* 46: 874–880.

LAMBRECHTS, M.G. and I.S. PRETORIUS. 2000. Yeast and its importance to wine aroma—a review. *S. Afr. J. Enol. Vitic.* 21: 97–129 (special issue).

LARSEN, J.T., J.C. NIELSEN, B. KRAMP, M. RICHELIEU, M.J. RIISAGER, N. ARNEBORG, and C.G. EDWARDS. 2003. Impact of different strains of *Saccharomyces cerevisiae* on malolactic fermentation by *Oenococcus oeni. Am. J. Enol. Vitic.* 54: 246–251.

LAURENT, M.-H., T. HENICK-KLING, and T.E ACREE. 1994. Changes in the aroma and odor of Chardonnay wine due to malolactic fermentation. *Wein-Wiss.* 49: 2–9.

LAVALLÉE, F., Y. SALVAS, S. LAMY, D.Y. THOMAS, R. DEGRÉ, and L. DULAU. 1994. PCR and DNA fingerprinting used as quality control in production of wine yeast strains. *Am. J. Enol. Vitic.* 45: 86–91.

LAWRENCE, N.L., D.C. WILSON, and C.S. PEDERSON. 1959. The growth of yeasts in grape juice stored at low temperatures. II. The types of yeast and their growth in pure culture. *Appl. Microbiol.* 7: 7–11.

LAY, H. 2003. Untersuchungen ueber die Entstehung des "Maeuseltons" in Wein und Modellloesungen. *Mitt. Klosterneuburg* 53: 243–250.

LEE, S.-J., D. RATHBONE, S. ASIMONT, R. ADDEN, and S.E. EBELER. 2004. Dynamic changes in ester formation during Chardonnay juice fermentations with different yeast inoculation and initial Brix conditions. *Am. J. Enol. Vitic.* 55: 346–354.

LEE, T.H. and R.F. SIMPSON. 1993. Microbiology and chemistry of cork taints in wine. In: *Wine Microbiology and Biotechnology.* G.H. Fleet (Ed.), Chapter 12, pp. 353–372. Harwood Academic Publishers, Chur, Switzerland.

LEMA, C., C. GARCIA-JARES, I. ORRIOLS, and L. ANGULO. 1996. Contribution of *Saccharomyces* and non-*Saccharomyces* populations to the production of some components of Albariño wine aroma. *Am. J. Enol. Vitic.* 47: 206–216.

LEMARESQUIER, H. 1987. Inter-relationships between strains of *Saccharomyces cerevisiae* from the Champagne area and lactic acid bacteria. *Lett. Appl. Microbiol.* 4: 91–94.

LEROI, F. and M. PIDOUX. 1993. Characterization of interactions between *Lactobacillus hilgardii* and *Saccharomyces florentinus* isolated from sugary kefir grains. *J. Appl. Bacteriol.* 74: 54–60.

LICKER, J.L., T.E. ACREE, and T. HENICK-KLING. 1999. What is "Brett" (*Brettanomyces*) flavor? A preliminary investigation. In: *Chemistry of Wine Flavor.* A.L. Waterhouse and S.E. Ebeler (Eds.), pp. 96–115. American Chemical Society, Washington, DC.

LIU, J.-W.R. and J.F. GALLANDER. 1982. Effect of insoluble solids on the sulfur dioxide content and rate of malolactic fermentation in white table wines. *Am. J. Enol. Vitic.* 33: 194–197.

LIU, J. and J.F. GALLANDER. 1983. Effect of pH and sulfur dioxide on the rate of malolactic fermentation in red table wines. *Am. J. Enol. Vitic.* 34: 44–46.

LIU, S.-Q. 2002. A review. Malolactic fermentation in wine—beyond deacidification. *J. Appl. Microbiol.* 92: 589–601.

LIU, S.-Q. and C.R. DAVIS. 1994. Analysis of wine carbohydrates using capillary gas liquid chromatography. *Am. J. Enol. Vitic.* 45: 229–234.

LIU, S.-Q. and G.J. PILONE. 2000. An overview of formation and roles of acetaldehyde in winemaking with emphasis on microbiological implications. *Int. J. Food Sci. Technol.* 35: 49–61.

LIU, S.-Q., G.G. PRITCHARD, M.J. HARDMAN, and G.J. PILONE. 1994. Citrulline production and ethyl carbamate (urethane) precursor formation from arginine degradation by wine lactic acid bacteria *Leuconostoc oenos* and *Lactobacillus buchneri*. *Am. J. Enol. Vitic.* 45: 235–242.

LIU, S.-Q., C.R. DAVIS, and J.D. BROOKS. 1995a. Growth and metabolism of selected lactic acid bacteria in synthetic wine. *Am. J. Enol. Vitic.* 46: 166–174.

LIU, S.-Q., G.G. PRITCHARD, M.J. HARDMAN, and G.J. PILONE. 1995b. Occurrence of arginine deiminase pathway enzymes in arginine catabolism by wine lactic acid bacteria. *Appl. Environ. Microbiol.* 61: 310–316.

LLAUBÉRES, R.M., B. RICHARD, A. LONVAUD, D. DUBOURDIEU, and B. FOURNET. 1990. Structure of an exocellular β-D-glucan from *Pediococcus* sp., a wine lactic bacteria. *Carb. Res.* 203: 103–107.

LLAURADÓ, J.M., N. ROZÈS, M. CONSTANTÍ, and A. MAS. 2005. Study of some *Saccharomcyes cerevisiae* strains for winemaking after preadaptation at low temperatures. *J. Agric. Food Chem.* 53: 1003–1011.

LONGO, E., J. CANSADO, D. AGRELO, and T.G. VILLA. 1991. Effect of climatic conditions on yeast diversity in grape musts from northwest Spain. *Am. J. Enol. Vitic.* 42: 141–144.

LONVAUD-FUNEL, A. 1999. Lactic acid bacteria in the quality improvement and depreciation of wine. *Antonie van Leeuwenhoek* 76: 317–331.

LONVAUD-FUNEL, A. 2001. Biogenic amines in wine: role of lactic acid bacteria. *FEMS Microbiol. Lett.* 199: 9–13.

LONVAUD-FUNEL, A. and A. JOYEUX. 1988. A bacterial disease causing ropiness of wines. *Sci. Aliments* 8: 33–50.

LONVAUD-FUNEL, A. and A. JOYEUX. 1993. Antagonism between lactic acid bacteria of wines: inhibition of *Leuconostoc oenos* by *Lactobacillus plantarum* and *Pediococcus pentosaceus*. *Food Microbiol.* 10: 411–419.

LONVAUD-FUNEL, A. and A. JOYEUX. 1994. Histamine production by wine lactic acid bacteria: isolation of a histamine-producing strain of *Leuconostoc oenos*. *J. Appl. Bacteriol.* 77: 401–407.

LONVAUD-FUNEL, A., C. DESENS, and A. JOYEUX. 1985. Stimulation de la fermentation malolactique par l'addition au vin d'enveloppes cellularies de levure adjuvants de nature polysaccharidique et azotee. *Conn. Vigne Vin* 19: 229–240.

LONVAUD-FUNEL, A., A. JOYEUX, and C. DESENS. 1988. Inhibition of malolactic fermentation of wines by products of yeast metabolism. *J. Sci. Food Agric.* 44: 183–191.

LONVAUD-FUNEL, A., A. JOYEUX, and O. LEDOUX. 1991. Specific enumeration of lactic acid bacteria in fermenting grape must and wine by colony hybridization with non-isotopic DNA probes. *J. Appl. Bacteriol.* 71: 501–508.

LOUREIRO, V. 2000. Spoilage yeasts in foods and beverages: characterisation and ecology for improved diagnosis and control. *Food Res. Int.* 33: 247–256.

LOUREIRO, V. and A. QUEROL. 1999. The prevalence and control of spoilage yeasts in foods and beverages. *Trends Food Sci. Tech.* 10: 356–365.

LOUREIRO, V. and M. MALFEITO-FERREIRA. 2003. Spoilage yeasts in the wine industry. *Int. J. Food Microbiol.* 86: 23–50.

LOUW, A. 2001. The occurrence of bitterness in wine: an overview. *Wynboer* 149: 95–100.

LUBBERS, S., C. CHARPENTIER, M. FEUILLAT, and A. VOILLEY. 1994. Influence of yeast cell walls on the behavior of aroma compounds in a model wine. *Am. J. Enol. Vitic.* 45: 29–33.

LÜTHI, H. and U. VETSCH. 1959. Contributions to the knowledge of the malo-lactic fermentation in wines and ciders. II. The growth promoting effect of yeast extract on lactic acid bacteria causing malo-lactic fermentation in wines. *J. Appl. Bacteriol.* 22: 384–391.

MACRIS, B.J. and P. MARKAKIS. 1974. Transport and toxicity of sulphur dioxide in *Saccharomyces cerevisiae* var. *ellipsoideus. J. Sci. Food Agric.* 25: 21–29.

MAICAS, S., J.V. GIL, I. PARDO, and S. FERRER. 1999. Improvement of volatile composition of wines by controlled addition of malolactic bacteria. *Food Res. Int.* 32: 491–496.

MAKDESI, A.K. and L.R. BEUCHAT. 1996. Improved selective medium for enumeration of benzoate-resistant, heat-stressed *Zygosaccharomyces bailii. Food Microbiol.* 13: 281–290.

MALFEITO-FERREIRA, M., A. ST. AUBYN, and V. LOUREIRO. 1989. Long-chain fatty acid composition as a tool for differentiating spoilage wine yeasts. *Mycotaxon.* 36: 35–42.

MALFEITO-FERREIRA, M., M. TARECO, and V. LOUREIRO. 1997. Fatty acid profiling: a feasible typing system to trace yeast contamination in wine bottling plants. *Int. J. Food Microbiol.* 38: 143–155.

MALLETROIT, V., J.-X. GUINARD, R.E. KUNKEE, and M.J. LEWIS. 1991. Effect of pasteurization on microbiological and sensory quality of white grape juice and wine. *J. Food Proc. Preserv.* 15: 19–29.

MAMEDE, M.E.O., H.M.A.B. CARDELLO, and G.M. PASTORE. 2005. Evaluation of an aroma similar to that of sparkling wine: sensory and gas chromatography analyses of fermented grape musts. *Food Chem.* 89: 63–68.

MANCA DE NADRA, M.C. and A.M. STRASSER DE SAAD. 1995. Polysaccharide production by *Pediococcus pentosaceus* from wine. *Int. J. Food Microbiol.* 27: 101–106.

MANGINOT, C., J.L. ROUSTAN, and J.M. SABLAYROLLES. 1998. Nitrogen demand of different yeast strains during alcoholic fermentation. Importance of the stationary phase. *Enz. Microbial Technol.* 23: 511–517.

MANSFIELD, A.K., B.W. ZOECKLEIN, and R.S. WHITON. 2002. Quantification of glycosidase activity in selected strains of *Brettanomyces bruxellensis* and *Oenococcus oeni. Am. J. Enol. Vitic.* 53: 303–307.

MARA, P.A. and L.F. BISSON. 2005. Bacterial causes of winery chloroanisole contamination. Abstr. 56th American Society for Enology Viticulture Annual Meeting, Seattle, WA. *Am. J. Enol. Vitic.* 56: 298A.

MARET, R. and T. SOZZI. 1977. Flore malolactique de moûts et de vins du Canton du Valais (Suisse). I. *Lactobacillus* et *Pédiococcus. Ann. Tech. Agric.* 27: 255–273.

MARET, R. and T. SOZZI. 1979. Flore malolactique de mouts et de vins du Canton du Valais (Suisse). II. Evolution des populations de lactobacilles et de pédiocoques au cours de la vinification d'un vin blanc (un Fendant) et d'un vin rouge (une Dole). *Ann. Tech. Agric.* 28: 31–40.

MARGALIT, Y. 2004. *Concepts in Wine Technology.* The Wine Appreciation Guild, San Francisco, CA.

MARGALITH, P.Z. 1981. *Flavour Microbiology.* Charles C. Thomas Publishers, Springfield, IL.

MAROIS, J.J., A.M. BLEDSOE, R.W. RICKER, and R.M. BOSTOCK, 1993. Sampling for *Botrytis cinerea* in harvested grape berries. *Am. J. Enol. Vitic.* 44: 261–265.

MARTINEAU, B. and T. HENICK-KLING. 1995a. Formation and degradation of diacetyl in wine during alcoholic fermentation with *Saccharomyces cerevisiae* strain EC1118 and malolactic fermentation with *Leuconostoc oenos* strain MCW. *Am. J. Enol. Vitic.* 46: 442–448.

MARTINEAU, B. and T. HENICK-KLING. 1995b. Performance and diacetyl production of commercial strains of malolactic bacteria in wine. *J. Appl. Bacteriol.* 78: 526–536.

MARTINEAU, B., T. ACREE, and T. HENICK-KLING. 1995. Effect of wine type on the detection threshold for diacetyl. *Food Res. Int.* 28: 139–143.

MARTINEZ-MURCIA, A.J., N.M. HARLAND, and M.D. COLLINS. 1993. Phylogenetic analysis of some leuconostocs and related organisms as determined from large-subunit rRNA gene sequences: assessment of congruence of small- and large-subunit rRNA derived trees. *J. Appl. Bacteriol.* 74: 532–541.

MARTINEZ-RODRIGUEZ, A.J., A.V. CARRASCOSA, and M.C. POLO. 2001. Release of nitrogen compounds to the extracellular medium by three strains of *Saccharomyces cerevisiae* during induced autolysis in a model wine system. *Int. J. Food Microbiol.* 68: 155–160.

MASLOW, J.N., A.M. SLUTSKY, and R.D. ARBEIT. 1993. Application of pulsed-field gel electrophoresis to molecular epidemiology. In: *Diagnostic Molecular Microbiology: Principles and Applications.* D.H. Persing, T.F. Smith, F.C. Tenover, and T.J. White (Eds.), pp. 563–572. American Society for Microbiology, Washington, DC.

MASON, A.B. and J.-P. DUFOUR. 2000. Alcohol acetyltransferases and the significance of ester synthesis in yeast. *Yeast* 16: 1287–1298.

MATEO, J.J., M. JIMENEZ, T. HUERTA, and A. PASTOR. 1992. Comparison of volatiles produced by four *Saccharomyces cerevisiae* strains isolated form Monastrell musts. *Am. J. Enol. Vitic.* 43: 206–209.

MAURICIO, J.C., S. GUIJO, and J.M. ORTEGA. 1991. Relationship between phospholipid and sterol contents in *Saccharomyces cerevisiae* and *Torulaspora delbrueckii* and their fermentation activity in grape musts. *Am. J. Enol. Vitic.* 42: 301–308.

MAURICIO, J.C., J. MORENO, L. ZEA, J.M. ORTEGA, and M. MEDINA. 1997. The effects of grape must fermentation conditions on volatile alcohols and esters formed by *Saccharomyces cerevisiae. J. Sci. Food Agric.* 75: 155–160.

MAYER, A.M. and E. HAREL. 1979. Polyphenol oxidases in plants. *Phytochem.* 18: 193–215.

MAYER, A.M. and R.C. STAPLES. 2002. Laccase: new functions for an old enzyme. *Phytochem.* 60: 551–565.

McDANIEL, M., L.A. HENDERSON, B.T. WATSON, and D. HEATHERBELL. 1987. Sensory panel training and screening for descriptive analysis of the aroma of

Pinot noir wines fermented by several strains of malolactic bacteria. *J. Sensory Studies* 2: 149–167.

McDONALD, V.R. 1963. Direct microscopic technique to detect viable yeast cells in pasteurized orange drink. *J. Food Sci.* 28: 135–139.

McEVILY, A.J., R. IYENGAR, and W.S. OTWELL. 1992. Inhibition of enzymatic browning in foods and beverages. *Crit. Rev. Food Sci. Nutr.* 32: 253–273.

McMAHON, H., B.W. ZOECKLEIN, K.C. FUGELSANG, and Y. JASINSKI. 1999. Quantification of glycosidase activities in selected yeasts and lactic acid bacteria. *J. Indust. Microbiol. Biotechnol.* 23: 198–203.

MEIDELL, J. 1987. Unsuitability of fluorescence microscopy for the rapid detection of small numbers of yeast cells on a membrane filter. *Am. J. Enol. Vitic.* 38: 159–160.

MENDES FERREIRA, A., M.C. CLÍMACO, and A. MENDES FAIA. 2001. The role of non-*Saccharomyces* species in releasing glycosidic bound fraction of grape aroma components—a preliminary study. *J. Appl. Microbiol.* 91: 67–71.

MEYER, S.A., R.W. PAYNE, and D. YARROW. 1998. *Candida* Berkhout. In: *The Yeasts*. C.P. Kurtzman and J.W. Fell (Eds.), 4th edition, Chapter 64, pp. 454–573. Elsevier, New York, NY.

MILLER, M.W. and H.J. PHAFF. 1998a. *Metschnikowia* Kamienski. In: *The Yeasts*. C. P. Kurtzman and J.W. Fell (Eds.), 4th edition, Chapter 39, pp. 256–267. Elsevier, New York, NY.

MILLER, M.W. and H.J. PHAFF. 1998b. *Saccharomycodes* E.C. Hansen. In: *The Yeasts*. C.P. Kurtzman and J.W. Fell (Eds.), 4th edition, Chapter 45, pp. 372–373. Elsevier, New York, NY.

MILLET, V. and A. LONVAUD-FUNEL. 2000. The viable but non-culturable state of wine micro-organisms during storage. *Lett. Appl. Microbiol.* 30: 136–141.

MILLS, D.A., E.A. JOHANNSEN, and L. COCOLIN. 2002. Yeast diversity and persistence in *Botrytis*-affected wine fermentations. *Appl. Environ. Microbiol.* 68: 4884–4893.

MILLS, J.M. 2001. *The Impact of Interactions between* Lactobacillus *and* Saccharomyces *spp. on Wine Fermentations*. Master of Science Thesis, Washington State University, Pullman, WA.

MIRA DE ORDUÑA, R., S.-Q. LIU, M.L. PATCHETT, and G.J. PILONE. 2000. Ethyl carbamate precursor citrulline formation from arginine degradation by malolactic wine lactic acid bacteria. *FEMS Microbiol. Lett.* 183: 31–35.

MIRA DE ORDUÑA, R., M.L. PATCHETT, S.Q. LIU, and G.J. PILONE. 2001. Growth and arginine metabolism of the wine lactic acid bacteria *Lactobacillus buchneri* and *Oenococcus oeni* at different pH values and arginine concentrations. *Appl. Environ. Microbiol.* 67: 1657–1662.

MISLIVEC, P.B., L.R. BEUCHAT, and M.A. COUSIN. 1992. Yeasts and molds. In: *Compendium of Microbiological Methods for the Examination of Foods*. C. Vanderzant and D.F. Splittstoesser (Eds.), 3rd edition, Chapter 16, pp. 239–249. American Public Health Association, Washington, DC.

MITRAKUL, C.M., T. HENICK-KLING, and C.M. EGLI. 1999. Discrimination of *Brettanomyces/Dekkera* yeast isolates from wine by using various DNA fingerprinting methods. *Food Microbiol.* 16: 3–14.

MOELLER, T., K. AKERSTRAND, and T. MASSOUD. 1997. Toxin-producing species of *Penicillium* and the development of mycotoxins in must and homemade wine. *Nat. Toxins* 5: 86–89.

MONK, P.R. 1986. Rehydration and propagation of active dry wine yeast. *Aust. Wine Ind. J.* 1: 3–5.

MONK, P.R. 1994. Nutrient requirements of wine yeast. In: *Proceedings of the New York Wine Industry Workshop.* T. Henick-Kling (Ed.), pp. 58–64. Geneva, NY.

MONTEIRO, F.F. and L.F. BISSON. 1992. Nitrogen supplementation of grape juice. I. Effect on amino acid utilization during fermentation. *Am. J. Enol. Vitic.* 43: 1–10.

MONTROCHER, R., M.C. VERNER, J. BRIOLAY, C. GAUTIER, and R. MARMEISSE. 1998. Phylogenetic analysis of the *Saccharomyces cerevisiae* group based on polymorphisms of rDNA spacer sequences. *Int. J. Sys. Bacteriol.* 48: 295–303.

MORA, J. and A. MULET. 1991. Effects of some treatments of grape juice on the population and growth of yeast species during fermentation. *Am. J. Enol. Vitic.* 42: 133–136.

MORA, J. and C. ROSSELLO. 1992. The growth and survival of *Pichia membranaefaciens* during fermentation of grape juice. *Am. J. Enol. Vitic.* 43: 329–332.

MORENO, J.J., C. MILLÁN, J.M. ORTEGA, and M. MEDINA. 1991. Analytical differentiation of wine fermentations using pure and mixed yeast cultures. *J. Indust. Microbiol.* 7: 181–190.

MORENO-ARRIBAS, V., S. TORLOIS, A. JOYEUX, A. BERTRAND, and A. LONVAUD-FUNEL. 2000. Isolation, properties and behaviour of tyramine-producing lactic acid bacteria from wine. *J. Appl. Microbiol.* 88: 584–593.

MORENO-ARRIBAS, M.V., M.C. POLO, F. JORGANES, and R. MUNOZ. 2003. Screening of biogenic amine production by lactic acid bacteria isolated from grape must and wine. *Int. J. Food Microbiol.* 84: 117–123.

MORNEAU, A.D. and R. MIRA DE ORDUÑA. 2005. Reduction of wine acetaldehyde levels by lactic acid bacteria. Abstr. 56th American Society for Enology Viticulture Annual Meeting, Seattle, WA. *Am. J. Enol. Vitic.* 56: 297A–298A.

MORRIS, E.O. and A.A. EDDY. 1957. Method for the measurement of wild yeast infection in pitching yeast. *J. Inst. Brew.* 63: 34–35.

MOXON, E.R., P.B. RAINEY, M.A. NOWAK, and R.E. LENSKI. 1994. Adaptive evolution of highly mutable loci in pathogenic bacteria. *Curr. Biol.* 4: 24–33.

MULLER, C.J., K.C. FUGELSANG, and V.L. WAHLSTROM. 1993. Capture and use of volatile flavor constituents emitted during wine fermentations. In: *Beer and Wine Production: Analysis, Characterization and Technological Advances.* B.H. Gump (Ed.), pp. 219–234. American Chemical Society, Washington, DC.

MULLER, C.J., K.C. FUGELSANG, M.M. OSBORN, and B.H. GUMP. 1996. Effect of carbon monoxide on spoilage yeast. In: *Proceedings of the Wine Spoilage Microbiology Conference,* pp. 75–78. California State University, Fresno, CA.

MULLIS, K.B. 1990. The unusual origin of the polymerase chain reaction. *Sci. Am.* 262: 56–61, 64–65.

MUNOZ, E. and W.M. INGLEDEW. 1989a. Effect of yeast hulls on stuck and sluggish wine fermentations: importance of the lipid component. *Appl. Environ. Microbiol.* 55: 1560–1564.

Munoz, E. and W.M. Ingledew. 1989b. An additional explanation for the promotion of more rapid, complete fermentation by yeast hulls. *Am. J. Enol. Vitic.* 40: 61–64.

Murrell, W.G. and B.C. Rankine. 1979. Isolation and identification of a sporing *Bacillus* from bottled brandy. *Am. J. Enol. Vitic.* 30: 247–249.

Musters, W., R.J. Planta, H. van Heerikhuizen, and H.A. Raué. 1990. Functional analysis of the transcribed spacers of *Saccharomyces cerevisiae* ribosomal DNA: it takes a precursor to form a ribosome. In: *The Ribosome: Structure, Function and Evolution.* W.E. Hill, A. Dahlberg, R.A. Garrett, P.B. Moore, D. Schlessinger, and J.R. Warner (Eds.), pp. 435–442. American Society for Microbiology, Washington, DC.

Nadal, D., B. Colomer, and B. Pina. 1996. Molecular polymorphism distribution in phenotypically distinct populations of wine yeast strains. *Appl. Environ. Microbiol.* 62: 1944–1950.

Nault, I., V. Gerbaux, J.P. Larpent, and Y. Vayssier. 1995. Influence of pre-culture conditions on the ability of *Leuconostoc oenos* to conduct malolactic fermentation in wine. *Am. J. Enol. Vitic.* 46: 357–362.

Navarro, L., M. Zarazaga, J. Saenz, F. Ruiz-Larrea, and C. Torres. 2000. Bacteriocin production by lactic acid bacteria isolated from Rioja red wines. *J. Appl. Microbiol.* 88: 44–51.

Nel, H.A., R. Bauer, G.M. Wolfaardt, and L.M.T. Dicks. 2002. Effect of bacteriocins Pediocin PD-1, Plantaricin 423, and nisin on biofilms of *Oenococcus oeni* on a stainless steel surface. *Am. J. Enol. Vitic.* 53: 191–196.

Nes, I.F., D.B. Diep, L.S. Havarstein, M.B. Brurberg, V. Eijsink, and H. Holo. 1996. Biosynthesis of bacteriocins in lactic acid bacteria. *Antonie van Leeuwenhoek* 70: 113–128.

Ng, W., M. Mankotia, P. Pantazopoulos, R.J. Neil, and P.M. Scott. 2004. Ochratoxin A in wine and grape juice sold in Canada. *Food Add. Contam.* 21: 971–981.

Nielsen, J.C. and M. Richelieu. 1999. Control of flavor development in wine during and after malolactic fermentation by *Oenococcus oeni. Appl. Environ. Microbiol.* 65: 740–745.

Nielsen, J.C., C. Prahl, and A. Lonvaud-Funel. 1996. Malolactic fermentation in wine by direct inoculation with freeze-dried *Leuconostoc oenos* cultures. *Am. J. Enol. Vitic.* 47: 42–48.

Nieuwoudt, H.H., B.A. Prior, I.S. Pretorius, and F.F. Bauer. 2002. Glycerol in South African tables wines: an assessment of its relationship to wine quality. *S. Afr. J. Enol. Vitic.* 23: 22–30.

Nigam, P. 2000. Wines. Specific aspects of oenology. In: *Encyclopedia of Food Microbiology.* R.K. Robinson, C.A. Batt, and P.D. Patel (Eds.), Volume 3, pp. 2316–2322. Academic Press, New York, NY.

Nishino, H., S. Miyazaki, and K. Tohjo. 1985. Effect of osmotic pressure on the growth rate and fermentation activity of wine yeasts. *Am. J. Enol. Vitic.* 36: 170–174.

Nissen, P. and N. Arneborg. 2003. Characterization of early deaths of non-*Saccharomyces* yeasts in mixed culture with *Saccharomyces cerevisiae. Arch. Microbiol.* 180: 257–263.

NOBLE, A.C. and G.F. BURSICK. 1984. The contribution of glycerol to perceived viscosity and sweetness in white wine. *Am. J. Enol. Vitic.* 35: 110–112.

NONOMURA, H. 1983. *Lactobacillus yamanashiensis* subsp. *yamanashiensis* and *Lactobacillus yamanashiensis* subsp. *mali* sp. and subsp. nov., nom. rev. *Int. J. Sys. Bacteriol.* 33: 406–407.

NYGAARD, M., H. GELFMAN, C. CATANIA, and C. PRAHL. 1998. Malolactic fermentation in Chardonnay: timing of inoculation, interactions with yeast and flavour development. Abstr. 49th American Society of Enology and Viticulture Annual Meeting, Sacramento, CA. *Am. J. Enol. Vitic.* 49: 457–458.

NYGAARD, M., L. PETERSEN, E. PILATTE, and G. LAGARDE. 2002. Prophylactic use of lysozyme to control indigenous lactic acid bacteria during alcoholic fermentation. Abstr. 53rd American Society of Enology and Viticulture Annual Meeting, Portland, OR. *Am. J. Enol. Vitic.* 53: 240A.

NYKÄNEN, L. 1986. Formation and occurrence of flavor compounds in wine and distilled alcoholic beverages. *Am. J. Enol. Vitic.* 37: 84–96.

NYKÄNEN, L. and I. NYKÄNEN. 1977. Production of esters by different yeast strains in sugar fermentations. *J. Inst. Brew.* 83: 30–31.

O'CONNOR-COX, E., F.M. MOCHABA, E.J. LODOLO, M. MAJARA, and B. AXCELL. 1997. Methylene blue staining: use at your own risk. *Tech. Quart. Master Brew. Assoc. Am.* 34: 306–312.

OLIJVE, W. and J.J. KOK. 1979. Analysis of growth of *Gluconobacter oxydans* in glucose containing media. *Arch. Microbiol.* 121: 283–290.

OLIVE, D.M. and P. BEAN. 1999. Principles and applications of methods for DNA-based typing of microbial organisms. *J. Clin. Microbiol.* 37: 1661–1669.

OLIVER, J.D. 2005. The viable but nonculturable state in bacteria. *J. Microbiol.* 43: 93–100.

OLSEN, E.B. 1994. The use of ML starter cultures in the winery. In: *Proceedings of the New York Wine Industry Workshop.* T. Henick-Kling (Ed.), pp. 116–119. Geneva, NY.

OSBORNE, J.P. 2005. *Inhibition of the malolactic fermentation by* Saccharomyces cerevisiae *during the alcoholic fermentation.* Ph.D. Dissertation. Washington State University, Pullman, WA.

OSBORNE, J.P., R. MIRA DE ORDUÑA, G.J. PILONE, and S.-Q. LIU. 2000. Acetaldehyde metabolism by wine lactic acid bacteria. *FEMS Microbiol. Lett.* 191: 51–55.

OUGH, C.S. 1964. Fermentation rates of grape juice. I. Effects of temperature and composition on white juice fermentation rates. *Am. J. Enol. Vitic.* 15: 167–177.

OUGH, C.S. 1966. Fermentation rates of grape juice. II. Effect of initial °Brix, pH, and fermentation temperature. *Am. J. Enol. Vitic.* 17: 20–26.

OUGH, C.S. 1971. Measurement of histamine in California wines. *J. Agric. Food Chem.* 19: 241–244.

OUGH, C.S. 1976. Ethyl carbamate in fermented beverages and food. I. Naturally occurring ethyl carbamate. *J. Agric. Food Chem.* 24: 323–328.

OUGH, C.S. 1993a. Dimethyl dicarbonate and diethyl dicarbonate. In: *Antimicrobials in Foods.* P.M. Davidson and A.L. Branen (Eds.), 2nd edition, Chapter 9, pp. 343–368. Marcel Dekker, Inc., New York, NY.

OUGH, C.S. 1993b. Sulfur dioxide and sulfites. In: *Antimicrobials in Foods*. P.M. Davidson and A.L. Branen (Eds.), 2nd edition, Chapter 5, pp. 137–190. Marcel Dekker, Inc., New York, NY.

OUGH, C.S. and J.L. INGRAHAM. 1960. Use of sorbic acid and sulfur dioxide in sweet table wines. *Am. J. Enol. Vitic.* 11: 117–122.

OUGH, C.S. and M.A. AMERINE. 1966. Effects of temperature on wine making. California Agricultural Experiment Station Bulletin 827. University of California, Davis, CA.

OUGH, C.S. and R.E. KUNKEE. 1974. The effect of fumaric acid on malo-lactic fermentation in wine from warm areas. *Am. J. Enol. Vitic.* 25: 188–190.

OUGH, C.S. and M.L. GROAT. 1978. Particle nature, yeast strain, and temperature interactions on the fermentation rates of grape juice. *Appl. Environ. Microbiol.* 35: 881–885.

OUGH, C.S. and A.A. BELL. 1980. Effects of nitrogen fertilization of grapevines on amino acids metabolism and higher-alcohol formation during grape juice fermentation. *Am. J. Enol. Vitic.* 31: 122–123.

OUGH, C.S. and E.A. CROWELL. 1987. Use of sulfur dioxide in winemaking. *J. Food Sci.* 52: 386–388.

OUGH, C.S. and G. TRIOLI. 1988. Urea removal from wine by an acid urease. *Am. J. Enol. Vitic.* 39: 303–307.

OUGH, C.S. and M.A. AMERINE. 1988. *Methods for Analysis of Musts and Wines*, 2nd edition. John Wiley & Sons, New York, NY.

OUGH, C.S., E.A. CROWELL, R.E. KUNKEE, M.R. VILAS, and S. LAGIER. 1987. A study of histamine production by various wine bacteria in model solutions and in wine. *J. Food Process. Preserv.* 12: 63–70.

OUGH, C.S., R.E. KUNKEE, M.R. VILAS, E. BORDEU, and M.-C. HUANG. 1988a. The interactions of sulfur dioxide, pH, and dimethyl dicarbonate on the growth of *Saccharomyces cerevisiae* Montrachet and *Leuconostoc oenos* MCW. *Am. J. Enol. Vitic.* 39: 279–282.

OUGH, C.S., E.A. CROWELL, and B.R. GUTLOVE. 1988b. Carbamyl compound reactions with ethanol. *Am. J. Enol. Vitic.* 39: 239–242.

OUGH, C.S., E.A. CROWELL, and L.A. MOONEY. 1988c. Formation of ethyl carbamate precursors during grape juice (Chardonnay) fermentation. I. Addition of amino acids, urea, and ammonia: effects of fortification on intracellular and extracellular precursors. *Am. J. Enol. Vitic.* 39: 243–249.

OUGH, C.S., M. DAVENPORT, and K. JOSEPH. 1989a. Effect of certain vitamins on growth and fermentation rate of several commercial active dry wine yeasts. *Am. J. Enol. Vitic.* 40: 208–213.

OUGH, C.S., D. STEVENS, and J. ALMY. 1989b. Preliminary comments on effects of grape vineyard nitrogen fertilization on the subsequent ethyl carbamate formation in wines. *Am. J. Enol. Vitic.* 40: 219–220.

OUGH, C.S., Z. HUANG, D. AN, and D. STEVENS. 1991. Amino acid uptake by four commercial yeasts at two different temperatures of growth and fermentation: effects of urea excretion and readsorption. *Am. J. Enol. Vitic.* 42: 26–40.

OURA, E. 1977. Reaction productions of yeast fermentations. *Process. Biochem.* 12: 19–21.

PALLMAN, C.L., J.A. BROWN, T.L. OLINEKA, L. COCOLIN, D.A. MILLS, and L.F. BISSON. 2001. Use of WL medium to profile native flora fermentations. *Am. J. Enol. Vitic.* 52: 198–203.

PAMPULHA, M.E. and V. LOUREIRO. 1989. Interaction of the effects of acetic acid and ethanol on inhibition of fermentation in *Saccharomyces cerevisiae. Biotech. Lett.* 11: 269–274.

PARDO, I., M.J. GARCÍA, M. ZÚÑIGA, and F. URUBURU. 1988. Evaluation of the API 50 CHL system for identification of *Leuconostoc oenos. Am. J. Enol. Vitic.* 39: 347–350.

PARISH, M.E. and D.E. CARROLL. 1985. Indigenous yeasts associated with Muscadine (*Vitis rotundifolia*) grapes and musts. *Am. J. Enol. Vitic.* 36: 165–169.

PARISH, M.E. and D.E. CARROLL. 1988. Effects of combined antimicrobial agents on fermentation initiation by *Saccharomyces cerevisiae* in a model broth system. *J. Food Sci.* 53: 240–242.

PARISH, M.E. and P.M. DAVIDSON. 1993. Methods of evaluation. In: *Antimicrobials in Foods.* P.M. Davidson and A.L. Branen (Eds.), 2nd edition, Chapter 17, pp. 597–615. Marcel Dekker, Inc., New York, NY.

PARK, S.K., R.B. BOULTON, and A.C. NOBLE. 2000. Formation of hydrogen sulfide and glutathione during fermentation of white grape musts. *Am. J. Enol. Vitic.* 51: 91–97.

PARKAR, S.G., S.H. FLINT, and J.D. BROOKS. 2004. Evaluation of the effect of cleaning regimes on biofilms of thermophilic bacilli on stainless steel. *J. App. Microbiol.* 96: 110–116.

PASTERIS, S.E. and A.M. STRASSER DE SAAD. 2005. Aerobic glycerol catabolism by *Pediococcus pentosaceus* isolated from wine. *Food Microbiol.* 22: 399–407.

PATARATA, L., M.S. PIMENTEL, B. POT, K. KERSTERS, and A.M. FAIA. 1994. Identification of lactic acid bacteria isolated from Portuguese wines and musts by SDS-PAGE. *J. Appl. Bacteriol.* 76: 288–293.

PATYNOWSKI, R.J., V. JIRANEK, and A.J. MARKIDES. 2002. Yeast viability during fermentation and *sur lie* ageing of a defined medium and subsequent growth of *Oenococcus oeni. Aust. J. Grape Wine Res.* 8: 62–69.

PAWSEY, R.H. 1974. *Techniques with Bacteria.* Hutchinson Publishers, London.

PEDERSON, C.S., M.N. ALBURY, and M.D. CHRISTENSEN. 1961. The growth of yeasts in grape juice stored at low temperatures. IV. Fungistatic effects of organic acids. *Appl. Microbiol.* 9: 162–167.

PEDERSON, C.S., M.N. ALBURY, D.C. WILSON, and N.L. LAWRENCE. 1959. The growth of yeasts in grape juice stored at low temperatures. I. Control of yeast growth in a commercial operation. *Appl. Microbiol.* 7: 1–6.

PEINADO, R.A., J.A. MORENO, D. MUNOZ, M. MEDINA, and J. MORENO. 2004. Gas chromatographic quantification of major volatile compounds and polyols in wine by direct injection. *J. Agric. Food Chem.* 52: 6389–6393.

PEÑA-NEIRA, A., B. FERNÁNDEZ DE SIMÓN, M.C. GARCÍA-VALLEJO, T. HERNÁNDEZ, E. CADAHÍA, and J.A. SUAREZ. 2000. Presence of cork-taint responsible compounds in wines and their cork stoppers. *Eur. Food Res. Tech.* 211: 257–261.

PEREZ, M.A., F.J. GALLEGO, I. MARTINEZ, and P. HIDALGO. 2001. Detection, distribution and selection of microsatellites (SSRs) in the genome of the yeast *Saccharomyces cerevisiae* as molecular markets. *Lett. Appl. Microbiol.* 33: 461–466.

PERI, C., M. RIVA, and P. DECIO. 1988. Crossflow membrane filtration of wines: comparison of performance of ultrafiltration, microfiltration, and intermediate cut-off membranes. *Am. J. Enol. Vitic.* 39: 162–168.

PEYNAUD, E. 1984. *Knowing and Making Wine.* John Wiley & Sons, New York, NY.

PFAFF, H.J., M.N. MILLER, and E.M. MRAK. 1978. *The Life of Yeasts*, 2nd edition. Harvard University Press, Cambridge, MA.

PILATTE, E. and C. PRAHL. 1997. Biological deacidification of acid grape varieties by inoculation on must with a freeze-dried culture of *Lactobacillus plantarum*. Abstr. 48th American Society of Enology and Viticulture Annual Meeting, San Diego, CA. *Am. J. Enol. Vitic.* 48: 386.

PILONE, D.A., G.J. PILONE, and B.C. RANKINE. 1973. Influence of yeast strain, pH, and temperature on degradation of fumaric acid in grape juice fermentation. *Am. J. Enol. Vitic.* 24: 97–102.

PILONE, G.J. 1986. Effect of triadimenol fungicide on yeast fermentations. *Am. J. Enol. Vitic.* 37: 304–305.

PILONE, G.J. and C. PRAHL. 1990. The Viniflora concept—a new approach to malolactic fermentation. In: *Proceedings of the New Zealand Grape and Wine Symposium.* pp. 79–93. New Zealand Society for Viticulture and Oenology, Auckland, NZ.

PILONE, G.J. and R.E. KUNKEE. 1965. Sensory characterization of wines fermented with several malo-lactic strains of bacteria. *Am. J. Enol. Vitic.* 16: 224–230.

PILONE, G.J., R.E. KUNKEE, and A.D. WEBB. 1966. Chemical characterization of wines fermented with various malolactic bacteria. *Appl. Microbiol.* 14: 608–615.

PILONE, G.J. and R.E. KUNKEE. 1972. Characterization and energetics of *Leuconostoc oenos* ML 34. *Am. J. Enol. Vitic.* 23: 61–70.

PILONE, G.J., B.C. RANKINE, and D.A. PILONE. 1974. Inhibiting malo-lactic fermentation in Australian dry red wines by adding fumaric acid. *Am. J. Enol. Vitic.* 25: 99–107.

PILONE, G.J., M.G. CLAYTON, and R.J. VAN DUIVENBODEN. 1991. Characterization of wine lactic acid bacteria: single broth culture for tests of heterofermentation, mannitol from fructose and ammonia from arginine. *Am. J. Enol. Vitic.* 42: 153–157.

PIMENTEL, M.S., M.H. SILVA, I. CORTÊS, and A.M. FAIA. 1994. Growth and metabolism of sugar and acids of *Leuconostoc oenos* under different conditions of temperature and pH. *J. Appl. Bacteriol.* 76: 42–48.

PINA, C., C. SANTOS, J.A. COUTO, and T. HOGG. 2004. Ethanol tolerance of five non-*Saccharomyces* wine yeasts in comparison with a strain of *Saccharomyces cerevisiae*—influence of different culture conditions. *Food Microbiol.* 21: 439–447.

PITT, J.I. 2000. *Penicillium.* In: *Encyclopedia of Food Microbiology.* R.K. Robinson, C.A. Batt, and P.D. Patel (Eds.), Volume 3, pp. 1647–1655. Academic Press, New York, NY.

PLATA, C., C. MILLÁN, J.C. MAURICIO, and J.M. ORTEGA. 2003. Formation of ethyl acetate and isoamyl acetate by various species of wine yeasts. *Food Microbiol.* 20: 217–224.

POMPILIO, R. 1993. Malolactic fermentation . . . who's doing what—and why? *Vineyard Winery Manage.* 19 (Nov/Dec): 44–47.

POOLMAN, B., D. MOLENAAR, E.J. SMID, T. UBBINK, T. ABEE, P.P. RENAULT, and W.N. KONINGS. 1991. Malolactic fermentation: electrogenic malate uptake and malate/lactate antiport generate metabolic energy. *J. Bacteriol.* 173: 6030–6037.

PORTER, L.J. and C.S. OUGH. 1982. The effects of ethanol, temperature, and dimethyl dicarbonate on viability of *Saccharomyces cerevisiae* Montrachet No. 522 in wine. *Am. J. Enol. Vitic.* 33: 222–225.

PRAHL, C. and J.C. NIELSEN. 1995. Development of *Leuconostoc oenos* malolactic cultures for direct inoculation. Presented at the Fifth International Symposium on Enology. Bordeaux, France (June 15–17).

PRAPHAILONG, W., M. VAN GESTEL, G.H. FLEET, and G.M. HEARD. 1997. Evaluation of the Biolog system for identification of food and beverage yeasts. *Lett. Appl. Microbiol.* 24: 455–459.

PRIPIS-NICOLAU, L., G. DE REVEL, A. BERTRAND, and A. LONVAUD-FUNEL. 2004. Methionine catabolism and production of volatile sulphur compounds by *Oenococcus oeni. J. Appl. Microbiol.* 96: 1176–1184.

PUIG-DEU, M., E. LÓPEZ-TAMAMES, S. BUXADERAS, and M.C. TORRE-BORONAT. 1999. Quality of base and sparkling wines as influenced by the type of fining agent added pre-fermentation. *Food Chem.* 66: 35–42.

PUPKO, T. and D. GRAUR. 1999. Evolution of microsatellites in the yeast *Saccharomyces cerevisiae*: role of length and number of repeated units. *J. Mol. Evol.* 48: 313–316.

QUESADA, M.P. and J.L. CENIS. 1995. Use of random amplified polymorphic DNA (RAPD-PCR) in the characterization of wine yeasts. *Am. J. Enol. Vitic.* 46: 204–208.

QUINSLAND, D. 1978. Identification of common sediments in wine. *Am. J. Enol. Vitic.* 29: 70–71.

RACCACH, M. 1987. Pediococci and biotechnology. *CRC Crit. Rev. Microbiol.* 14: 291–309.

RADLER, F. 1990a. Possible use of nisin in winemaking. I. Action of nisin against lactic acid bacteria and wine yeasts in solid and liquid media. *Am. J. Enol. Vitic.* 41: 1–6.

RADLER, F. 1990b. Possible use of nisin in winemaking. II. Experiments to control lactic acid bacteria in the production of wine. *Am. J. Enol. Vitic.* 41: 7–11.

RADLER, F. and B. GERWARTH. 1971. Über die bildung von flüchtigen gärungsnebenprodukten durch milchsäurebakterien. *Arch. Mikrobiol.* 76: 299–307.

RADLER, F. and C. YANNISSIS. 1972. Weinsäureabbau bei Milchsäurebakterien. *Arch. Microbiol.* 82: 219–239.

RADLER, F., P. PFEIFFER, and M. DENNERT. 1985. Killer toxin in new isolates of the yeasts *Hanseniaspora uvarum* and *Pichia kluyveri. FEMS Microbiol. Lett.* 29: 269–272.

RAMOS, M.T. and A. MADEIRA-LOPES. 1990. Effects of acetic acid on the temperature profile of ethanol tolerance in *Saccharomyces cerevisiae. Biotech. Lett.* 12: 229–234.

RAMOS, A., J.S. LOLKEMA, W.N. KONINGS, and H. SANTOS. 1995. Enzyme basis for pH regulatin of citrate and pyruvate metabolism by *Leuconostoc oenos. Appl. Environ. Microbiol.* 61: 1303–1310.

RANKINE, B.C. 1967. Formation of higher alcohols by wine yeasts, and relationship to taste thresholds. *J. Sci. Food Agric.* 18: 583–589.

RANKINE, B.C. 1972. Influence of yeast strain and malo-lactic fermentation on composition and quality of table wines. *Am. J. Enol. Vitic.* 23: 152–158.

RANKINE, B.C. and D.A. PILONE. 1973. *Saccharomyces bailii*, a resistant yeast causing serious spoilage of bottled table wine. *Am. J. Enol. Vitic.* 24: 55–58.

RANKINE, B.C., J.C.M. FORNACHON, and D.A. BRIDSON. 1969. Diacetyl in Australian dry red wines and its significance in wine quality. *Vitis* 8: 129–134.

RAPP, A. and H. MANDERY. 1986. Wine aroma. *Experientia* 42: 873–884.

RASMUSSEN, J.E., E. SCHULTZ, R.E. SNYDER, R.S. JONES, and C.R. SMITH. 1995. Acetic acid as a causative agent in producing stuck fermentations. *Am. J. Enol. Vitic.* 46: 278–280.

RAUHUT, D. 1993. Yeasts-production of sulfur compounds. In: *Wine Microbiology and Biotechnology*. G.H. Fleet (Ed.), Chapter 6, pp. 183–223. Harwood Academic Publishers, Chur, Switzerland.

RAVJI, R.G., S.B. RODRIGUEZ, and R.J. THORNTON. 1988. Glycerol production by four common grape molds. *Am. J. Enol. Vitic.* 39: 77–82.

RENEE TERRELL, F., J.R. MORRIS, M.G. JOHNSON, E.E. GBUR, and D.J. MAKUS. 1993. Yeast inhibition in grape juice containing sulfur dioxide, sorbic acid and dimethyldicarbonate. *J. Food Sci.* 58: 1132–1134.

RIBÉREAU-GAYON, P. 1985. New developments in wine microbiology. *Am. J. Enol. Vitic.* 36: 1–10.

RIBÉREAU-GAYON, P., D. DUBOURDIEU, B. DONÉCHE, and A. LONVAUD. 2000. *Handbook of Enology. Volume 1. The Microbiology of Wine and Vinifications.* John Wiley & Sons, New York, NY.

RICHMOND, J.Y. and R.W. McKINNEY. 1993. *Biosafety in Microbiological and Biomedical Laboratories*, 3rd edition. Public Health Service, U.S. Department of Health and Human Services, Washington, DC.

RIESEN, R. 1992. Undesirable fermentation aromas. In: *Proceedings of the ASEV/ES Workshop: Wine Aroma Defects*. T. Henick–Kling (Ed.), pp. 1–43. American Society of Enology and Viticulture (Eastern Section), Corning, NY.

RIVAS-GONZALO, J.C., J.F. SANTOS-HERNANDEZ, and A. MARINÉ-FONT. 1983. Study of the evolution of tyramine content during the vinification. *J. Food Sci.* 48: 417–418.

RODRIGUES, N., G. GONCALVES, S. PEREIRA-DA-SILVA, M. MALFEITO-FERREIRA, and V. LOUREIRO. 2001. Development and use of a new medium to detect yeasts of the genera *Dekkera/Brettanomyces. J. Appl. Microbiol.* 90: 588–599.

RODRIGUEZ, A.V. and M.C. MANCA DE NADRA. 1995a. Effect of pH and hydrogen peroxide produced by *Lactobacillus hilgardii* on *Pediococcus pentosaceus* growth. *FEMS Microbiol. Lett.* 128: 59–62.

RODRIGUEZ, A.V. and M.C. MANCA DE NADRE. 1995b. Production of hydrogen peroxide by *Lactobacillus hilgardii* and its effect on *Leuconostoc oenos* growth. *Curr. Microbiol.* 30: 23–25.

RODRÍGUEZ, M.E., C.A. LOPES, M. VAN BROOCK, S. VALLES, D. RAMÓN, and A.C. CABALLERO. 2004. Screening and typing of Patagonian wine yeasts for glycosidase activities. *J. Appl. Microbiol.* 96: 84–95.

RODRIGUEZ, S.B., E. AMBER, R.J. THORNTON, and M.R. McLELLAN. 1990. Malolactic fermentation in Chardonnay: growth and sensory effects of commercial strains of *Leuconostoc oenos*. *J. Appl. Bacteriol.* 68: 139–144.

ROHN, S., H.M. RAWEL, and J. KROLL. 2002. Inhibitory effects of plant phenols on the activity of selected enzymes. *J. Agric. Food Chem.* 50: 3566–3571.

ROJAS, V., J.V. GIL, F. PIÑAGA, and P. MANZANARES. 2001. Studies on acetate ester production by non-*Saccharomyces* wine yeasts. *Int. J. Food Microbiol.* 70: 283–289.

ROJAS, V., J.V. GIL, F. PIÑAGA, and P. MANZANARES. 2003. Acetate ester formation in wine by mixed cultures in laboratory fermentations. *Int. J. Food Microbiol.* 86: 181–188.

ROMANO, P. and G. SUZZI. 1993. Sulphur dioxide and wine microorganisms. In: *Wine Microbiology and Biotechnology.* G.H. Fleet (Ed.), Chapter 13, pp. 373–393. Harwood Academic Publishers, Chur, Switzerland.

ROMANO, P., G. SUZZI, P. DOMIZIO, and F. FATICHENTI. 1997. Secondary products formation as a tool for discriminating non-*Saccharomyces* wine strains. *Antonie van Leeuwenhoek* 71: 239–242.

ROSA, C.A. DA ROCHA, V. PALACIOS, M. COMBINA, M.E. FRAGA, A. DE OLIVEIRA REKSON, C.E. MAGNOLI, and A.M. DALCERO. 2002. Potential ochratoxin A producers from wine grapes in Argentina and Brazil. *Food Add. Contam.* 19: 408–414.

ROSENQUIST, J.K. and J.C. MORRISON. 1989. Some factors affecting cuticle and wax accumulation on grape berries. *Am. J. Enol. Vitic.* 40: 241–244.

ROZES, N., C. GARCIA-JARES, F. LARUE, and A. LONVAUD-FUNEL. 1992. Differentiation between fermenting and spoilage yeast in wine by total free fatty acid analysis. *J. Sci. Food Agric.* 59: 351–357.

RUIZ, A., M. POBLET, A. MAS, and J.M. GUILLAMON. 2000. Identification of acetic acid bacteria by RFLP of PCR-amplified 16S rDNA and 16S-23S rDNA intergenic spacer. *Int. J. Syst. Microbiol.* 50: 1981–1987.

SABATE, J., J. CANO, B. ESTEVE-ZARZOSO, and J.M. GUILLAMÓN. 2002. Isolation and identification of yeasts associated with vineyard and winery by RFLP analysis of ribosomal genes and mitochondrial DNA. *Microbiol. Res.* 157: 267–274.

SABLAYROLLES, J.-M., C. DUBOIS, C. MANGINOT, J.-L. ROUSTAN, and P. BARRE. 1996. Effectiveness of combined ammoniacal nitrogen and oxygen additions for completion of sluggish and stuck wine fermentations. *J. Ferm. Bioeng.* 82: 377–381.

SAEKI, A., M. TANIGUCHI, K. MATSUSHITA, H. TOYAMA, G. THEERAGOOL, G. LOTONG, and O. ADACHI. 1997. Microbiological aspects of acetate oxidation by acetic acid bacteria, unfavorable phenomena in vinegar fermentation. *Biosci. Biotechnol. Biochem.* 61: 317–323.

SAGE, L., D. GARON, and F. SEIGLE-MURANDI. 2004. Fungal microflora and ochratoxin A risk in French vineyards. *J. Agric. Food Chem.* 52: 5764–5768.

SALEMA, M., B. POOLMAN, J.S. LOLKEMA, M.C. DIAS, and W.N. KONINGS. 1994. Uniport of monoanionic L-malate in membrane vesicles from *Leuconostoc oenos. Eur. J. Biochem.* 225: 289–295.

SALEMA, M., J.S. LOLKEMA, M.V. SAN RAMAO, and M.C. LOUREIRO-DIAS. 1996. The proton motive force generated in *Leuconostoc oenos* by L-malate fermentation. *J. Bacteriol.* 178: 3127–3132.

SANCHO, T., G. GIMENEZ-JURADO, M. MALFEITO-FERREIRA, and V. LOUREIRO. 2000. Zymological indicators: a new concept applied to the detection of potential spoilage yeast species associated with fruit pulps and concentrates. *Food Microbiol.* 17: 613–624.

SAUVAGEOT, F. and P. VIVIER. 1997. Effects of malolactic fermentation of sensory properties of four Burgundy wines. *Am. J. Enol. Vitic.* 48: 187–192.

SAUVAGEOT, N., K. GOUFFI, J.-M. LAPACE, and Y. AUFFRAY. 2000. Glycerol metabolism in *Lactobacillus collinoides*: production of 3-hydroxypropionaldehyde, a precursor of acrolein. *Int. J. Food Microbiol.* 55: 167–170.

SCHMITT, M.J. and F. NEUHAUSEN. 1994. Killer toxin-secreting double-stranded RNA mycoviruses in the yeasts *Hanseniaspora uvarum* and *Zygosaccharomyces bailii. J. Virol.* 68: 1765–1772.

SCHMITTHENNER, J. 1950. Die Werkung der Kohlensaure aus Hefen and Bakererein. Bad Kreuznach: Seitz-Werke.

SCHOEMAN, H., M. VIVIER, M. DU TOIT, L.M.T. DICKS, and I.S. PRETORIUS. 1999. The development of bactericidal yeast strains by expressing the *Pediococcus acidilactici* pediocin gene (pedA) in *Saccharomyces cerevisiae. Yeast* 15: 647–656.

SCHREIER, P. 1979. Flavor composition of wines: a review. *CRC Crit. Rev. Food Sci. Nutr.* 12: 59–111.

SCHÜLTZ, H. and F. RADLER 1984. Anaerobic reduction of glycerol to propandiol-1,3 by *Lactobacillus brevis* and *Lactobacillus buchneri. Sys. Appl. Microbiol.* 5: 169–178.

SCHÜTZ, M. and J. GAFNER. 1995. Lower fructose uptake capacity of genetically characterized strains of *Saccharomyces bayanus* compared to strains of *Saccharomyces cerevisiae*: a likely cause of reduced alcoholic fermentation activity. *Am. J. Enol. Vitic.* 46: 175–180.

SCOTT, P.M., T. FULEKI, and J. HARWIG. 1977. Patulin content of juice and wine produced from moldy grapes. *J. Agric. Food Chem.* 25: 434–437.

SEA, K., C. BUTZKE, and R. BOULTON. 1998. Seasonal variation in the production of hydrogen sulfide during wine fermentations. In: *Chemistry of Wine Flavour.* A.L. Waterhouse and S.E. Ebeler (Eds.), pp. 81–95. American Chemical Society, Washington, DC.

SEISKARI, P., Y.-Y. LINKO, and P. LINKO. 1985. Continuous production of gluconic acid by immobilized *Gluconobacter oxydans* cell bioreactor. *Appl. Microbiol. Biotechnol.* 21: 356–360.

SEMON, M.J., C.G. EDWARDS, D. FORSYTH, and C. DINN. 2001. Inducing malolactic fermentation in Chardonnay musts and wines using different strains of *Oenococcus oeni. Aust. J. Grape Wine Res.* 7: 52–59.

SEÑIRES, A.Z. and N.V. ALEGADO. 2005. Effective stratagies in implementing HACCP in San Miguel breweries. *Tech. Quart. Master Brew. Assoc. Am.* 42: 16–20.

SERRA, R., L. ABRUNHOSA, Z. KOZAKIEWICZ, and A. VENÂNCIO. 2003. Black *Aspergillus* species as ochratoxin A producers in Portuguese wine grapes. *Int. J. Food Microbiol.* 88: 63–68.

SHARF, R. and P. MARGALITH. 1983. The effect of temperature on spontaneous wine fermentation. *Eur. J. Appl. Microbiol. Technol.* 17: 311–313.

SHARPE, M.E., E.I. GARVIE, and R.H. TILBURY. 1972. Some slime-forming heterofermentative species of the genus *Lactobacillus. Appl. Microbiol.* 23: 389–397.

SHIHATA, A.M. and E.M. MRAK. 1951. The fate of yeast in the digestive tract of *Drosophila. Am. Natural.* 85(825): 381–383.

SHIMAZU, Y. and M. WATANABE. 1981. Effects of yeast strains and environmental conditions on formation of organic acids in must during fermentation. *J. Ferment. Technol.* 59: 27–32.

SHIMIZU, K. 1993. Killer yeasts. In: *Wine Microbiology and Biotechnology.* G.H. Fleet (Ed.), Chapter 8, pp. 243–264. Harwood Academic Publishers, Chur, Switzerland.

SHINOHARA, T., S. KUBODERA, and F. YANAGIDA. 2000. Distribution of phenolic yeasts and production of phenolic off-flavors in wine fermentation. *J. Biosci. Bioeng.* 90: 90–97.

SIA, E.A., S. JINKS-ROBERTSON, and T.D. PETES. 1997. Genetic control of microsatellite instability. *Mutat. Res.* 383: 61–70.

SIANTAR, D.P., C.A. HALVERSON, C. KIRMIZ, G.F. PETERSON, N.R. HILL, and S.M. DUGAR. 2003. Ochratoxin A in wine: survey by antibody- and polymeric-based SPE columns using HPLC/fluorescence detection. *Am. J. Enol. Vitic.* 54: 170–177.

SIEIRO, C., J. CANSADO, D. AGRELO, J.B. VELÁZQUEZ, and T.G. VILLA. 1990. Isolation and enological characterization of malolactic bacteria from the vineyards of northwestern Spain. *Appl. Environ. Microbiol.* 56: 2936–2938.

SILLA SANTOS, M.H. 1996. Biogenic amines: their importance in food. *Int. J. Food Microbiol.* 29: 213–231.

SILVA PEREIRA, C., J.J. FIGUEIREDO MARQUES, and M.V. SAN ROMÃO. 2000. Cork taint in wine: scientific knowledge and public perception—a critical review. *CRC Crit. Rev. Microbiol.* 26: 147–162.

SILVA, P., H. CARDOSO, and H. GERÓS. 2004. Studies on the wine spoilage capacity of *Brettanomyces/Dekkera* spp. *Am. J. Enol. Vitic.* 55: 65–72.

SILVA, S., F. RAMÓN-PORTUGAL, P. ANDRADE, S. ABREU, M. DE FATIMA TEXEIRA, and P. STREHAIANO. 2003. Malic acid consumption by dry immobilized cells of *Schizosaccharomyces pombe. Am. J. Enol. Vitic.* 54: 50–55.

SILVER, J. and T. LEIGHTON. 1981. Control of malolactic fermentation in wine. 2. Isolation and characterization of a new malolactic organism. *Am. J. Enol. Vitic.* 32: 64–72.

SIMPSON, R.F., J.M. AMON, and A.J. DAW. 1986. Off-flavour in wine caused by guaiacol. *Food Technol. Aust.* 38: 31–33.

SIMPSON, R.F., D.L. CAPONE, and M.A. SEFTON. 2004. Isolation and identification of 2-methoxy-3,5-dimethylpyrazine, a potent musty compound from wine corks. *J. Agric. Food Chem.* 52: 5425–5430.

SLAUGHTER, J.C. and G. McKERNAN. 1988. The influence of pantothenate concentration and inoculum size on the fermentation of a defined medium by *Saccharomyces cerevisiae. J. Inst. Brew.* 94: 14–18.

SLININGER, P.J., R.J. BOTHAST, and K.L. SMILEY. 1983. Production of 3-hydroxy-propionaldehyde from glycerol. *Appl. Environ. Microbiol.* 46: 62–67.

SMITH, C.R. 2002. Volatile acidity reduction and other winegrape maturity enhancement applications of flavour-proof membranes. In: *Proceedings of the 13th International Enology Symposium.* H. Trogus, J. Gafner, and A. Sütterlin (Eds.), pp. 509–522, Institut National de la Recherche Agronomique (INRA), Montpellier, France (June 9–12).

SMITH, M.TH. 1998a. *Dekkera* van der Walt. In: *The Yeasts.* C.P. Kurtzman and J.W. Fell (Eds.), 4th edition, Chapter 27, pp. 174–177. Elsevier, New York, NY.

SMITH, M.TH. 1998b. *Brettanomyces* Kufferath and van Laer. In: *The Yeasts.* C.P. Kurtzman and J.W. Fell (Eds.), 4th edition, Chapter 63, pp. 450–453. Elsevier, New York, NY.

SMITH, M.TH. 1998c. *Hanseniaspora* Zikes. In: *The Yeasts.* C.P. Kurtzman and J.W. Fell (Eds.), 4th edition, Chapter 34, pp. 214–220. Elsevier, New York, NY.

SNOW, P.G. and J.F. GALLANDER. 1979. Deacidification of white table wines through partial fermentation with *Schizosaccharomyces pombe. Am. J. Enol. Vitic.* 30: 45–48.

SOBOLOV, M. and K.L. SMILEY. 1960. Metabolism of glycerol by an acrolein-forming *Lactobacillus. J. Bacteriol.* 79: 261–266.

SOFOS, J.N. and F.F. BUSTA. 1993. Sorbic acid and sorbates. In: *Antimicrobials in Foods.* P.M. Davidson and A.L. Branen (Eds.), 2nd edition, Chapter 3, pp. 49–94. Marcel Dekker, Inc., New York, NY.

SOLES, R.M., C.S. OUGH, and R.E. KUNKEE. 1982. Ester concentration differences in wine fermented by various species and strains of yeasts. *Am. J. Enol. Vitic.* 33: 94–98.

SOMERS, T.C. 1984. *Botrytis cinerea*-consequences for red wines. *Aust. Grape. Wine.* 244: 80–85.

SOMMER, P., E. STOLPE, B. KRAMP, A. BUNTE, B. CAUCHY-ALVIN, E. PILATTE, J. AALLING, and J. HEINEMEYER. 2005. Flavor enhancement: mixed starter cultures of *Saccharomyces cerevisiae, Kluyveromyces thermotolerans,* and *Torulaspora delbrueckii.* Abstr. 56th American Society for Enology Viticulture Annual Meeting, Seattle, WA. *Am. J. Enol. Vitic.* 56: 309A.

SOUFLEROS, E., M.-L. BARRIOS, and A. BERTRAND. 1998. Correlation between the content of biogenic amines and other wine compounds. *Am. J. Enol. Vitic.* 49: 266–278.

SPAYD, S.E. and J. ANDERSEN-BAGGE. 1996. Free amino acid composition of grape juice from 12 *Vitis vinifera* cultivars in Washington. *Am. J. Enol. Vitic.* 47: 389–402.

SPAYD, S.E., R.L. WAMPLE, R.G. EVANS, R.G. STEVENS, B.J. SEYMOUR, and C.W. NAGEL. 1994. Nitrogen fertilization of White Riesling grapes in Washington. Must and wine composition. *Am. J. Enol. Vitic.* 45: 34–42.

SPAYD, S.E., C.W. NAGEL, and C.G. EDWARDS. 1995. Yeast growth in Riesling juice as affected by vineyard nitrogen fertilization. *Am. J. Enol. Vitic.* 46: 49–55.

SPIROPOULOS, A., J. TANAKA, I. FLERIANOS, and L.F. BISSON. 2000. Characterization of hydrogen sulfide formation in commercial and natural wine isolates of *Saccharomyces*. *Am. J. Enol. Vitic.* 51: 233–247.

SPLITTSTOESSER, D.F. 1978. Fruits and fruit products. In: *Food and Beverage Mycology*. L.R. Beuchat (Ed.), pp. 83–109. AVI Publishing Co., Westport, CT.

SPLITTSTOESSER, D.F. 1992. Direct microscopic count. In: *Compendium of Microbiological Methods for the Examination of Foods*. C. Vanderzant and D.F. Splittstoesser (Eds.), 3rd edition, Chapter 5, pp. 97–104. American Public Health Association, Washington, DC.

SPLITTSTOESSER, D.F. and B.O. STOYLA. 1987. Lactic-acid spoilage in wine. *Wines Vines* 68: 65–66.

SPLITTSTOESSER, D.F. and J.J. CHURNEY. 1992. The incidence of sorbic acid-resistant gluconobacters and yeasts on grapes grown in New York State. *Am. J. Enol. Vitic.* 43: 290–293.

SPONHOLZ, W.R. 1991. Nitrogen compounds in grapes, must, and wine. In: *Proceedings of the International Symposium on Nitrogen in Grapes and Wine*. J. Rantz (Ed.), pp. 196–199. American Society for Enology and Viticulture, Davis, CA.

SPONHOLZ, W.R. 1993. Wine spoilage by microorganisms. In: *Wine Microbiology and Biotechnology*. G.H. Fleet (Ed.), Chapter 14, pp. 395–420. Harwood Academic Publishers, Chur, Switzerland.

SPONHOLZ, W.R. and H.H. DITTRICH. 1985. Über die Herkunft von Gluconsäure, 2- und 5-oxo Gluconsäure sowie Glucuron- und Galacturonsäure in Mosten und Weinen. *Vitis* 24: 51–58.

STANDER, M.A. and P.S. STEYN. 2002. Survey of ochratoxin A in South African wines. *S. Afr. J. Enol. Vitic.* 23: 9–13.

STENDER, H., A.J. BROOMER, K. OLIVEIRA, H. PERRY-O'KEEFE, J.J. HYLDIG-NIELSEN, A. SAGE, and J. COULL. 2001a. Rapid detection, identification, and enumeration of *Escherichia coli* cells in municipal water by chemiluminescent *in situ* hybridization. *Appl. Environ. Microbiol.* 67: 142–147.

STENDER, H., C. KURTZMAN, J.J. HYLDIG-NIELSEN, D. SØRENSEN, A. BROOMER, K. OLIVEIRA, H. PERRY-O'KEEFE, A. SAGE, B. YOUNG, and J. COULL. 2001b. Identification of *Dekkera bruxellensis* (*Brettanomyces*) from wine by fluorescence *in situ* hybridization using peptide nucleic acid probes. *Appl. Environ. Microbiol.* 67: 938–941.

STENDER, H., A. SAGE, K. OLIVEIRA, A.J. BROOMER, B. YOUNG, and J. COULL. 2001c. Combination of ATP bioluminescence and PNA probes allows rapid total counts and identification of specific microorganisms in mixed populations. *J. Microbiol. Meth.* 46: 69–75.

STEVENS, D.F. and C.S. OUGH. 1993. Ethyl carbamate formation: reaction of urea and citrulline with ethanol in wine under low to normal temperature conditions. *Am. J. Enol. Vitic.* 44: 309–312.

STINES, A.P., J. GRUBB, H. GOCKOWIAK, P.A. HENSCHKE, P.B. HØJ, and R. VAN HEESWIJCK. 2000. Proline and arginine accumulation in developing berries of *Vitis vinifera* L. in Australian vineyards: influence of vine cultivar, berry maturity and tissue type. *Aust. J. Grape Wine Res.* 6: 150–158.

STRASSER DE SAAD, A.M. and M.C. MANCA DE NADRA. 1993. Characterization of bacteriocin produced by *Pediococcus pentosaceus* from wine. *J. Appl. Bacteriol.* 74: 406–410.

STRATFORD, M., P. MORGAN, and A.H. ROSE. 1987. Sulphur dioxide resistance in *Saccharomyces cerevisiae* and *Saccharomycodes ludwigii. J. Gen. Microbiol.* 133: 2173–2179.

STRATTON, J.E., R.W. HUTKINS, and S.L. TAYLOR. 1991. Biogenic amines in cheese and other fermented foods: a review. *J. Food Prot.* 54: 460–470.

SUSLOW, T.V., M.N. SCHROTH, and M. ISAKA. 1982. Application of a rapid method for Gram differentiation of plant pathogenic and saprophytic bacteria without staining. *Phytopathol.* 72: 917–918.

SUZUKI, K. 1993. Cell envelopes and classification. In: *Handbook of New Bacterial Systematics.* A.G. O'Donnell and M. Goodfellow (Eds.), pp. 195–250. Academic Press, London.

SUZZI, G., P. ROMANO, and C. ZAMBONELLI. 1985. *Saccharomyces* strain selection in minimizing SO_2 requirement during vinification. *Am. J. Enol. Vitic.* 36: 199–202.

SVEUM, W.H., L.J. MOBERG, R.A. RUDE, and J.F. FRANK. 1992. Microbiological monitoring of the food processing environment. In: *Compendium of Microbiological Methods for the Examination of Foods.* C. Vanderzant and D.F. Splittstoesser (Eds.), 3rd edition, Chapter 3, pp. 51–74. American Public Health Association, Washington, DC.

SWANSON, K.M.J., F.F. BUSTA, E.H. PETERSON, and M.G. JOHNSON. 1992. Colony count methods. In: *Compendium of Microbiological Methods for the Examination of Foods.* C. Vanderzant and D.F. Splittstoesser (Eds.), 3rd edition, Chapter 4, pp. 75–95. American Public Health Association, Washington, DC.

SWINGS, J. 1992. The genera *Acetobacter* and *Gluconobacter.* In: *The Prokaryotes.* A. Balows, H.G. Trüper, M. Dworkin, W. Harder, and K.-H. Schleifer (Eds.), 2nd edition, Volume III, Chapter 111, pp. 2268–2286. Springer-Verlag, New York, NY.

TAMACHKIAROW, A. and H.-C. FLEMMING. 2003. On-line monitoring of biofilm formation in a brewery water pipeline system with a fibre optical device. *Water Sci. Tech.* 47: 19–24.

TAMAYO, C., J. UBEDA, and A. BRIONES. 1999. Relationship between H_2S-producing strains of wine yeast and different fermentation conditions. *Can. J. Microbiol.* 45: 343–346.

TAUTZ, D. 1989. Hypervariability of simple sequences as a general source of polymorphic DNA markers. *Nucleic Acids Res.* 17: 6463–6471.

TAYAMA, K., H. MINAKAMI, S. FUJIYAMA, H. MASSAI, and A. MISAKI. 1986. Structure of an acidic polysaccharide elaborated by *Acetobacter* sp. NDI-1005. *Agric. Biol. Chem.* 50: 1271–1278.

TEGMO-LARSSON, I.M., T.D. SPITTLER, and S.B. RODRIGUEZ. 1989. Effect of malolactic fermentation on ethyl carbamate formation in Chardonnay wine. *Am. J. Enol. Vitic.* 40: 106–108.

TENOVER, F.C., R.D. ARBEIT, R.V. GOERING, P.A. MICKELSEN, B.E. MURRAY, D.H. PERSING, and B. SWAMINATHAN. 1995. Interpreting chromosomal DNA restriction patterns produced by pulsed-field gel electrophoresis: criteria for bacterial strain typing. *J. Clin. Microbiol.* 33: 2233–2239.

THELWELL, N., S. MILLINGTON, A. SOLINAS, J. BOOTH, and T. BROWN. 2000. Mode of action and application of Scorpion primers to mutation detection. *Nucleic Acids Res.* 28: 3752–3761.

THOMAS, C.S., R.B. BOULTON, M.W. SILACCI, and W.D. GUBLER. 1993. The effect of elemental sulfur, yeast strain and fermentation medium on hydrogen sulfide production during fermentation. *Am. J. Enol. Vitic.* 44: 211–216.

THOMAS, D.S. 1993. Yeasts as spoilage organisms in beverages. In: *The Yeasts.* A.H. Rose and J.S. Harrison (Eds.), 2nd edition, Volume 5, pp. 517–561. Academic Press, New York, NY.

THOMAS, D.S. and R.R. DAVENPORT. 1985. *Zygosaccharomyces bailii*—a profile of characteristics and spoilage activities. *Food Microbiol.* 2: 157–169.

THOMPSON, A. 2000. ATP bioluminescence. Application in beverage microbiology. In: *Encyclopedia of Food Microbiology.* R.K. Robinson, C.A. Batt, and P.D. Patel (Eds.), Volume 1, pp. 101–109. Academic Press, New York, NY.

THORNTON, R.J. and S.B. RODRIGUEZ. 1996. Deacidification of red and white wines by a mutant of *Schizosaccharomyces malidevorans* under commercial winemaking conditions. *Food Microbiol.* 13: 475–482.

TIMBERLAKE, C.F. and P. BRIDLE. 1976. Interactions between anthocyanins, phenolic compounds, and acetaldehyde and their significance in red wines. *Am. J. Enol. Vitic.* 27: 97–105.

TONON, T. and A. LONVAUD-FUNEL. 2002. Arginine metabolism by wine lactobacilli isolated from wine. *Food Microbiol.* 19: 451–461.

TORO, M.E. and F. VAZQUEZ. 2002. Fermentation behaviour of controlled mixed and sequential cultures of *Candida cantarellii* and *Saccharomyces cerevisiae* wine yeasts. *World J. Microbiol. Biotechnol.* 18: 347–354.

TRACEY, R.P. and T.J. BRITZ. 1987. A numerical taxonomy study of *Leuconostoc oenos* strains from wine. *J. Appl. Bacteriol.* 63: 525–532.

TRACEY, R.P. and T.J. BRITZ. 1989. Freon 11 extraction of volatile metabolites formed by certain lactic acid bacteria. *Appl. Environ. Microbiol.* 55: 1617–1623.

TRAVERSO-RUEDA, S. and V.L. SINGLETON. 1973. Catecholase activity in grape juice and its implications in winemaking. *Am. J. Enol. Vitic.* 24: 103–109.

TREDOUX, H.G., J.L.F. KOCK, P.M. LATEGAN, and H.B. MULLER. 1987. A rapid identification technique to differentiate between *Saccharomyces cerevisiae* strains and other yeast species in the wine industry. *Am. J. Enol. Vitic.* 38: 161–164.

UGARTE, P., E. AGOSIN, E. BORDEU, and J.I. VILLALOBOS. 2005. Reduction of 4-ethylphenol and 4-ethylguaiacol concentration in red wines using reverse osmosis and adsorption. *Am. J. Enol. Vitic.* 56: 30–36.

UGLIANO, M., A. GENOVESE, and L. MOIO. 2003. Hydrolysis of wine aroma precursors during malolactic fermentation with four commercial starter cultures of *Oenococcus oeni. J. Agric. Food Chem.* 51: 5073–5078.

UTHURRY, C.A., J.A. SUÁREZ LEPE, J. LOMBARDERO, and J.R. GARCÍA DEL HIERRO. 2006. Ethyl carbamate production by selected yeasts and lactic acid bacteria in red wine. *Food Chem.* 94: 262–270.

VALERO, E., L. MOYANO, M.C. MILLAN, M. MEDINA, and J.M. ORTEGA. 2002. Higher alcohols and esters production by *Saccharomyces cerevisiae*. Influence of the initial oxygenation of the grape must. *Food Chem.* 78: 57–61.

VAN DER WALT, J.P. and A.E. VAN KERKEN. 1959. The wine yeasts of the Cape. Part II. The occurrence of *Brettanomyces intermedius* and *Brettanomyces schanderlii* in South African table wines. *Antonie van Leeuwenhoek* 25: 145–151.

VAN DER WALT, J.P. and A.E. VAN KERKEN. 1961. The wine yeasts of the Cape. Part V. Studies on the occurrence of *Brettanomyces intermedius* and *Brettanomyces schanderlii. Antonie van Leeuwenhoek* 27: 81–90.

VAN KEULEN, H., D.G. LINDMARK, K.E. ZEMAN, and W. GERLOSKY. 2003. Yeasts present during spontaneous fermentation of Lake Erie Chardonnay, Pinot Gris and Riesling. *Antonie van Leeuwenhoek* 83: 149–154.

VAN RENSBURG, P. and I.S. PRETORIUS. 2000. Enzymes in winemaking: harnessing natural catalysts for efficient biotransformations—a review. *S. Afr. J. Enol. Vitic.* 21: 52–73.

VAN VUUREN, H.J.J. and C.J. JACOBS. 1992. Killer yeasts in the wine industry: a review. *Am. J. Enol. Vitic.* 43: 119–128.

VAN VUUREN, H.J.J. and L.M.T. DICKS. 1993. *Leuconostoc oenos*: a review. *Am. J. Enol. Vitic.* 44: 99–112.

VARELA, F., F. CALDERÓN, M.C. GONZÁLES, B. COLOMO, and J.A. SUÁREZ. 1999. Effect of clarification on the fatty acid composition of grape must and the fermentation kinetics of white wines. *Eur. Food Res. Tech.* 209: 439–444.

VAUGHAN-MARTINI, A. and A. MARTINI. 1995. Facts, myths and legends on the prime industrial microorganism. *J. Indust. Microbiol.* 14: 514–522.

VAUGHAN-MARTINI, A. and A. MARTINI. 1998a. *Saccharomyces* Meyen ex Reess. In: *The Yeasts.* C.P. Kurtzman and J.W. Fell (Eds.), 4th edition, Chapter 44, pp. 358–371. Elsevier, New York, NY.

VAUGHAN-MARTINI, A. and A. MARTINI. 1998b. *Schizosaccharomyces* Lindner. In: *The Yeasts.* C.P. Kurtzman and J.W. Fell (Eds.), 4th edition, Chapter 48, pp. 391–394. Elsevier, New York, NY.

VAUGHN, R.H. 1955. Bacterial spoilage of wines with special reference to California conditions. *Adv. Food Res.* 6: 67–108.

VEIGA-DA-CUNHA, M., H. SANTOS, and E. VAN SCHAFTINGEN. 1993. Pathway and regulation of erythritol formation in *Leuconostoc oenos. J. Bacteriol.* 175: 3941–3948.

VERSALOVIC, J., T. KOEUTH, and J.R. LUPSKI. 1991. Distribution of repetitive DNA sequences in eubacteria and application to fingerprinting of bacterial genomes. *Nucleic Acids Res.* 19: 6823–6832.

VERSALOVIC, J., M. SCHNEIDER, F.J. DE BRUIJN, and J.R. LUPSKI. 1994. Genomic fingerprinting of bacteria using repetitive sequence-based polymerase chain reaction. *Meth. Mol. Cell Biol.* 5: 25–40.

VERSALOVIC, J., F.J. DE BRUIJN, and J.R. LUPSKI. 1998. Repetitive sequence-based PCR (rep-PCR) DNA fingerprinting of bacterial genomes. In: *Bacterial Genomes: Physical Structure and Analysis.* F.J. de Bruijn, J.R. Lupski, and G.M. Weinstock (Eds.), pp. 437–454. Chapman and Hall, New York, NY.

VERSTREPEN, K.J., G. DERDELINCKX, J.-P. DUFOUR, J. WINDERICKX, J.M. THEVELEIN, I.S. PRETORIUS, and F.R. DELVAUX. 2003. Flavor-active esters: adding fruitiness to beer. *J. Biosci. Bioeng.* 96: 110–118.

VIANNA, E. and S.E. EBELER. 2001. Monitoring ester formation in grape juice fermentations using solid phase microextraction coupled with gas chromatography-mass spectrometry. *J. Agric. Food Chem.* 49: 589–595.

VIDAL-CAROU, M.C., R. CODONY-SALCEDO, and A. MARINÉ-FONT. 1991. Changes in the concentration of histamine and tyramine during wine spoilage at various temperatures. *Am. J. Enol. Vitic.* 42: 145–149.

VILAS, M. 1993. Bottling line sampling and diagnostic techniques. *Vineyard Winery Manage.* 19 (Sept/Oct): 33–35.

VILLETTAZ, J.-C., D. STEINER, and H. TROGUS. 1984. The use of a beta glucanase as an enzyme in wine clarification and filtration. *Am. J. Enol. Vitic.* 35: 253–256.

VOS, P.J.A. and R.S. GRAY. 1979. The origin and control of hydrogen sulfide during fermentation of grape must. *Am. J. Enol. Vitic.* 30: 187–197.

WAHLSTROM, V.L. and K.C. FUGELSANG. 1988. Utilization of yeast hulls in winemaking. *Calif. Agric. Tech. Inst. Bull.* 880103.

WAINWRIGHT, T. 1970. Hydrogen sulphide production by yeast under conditions of methionine, pantothenate or vitamin B_6 deficiency. *J. Gen. Microbiol.* 61: 107–119.

WALKER, G.M. 1998. *Yeast. Physiology and Biotechnology.* John Wiley & Sons, New York, NY.

WALLING, E., M. DOLS-LAFARGUE, and A. LONVAUD-FUNEL. 2005. Glucose fermentation kinetics and exopolysaccharide production by ropy *Pediococcus damnosus* IOEB8801. *Food Microbiol.* 22: 71–78.

WANG, L.F. 1985. *Off-flavor Development in White Wine by* Brettanomyces *and* Dekkera. Master of Science Thesis, California State University, Fresno, CA.

WANG, X.D., J.C. BOHLSCHEID, and C.G. EDWARDS. 2003. Fermentative activity and production of volatile compounds by *Saccharomyces* grown in synthetic grape juice media deficient in assimilable nitrogen and/or pantothenic acid. *J. Appl. Microbiol.* 94: 1–11.

WARTH, A.D. 1977. Mechanism of resistance of *Saccharomyces bailii* to benzoic, sorbic and other weak acids used as food preservatives. *J. Appl. Bacteriol.* 43: 215–230.

WARTH, A.D. 1985. Resistance of yeast species to benzoic and sorbic acids and to sulfur dioxide. *J. Food Prot.* 48: 564–569.

WEBB, A.D. and J.L. INGRAHAM. 1960. Induced malo-lactic fermentations. *Am. J. Enol. Vitic.* 11: 59–63.

WEBSTER, D.R., C.G. EDWARDS, S.E. SPAYD, J.C. PETERSON, and B.J. SEYMOUR. 1993. Influence of vineyard nitrogen fertilization on the concentrations of monoterpenes, higher alcohols, and esters in aged Riesling wines. *Am. J. Enol. Vitic.* 44: 275–284.

WEENK, G., W. OLIJVE, and W. HARDER. 1984. Ketogluconate formation by *Gluconobacter* species. *Appl. Microbiol. Biotechnol.* 20: 400–405.

WEISS, N. 1992. The genera *Pediococcus* and *Aerococcus*. In: *The Prokaryotes*. A. Balows, H.G. Trüper, M. Dworkin, W. Harder, and K.-H. Schleifer (Eds.), 2nd edition, Volume II, Chapter 68, pp. 1502–1507. Springer-Verlag, New York, NY.

WEISS, N., V. SCHILLINGER, and O. KANDLER. 1983. *Lactobacillus trichodes* and *Lactobacillus heterohiochii*, subjective synonyms of *Lactobacillus fructivorans*. *Syst. Appl. Microbiol.* 4: 507–511.

WHITTENBURY, R. 1964. Hydrogen peroxide formation and catalase activity in the lactic acid bacteria. *J. Gen. Microbiol.* 35: 13–26.

WIBOWO, D., R. ESCHENBRUCH, C.R. DAVIS, G.H. FLEET, and T.H. LEE. 1985. Occurrence and growth of lactic acid bacteria in wine. A review. *Am. J. Enol. Vitic.* 36: 302–313.

WIBOWO, D., G.H. FLEET, T.H. LEE, and R.E. ESCHENBRUCH. 1988. Factors affecting the induction of malolactic fermentation in red wines with *Leuconostoc oenos*. *J. Appl. Bacteriol.* 64: 421–428.

WIJSMAN, M.R., J.P. VAN DIJKEN, B.H.A. VAN KLEEFF, and W.A. SCHEFFERS. 1984. Inhibition of fermentation and growth in batch cultures of the yeast *Brettanomyces intermedius* upon a shift from aerobic to anaerobic conditions (Custers effect). *Antonie van Leeuwenhoek* 50: 183–192.

WILKER, K.L. and M.R. DHARMADHIKARI. 1997. Treatment of barrel wood infected with acetic acid bacteria. *Am. J. Enol. Vitic.* 48: 516–520.

WILLIAMS, J.T., C.S. OUGH, and H.W. BERG. 1978. White wine composition and quality as influenced by method of must clarification. *Am. J. Enol. Vitic.* 29: 92–96.

WILLIAMS, J.G.K., A.R. KUBELIK, K.J. LIVAK, J.A. RAFALSKI, and S.V. TINGEY. 1990. DNA polymorphisms amplified by arbitrary primers are useful genetic markers. *Nucleic Acids Res.* 18: 6531–6535.

WILSON, B., C.R. STRAUSS, and P.J. WILLIAMS. 1986. The distribution of free and glycosidically-bound monoterpenes among skin, juice, and pulp fractions of some white grape varieties. *Am. J. Enol. Vitic.* 37: 107–114.

WIRTANEN, G. and S. SALO. 2003. Disinfection in food processing-efficacy testing of disinfectants. *Rev. Environ. Sci. Biotechnol.* 2: 293–306.

WISSELINK, H.W., R.A. WEUSTHUIS, G. EGGINK, J. HUGENHOLTZ, and G.J. GROBBEN. 2002. Mannitol production by lactic acid bacteria: a review. *Int. Dairy J.* 12: 151–161.

WOOLFORD, M.K. 1975. Microbiological screening of the straight chain fatty acids (C_1–C_{12}) as potential silage additives. *J. Sci. Food Agric.* 26: 219–228.

YAJIMA, M. and K. YOKOTSUKA. 2001. Volatile compound formation in white wines fermented using immobilized and free yeast. *Am. J. Enol. Vitic.* 52: 210–218.

YAMADA, S., K. NABE, M. IZUO, and I. CHIBATA. 1979. Enzymic production of dihydroxyacetone by *Acetobacter suboxydans* ATCC 621. *J. Ferment. Technol.* 57: 221–226.

YANAI, T. and N. SATO. 1999. Isolation and properties of β-glucosidase produced by *Debaryomyces hansenii* and its application in winemaking. *Am. J. Enol. Vitic.* 50: 231–235.

YANG, W.H. and E.C. PURCHASE. 1985. Adverse reactions to sulfites. *Can. Med. Assoc. J.* 133: 865–867, 880.

YARROW, D. 1998. Methods for the isolation, maintenance, and identification of yeasts. In: *The Yeasts.* C.P. Kurtzman and J.W. Fell (Eds.), 4th edition, Chapter 11, pp. 77–100. Elsevier, New York, NY.

YOKOTSUKA, K., A. OTAKI, A. NAITOH, and H. TANAKA. 1993. Controlled simultaneous deacidification and alcohol fermentation of a high-acid grape must using two immobilized yeasts, *Schizosaccharomyces pombe* and *Saccharomyces cerevisiae*. *Am. J. Enol. Vitic.* 44: 371–377.

YOKOTSUKA, K., M. YAJIMA, and T. MATSUDO. 1997. Production of bottle-fermented sparkling wine using yeast immobilized in double-layer gel beads or strands. *Am. J. Enol. Vitic.* 48: 471–481.

YOKOTSUKA, K., T. TAKAYANAGI, T. OKUDA, and M. YAJIMA. 2003. Production of sweet table wine by termination of alcohol fermentation using an antimicrobial substance from paprika seed. *Am. J. Enol. Vitic.* 54: 112–118.

YOUNG, E.T., J.S. SLOAN, and K. VAN RIPER. 2000. Trinucleotide repeats are clustered in regulatory genes in *Saccharomyces cerevisiae*. *Genetics* 154: 1053–1068.

YURDUGUL, S. and F. BOZOGLU. 2002. Studies on an inhibitor produced by lactic acid bacteria of wines on the control of malolactic fermentation. *Eur. Food Res. Technol.* 215: 38–41.

ZEE, J.A., R.E. SIMARD, L.L. HEUREUX, and J. TREMBLAY. 1983. Biogenic amines in wines. *Am. J. Enol. Vitic.* 34: 6–9.

ZEEMAN, W., J.P. SNYMAN, and C.J. VAN WYK. 1982. The influence of yeast strain and malolactic fermentation on some volatile bouquet substances and on quality of table wines. In: *Proceedings of the U.C.D. Grape and Wine Centennial.* A.D. Webb (Ed.), pp. 79–90. University of California, Davis, CA.

ZIMMERLI, B. and J. SCHLATTER. 1991. Ethyl carbamate: analytical methodology, occurrence, formation, biological activity and risk assessment. *Mutat. Res.* 259: 325–350.

ZOECKLEIN, B.W., K.C. FUGELSANG, B.H. GUMP, and F.S. NURY. 1995. *Wine Analysis and Production.* Chapman and Hall, New York, NY.

INDEX